Virtual Conference on Advances in Electric Drives, Process Control and Automation (VCADPCA 2019)

IOP Conference Series: Materials Science and Engineering
Volume 623

Online
4 June 2019

Editors:

V. Indragandhi
Raja Singh R
N.Kiran kumar
Sudha R
Prabhu K R

ISBN: 978-1-5108-9827-1
ISSN: 1757-8981

Printed from e-media with permission by:

Curran Associates, Inc.
57 Morehouse Lane
Red Hook, NY 12571

CURRAN ASSOCIATES INC. proceedings.com

Some format issues inherent in the e-media version may also appear in this print version.

Institute of Physics
Dirac House, Temple Back
Bristol BS1 6BE UK

Phone: 44 1 17 929 7481
Fax: 44 1 17 920 0979

techtracking@iop.org

Additional copies of this publication are available from:

Curran Associates, Inc.
57 Morehouse Lane
Red Hook, NY 12571 USA
Phone: 845-758-0400
Fax: 845-758-2633
Email: curran@proceedings.com
Web: www.proceedings.com

Table of contents

Volume 623

Virtual Conference on Advances in Electric Drives, Process Control and Automation

4 June 2019, India and Sweden

Accepted papers received: 23 August 2019
Published online: 18 October 2019

Preface

Papers

IOP Conf. Series: Materials Science and Engineering **623** (2019) 011001 doi:10.1088/1757-899X/623/1/011001

Preface

We would like to present, with great pleasure that the aim of the Virtual Conference on Advances in Electric Drives, Process Control and Automation (VCADPCA2019) is a platform for researchers, academicians as well as professionals from all over the world to present, discuss and promote advances in knowledge, research and practice in the field of Electric Drives, Process Control and Automation. VCADPCA2019 is offering a fantastic opportunity to attend a global scientific forum from the convenience of your desktop. No travelling, no hotel expenses, no time away from the office. The conference runs fully online, from paper submission, including reviewing, conference discussion and post conference processing. All papers will be referred to double tier approval process, single-blind peer-review and regular check. The online conference is a smart and affordable manner of presenting research results.

Editorial Board

IOP Conf. Series: Materials Science and Engineering **623** (2019) 011002 doi:10.1088/1757-899X/623/1/011002

Peer review statement

All papers published in this volume of *IOP Conference Series: Materials Science and Engineering* have been peer reviewed through processes administered by the proceedings Editors. Reviews were conducted by expert referees to the professional and scientific standards expected of a proceedings journal published by IOP Publishing.

Demand side EMS using artificial neural network-based LP method

N Loganathan[1], P S Mayurappriyan[2], K Lakshmi[3]

[1]Assistant Professor, Department of EEE, Sri Krishna College of Engineering and Technology, Coimbatore, India.(logukirsh@gmail.com)
[2]Professor, Department of Electronics and Instrumentation Engineering, Kumaraguru College of Technology, Coimbatore. (mayurappriyan.ps.eie@kct.ac.in)
[3]Professor and Head, Department of Electrical and Electronic Engineering, Sri Krishna College of Engineering and Technology, Coimbatore.(lakshmi01@gmail.com)

Abstract. This proposed system introduces advancement technique towards the interest sideways Energy Managing Structure (EMS) of a given concerning hourly power costs. This proposed system studies a bunch of interrelated cost receptive requests popular an educational institution. This requests canister stand provided concluded the fundamental framework as well as stochastic different power sources (for example, wind power plant & sun-based energy sources. What's more, the gathering of requests possesses a vitality storeroom. The proposed energy managing structure has a limit that every shopper can utilize their individual procedure to direct the current demand and costs trendy the energy dissemination framework. Towards tackle this energy managing structure issue then advancement calculation dependent on Linear Programming (LP) method takes remained executed. Notwithstanding LP calculation an ANN remained connected towards foresee the upcoming energy utilization the bunch of cost receptive requests. Main objective of the planned strategy remains toward boost the usage group of requests after it remains exposed towards an arrangement the compels. In this linear programming calculation enables bunch the interest to purchase, store and pitch vitality at reasonable occasions to change the hourly demand equal. Towards assess execution of this projected calculation of IEEE 14 transport framework remained measured. This outcome demonstrates the bunch of requests to vitality administration framework utilizing the proposed methodology expanding the effectiveness and limiting the misfortunes than the current strategies.

1. Introduction

The task of Energy Managing Structure (EMS) is essential aimed at the bunch of requests through their power source. In view of the energy request scope of a grid and energy price data, the energy managing structure ideally chooses the continually vitality utilization aimed at individually interest then decides the add up towards control utilization to the vitality bases. It takes measured three vitality bases, in particular, this fundamental network, solar then a breeze control station framework. The gathering of requirements claims a energy storing volume near stockpile strength in addition to use it on reasonable occasions after wanted. Energy the board issue with a little volume electric vitality framework utilizing Smart Grid (SG) innovation stayed measured in [1]. The storing room component, then the disseminated power capitals fill in by way of a simulated control station that purchases as well as supplies power trendy long periods of minimum power costs as well as moves vitality trendy long periods of maximum costs [2]. In [3] and [4] utilize the RO technique to game plan with the unit

IOP Conf. Series: Materials Science and Engineering **623** (2019) 012001 doi:10.1088/1757-899X/623/1/012001

responsibility issue, join similarly wind control and siphoned stockpiling hydro control plant. This prototypical of smart grid reflects a powerful framework through the reason toward consolidate the proportions of altogether clients with the end goal that every one of the generators and buyers must perform to productively disperse economically, financially savvy, and secure electrical power supply.
Extra to request reaction issues, Robust Advancement (RO) [5], [6] has been utilized in various power framework streamlining issues. Additionally, in [7],[8] RO approach built up a preservationist control maker. Vehicle to Grid (V2G) request reaction [9], [10], [11], the board of circulated age [12], and vitality capacity process [13], and vitality supply [14], [15], PI represents more wellsprings of uncertainty, counting prototypical misspecification besides clamour fluctuation [16], a additional straightforward an method near development of forecast interim aimed at wind plant control gauges produced via neural network methods. Wind control anticipating strategies can be comprehensively partitioned addicted to dual congregations: carnal plans then information serice (likewise called factual) strategies [17].
To deal with this EMS issue, this paper proposes a computation reliant on a Linear Programming (LP) show has been realized to increase the utility for the gathering of solicitation in regards to a game plan of goals, for instance, least consistently essentialness use, most noteworthy and least hourly stack levels, imperativeness amassing purposes of containment, and power availability from the essential grid and the DERs. Given the above setting, the responsibilities of this paper are fourfold:
Restricted gathering of reliable cost responsive requests, which can be provided completely through the principle network, stochastic DERs, and a energy store room. Results from a sensible contextual investigation demonstrate this effectiveness of the projected energy managing structure calculation. This article remains organized as pursues: Segment two gives the power monitoring framework. Segment three gives the execution of LP strategy toward resolve the power managing issue. Segment four exhibits the simulation outputs. Segment five gives the supposition.

2. Energy Managing Framework

Energy managing stays the optimistic, arranged as well as effective organization of acquirement, alteration, source then utilization of power get together provisions, appealing interested in record natural and monetary targets. Power the executives remains a nonstop capacity of power administrators. It remains imperative towards coordinate the power the board of association erection. It remains prudent towards set up a different authoritative element "power the executives" popular vast before power exhaustive organizations. Ability administration remains an imperative some portion vitality the board, in light of the fact that tremendous extent (normal 25 percent) of finish working expenses is vitality price.
This projected energy managing structure, solar cluster's principle work remains toward deliver 500KW power towards source the heaps. This proposed system of energy managing structure appears trendy figure1.

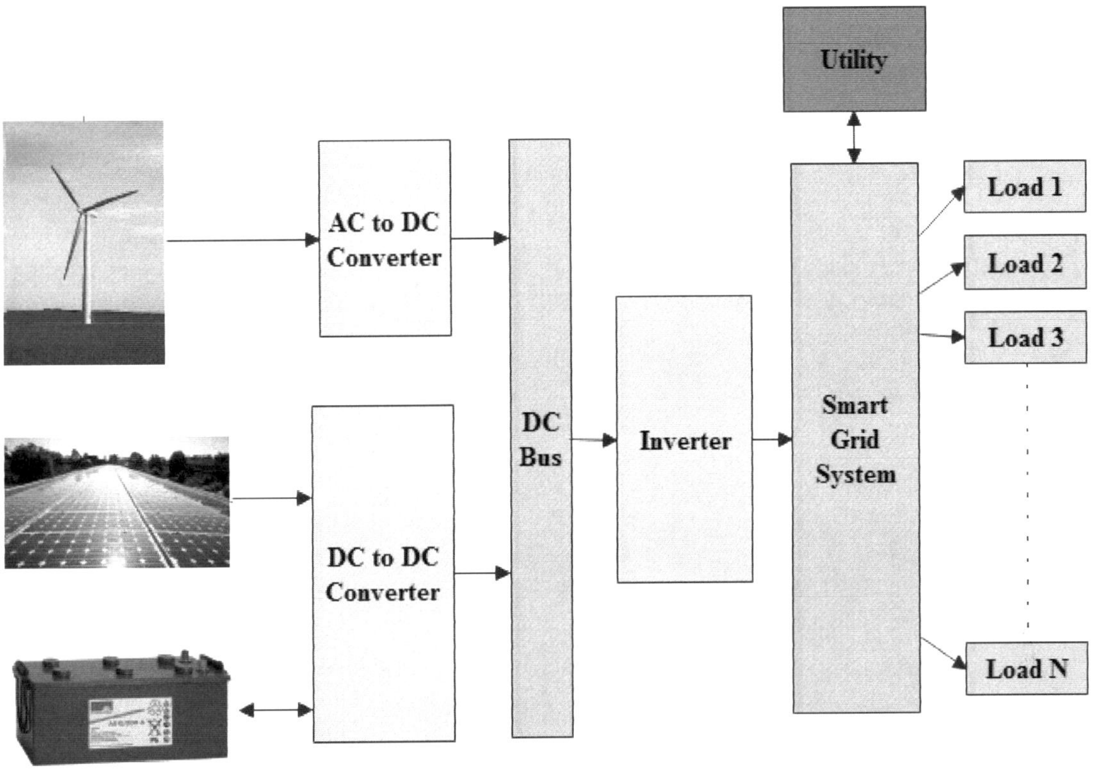

Figure.1 Proposed System of Energy Managing Structure

Wind develops produces 20 kW vitality supply to the interest. The converter goes about as a buck support converter to construct the yield voltage. From the DC converter the data banner of voltage and current is given to the MPPT controller. This controller finds the bungle banner and initiates the IGBT passage drive; this very decreases the amount of accentuations in the MPPT strategy.

Wind develops pitch controller has a working control structure that can vacillate the pitch edge of the turbine bleeding edges to decrease the torque made by the edges in a settled speed turbine additionally, to lessen the rotational speed in factor speed turbines. The variable speed action of wind electric structures yields higher yields for both low and high breeze speeds. The battery bank is used to store the DC vitality at the voltage of 115kW. From the battery, AC burdens are related. These AC burdens are sorted out as lighting load which considered light and fan for the structure considered. If the stack solicitation is lower than the DERs yield, the excess vitality will be used to ensnare the battery. The battery will be completely energized; the power can be inverter from DC to Air melding for the utilization of AC weights or excess vitality may be continued to the system. In case the DC supply does not exist or be figuratively speaking mostly available and the intrigue is on the dc stacks, the battery will supply the power explicitly. If the store requires to be upper more unmistakable than the battery yield, the AC structure will give vitality when it is available

3. Execution of LP Strategy Toward Resolve the Energy Managing Issue

The well-ordered methodology for scientific plan of straight programming technique to take care of vitality the board issue is as per the following.

Stage 1:
Information the interest factors for constant information and foreordained information utilizing Neural Network (NN) of the vitality the executive's framework

IOP Conf. Series: Materials Science and Engineering **623** (2019) 012001 doi:10.1088/1757-899X/623/1/012001

Stage 2:
Define the interest capacity to be upgraded (greatest or least) as a direct capacity of the various factors.

Stage 3:
Figure the requirements of vitality the executive's framework, for example, asset restrictions, advertise requests, between connection between various interest factors.

Stage 4:
From the considered contextual investigation thirteen distinct sorts of requests accessible and three unique kinds of vitality sources accessible. Let Mcd signify the quantity of units of vitality sources d in the unit of requests c, d =1, 2, 3: c = 1, 2, 3, 4, 5, 6, 7, 8, 9, 10, 11, 12, 13, 14, 15, 16. Give ed a chance to be the number of units devoured for interest. At that point the all-out number of units of requests C in the favored source.

$$\sum_{C=1}^{16}\sum_{j=1}^{3} m_{cd}e_d \quad (1)$$

Stage 5:
Give nc a chance to be the number of units of least day by day prerequisite of the interest I and it very well may be communicated as pursues

$$\sum_{c=1}^{16}\sum_{d=1}^{3} m_{cd}e_d \geq n_c \qquad (2)$$

Where c = 1, 2, 3 . . . 16

Stage 6:
For each source d, ed must be either positive or zero.

$$e_d \geq 0 \qquad (3)$$

Where d = 1, 2, 3

Stage 7:

Let fd be the vitality the board framework yield of vitality source d. Along these lines the absolute yield of vitality the board framework is given beneath

$$z = f1e1 + f2e2 + ... + f16e16 \qquad (4)$$

Stage 8:
The most significant normal for Prediction Intervals (PIs) is their inclusion likelihood. PI inclusion likelihood (PICP) is estimated by checking the quantity of target esteems secured by the developed PIs.

$$\mathrm{PICP} = \frac{1}{n}\sum_{c=1}^{n} i_o \ (5)$$

PICP is a proportion of legitimacy of PIs built with a related certainty level.

IOP Conf. Series: Materials Science and Engineering **623** (2019) 012001 doi:10.1088/1757-899X/623/1/012001

Stage 9:
PI standardized arrived at the midpoint of width (PINAW) surveys PIs from this angle and measures how wide they are

$$\text{PINAW} = \frac{1}{nT}\sum_{0=1}^{n}(V_o - S_o) \qquad (6)$$

Where Vo, So furthest point of confinement and lower farthest point of interest, T is the scope of the fundamental target characterized as the distinction between its base and most extreme qualities. PINAW is the normal width of PIs as a level of the basic target run.

4. Result and Conversation

The proposed LP strategy re-enactment were created utilizing MATLAB 7.10 programming bundle and the framework setup is Intel Core i5-2410M Processor with 2.90 GHz speed and 4 GB RAM. In proposed work three vitality sources, 13 requests and IEEE 14 transport framework considered as contextual investigation, over indicated time interims. The computational aftereffects of vitality the board issue achieved by the proposed direct programming system for the three vitality sources investigated.

4.1 Circumstance Learning – IEEE 14 Bus Systems

This examination is recognized away at the state of organizing, action, control, and monetarily conjectures. They exist of utilization in unequivocal the size and stage edge of burden transports, likewise, dynamic and open power stream more vital than conduction lines, and dynamic and responsive power to be implanted at the transports. For this work the straight programming the system is used for numerical investigation. The motivation behind this endeavor is to stretch out a MATLAB program to support the utilization of the gathering of solicitations when it is presented to a lot of requirements. Fig. 2 demonstrates the all-out requests and vitality sources. This LP count allows the bundle of solicitation to purchase, store and offer vitality at sensible occasions to change the hourly burden level to separate voltages, dynamic and responsive power control on each bus used for IEEE 14 transport structures. By basic IEEE 5 transport system is organized by using hand figuring and differentiated and MATLAB Program results and after that IEEE 14, transport structure MATLAB program is executed with the responsible data. This sort of examination is significant for handling the power stream issue in different control structures which will profitable to figure the dark amounts.

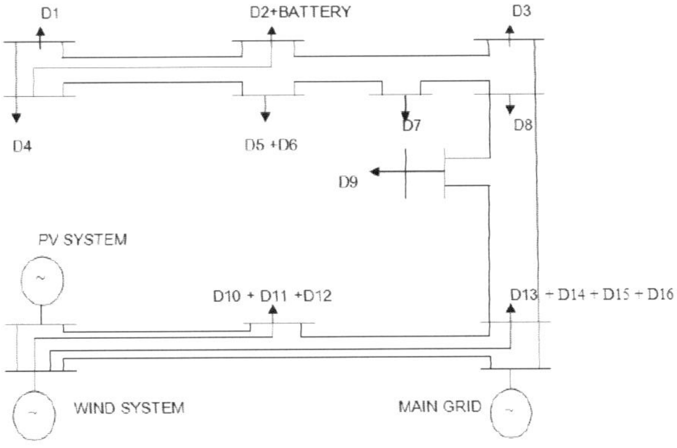

Figure 2. IEEE 14 Bus Scheme System

IOP Conf. Series: Materials Science and Engineering **623** (2019) 012001 doi:10.1088/1757-899X/623/1/012001

4.2 Demand requires statistics:

This paper considers the vitality the executives' framework requests situated in the school grounds. The conveyed vitality assets and a vitality storeroom are situated at bus14.The primary framework is associated into the bus3. The all out power provided from the disseminated vitality assets, fundamental lattice and vitality storeroom. The accessible interest side administration limit is taken as a small amount of the planned interest of the comparing hour. Also, the submitted interest side administration limit must be returned to the interest during that day so as to the conduct of vitality value delicate brilliant apparatuses. On the off chance that a high level of disseminated vitality assets age is considered, the fundamental commitment of interest side administration is levelling the heap and lessening the circulated vitality assets fluctuation. The hourly burden level for various requests and gathered the information on 16.10.2018 - 17.10.2018 at 12 PM - 11 AM from Tuesday to Wednesday in the college campus.

4.3 Simulation Prototypical for Projected Energy Management Strategy

This proposed expanded reenactment through a little period interim taking over six times to track. Load adjacent the executives utilize the corrupt prototypical of LP. This joined investigation objective needs every one of variables that remained additional trendy the previous aimed at every opportunity. The component permits this prototypical to confine distributed energy resources once its abundance stays away from accomplishing a plausible arrangement (age more noteworthy than request) and yet wind cut is stayed away from. Demand side management results in the decrease of misfortunes and appropriate use of the assets. This displaying indicates adaptability in age then demand stability that assesses the necessity seeing operative besides limit price. This change affect just scheduled diminished load on grounds that the main limitation on moved demand that influences a few moments of period stands aimed at every daytime, the base size that has been considered to make the gatherings. With the ultimate objective to survey the execution of the model and grasp the impact of moved interest side administration on the intrigue supply balance, this subsection displays a fundamental logical investigation for 24 hours. The intrigue is shown sinusoidal to give some sort of variety. The interest would not be adequate to achieve this outcome in a sensible application, external control techniques would be significant.

TABLE I. UTILIZATION OF POWER

Energy Supply	Estimated Volume in kW	Production Energy in kW
Photovoltaic	450	299.90
Wind Plant	25	15.00
Utility	250	86.01
Total	**670**	**400.91**
Total Energy Utilization in Demand		**399.51 kW**

The figure.3 demonstrates the all-out associated load, absolutely devoured control for the practical Linear Programming strategy for the timespan.

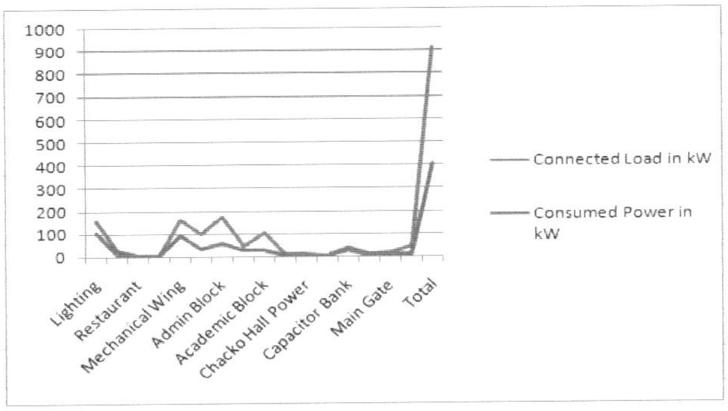

Figure 3. Total Consumed Energy

5. Conclusion

The proposed frameworks give the real-time frame checking just as control of interest side administration framework. It improves the presentation of association request to the point of dispersed vitality assets dissemination. The sun based and wind generation was planned to utilize MATLAB. The over-burden control which is created beginning sun based and wind power producing framework its moves to the circulated framework. The power usage of IEEE 14 transport framework has been resolved utilizing straight programming strategy in every one of the transports and requires is satisfied with the security framework for sun oriented and wind creating framework was actualized. This proposed framework fulfills the gathering of burden in the vitality the executive's framework and furthermore improves the association viability and limits the vitality misfortunes. The surplus power from circulated vitality assets can likewise be put away in the battery and it may be used by the interest while there is a heap of vitality.

6. References

[1] Smart grids, European Technology Platform for the Electricity Networks of the Future. [Online]. Available:http://www.smartgrids.eu.
[2] H. Pandic, J. M. Morales, A. J. conejo, and I. Kuzle 2013, "Offering model for a virtual power plant based on stochastic programming,"Appl.Energy, vol. 105, pp. 282-292.
[3] R. Jiang, J. Wang, and Y. Guan 2012, "Robust unit commitment through wind power and pumped storage hydro," IEEE Trans. Power Syst, vol. 27,no. 2, pp. 800–810.
[4] D. Bertsimas, E. Litvinov, X. A. Sun, J. Zhao, and T. Zheng 2013 , "Adaptive robust optimization for the security constrained unit commitment problem," IEEE Trans. Power Syst., vol. 28, no. 1, pp. 52–63.
[5] Diego Mejia – Giraldo2004 , "Adjustable Decisions for Reducing the Price of Robustness of Capacity Expansion Planning," IEEE Trans. Power Syst, vol.29, no. 4, pp. 1573–1582.
[6] Miroslav Baric and Francesco Borrelli 2003, "Decentralized Robust Control Invariance for a Network of Storage Devices," IEEE Trans. Power Syst, vol. 57, no. 4, pp. 1018–1024.
[7] L. Baringo and A. J. Conejo 2011, "Offering strategy via robust optimization, IEEE Trans. Power Syst, vol. 26, no. 3, pp. 1418–1425.
[8] Loganathan, K.Lakshmi and J.ArunVenkatesh 2016, "A Robust Energy Management System for Smart Grid", Springer in Advances in Intelligent Systems and Computing DOI 10.1007/978-981-10-1645-5_36 ISSN: 2194-5357, Vol. 467, No. 3, pp. 423-435
[9] N. Venkatesan, J. Solanki 2012, "Residential demand response model and impact on voltage profile and losses of an electric distribution network," Appl. Energy, vol. 96, pp. 84–91.
[10] S. Shao, M. Pipattanasomporn, and S. Rahman 2012, "Grid integration of electric vehicles and demand response with customer choice," IEEE Trans. Smart Grid, vol. 3, pp. 543–550.

[11] R. S. Ferreira, L. A. Barroso, and M. M. Carvalho 2012, "Demand response models with correlated price data: A robust optimization approach,"Appl. Energy, vol. 96, pp. 133–149.
[12] R.Sioshansi and W. Short 2009, "Evaluating the impacts of real-time pricing on the usage of wind generation," IEEE Trans. Power Syst, vol. 24, no.2, pp. 516–524.
[13] A. A. Thatte and L. Xie 2012, "Towards a unified operational value index of energy storage in smart grid environment," IEEE Trans. Smart Grid, vol. 3, no. 3, pp. 1418–1426.
[14] Danny Dudjianto, Marko Aunedi 2014, "Whole-Systems Assessment of the Value of Energy Storage in Low-Carbon Electricity Systems," IEEE Trans. Power Syst, vol. 5, no. 2, pp. 1098–1108.
[15] Y. He, B. Venkatesh, and L. Guan 2012, "Optimal scheduling for charging and discharging of electric vehicles," IEEE Trans. Smart Grid, vol. 3, no. 3, pp.1095 -1105.
[16] A. Khosravi, E. Mazloumi, S. Nahavandi, D. Creighton and J. Van Lint 2011, "A genetic algorithm-based method for improving quality of travel time prediction intervals ,"Transport. Res. Part C: Emerg. Technol, vol.19, no. 6,pp. 1364-1376.
[17] A. M. Foley, P. G. Leahy, , and E. J. McKeogh 2012, "Current methods and advances in forecasting of wind power generation," IEEE Trans. Renew. Energy, vol. 37, no. 1, pp. 1–8.

Design and fabrication of home automation

Vignesh T[1], Selvakumar D[2], Prasath C[3], ManikandanN[4] , Sri Viswanath R[5]

[1,3,4,5] Department of Mechatronics Engineering
[2] Department of Electronics and communication Engineering
Sri Krishna College of Engineering and technology, Coimbatore – 641008
vigneshtt@skcet.ac.in , selvakumard@skcet.ac.in, prasathc@skcet.ac.in,
manikandan@skcet.ac.in, 15eumt106@skcet.ac.in

Abstract. The objective of this project is to control all electronic device in the house, remotely and to automate some parts of the house like pipes, garage using Arduino or raspberry pi. The raspberry pi is a microprocessor which can be used for multiple purposes like getting input signal, based on the signal the output is calculated and sent to the actuators. Here we use raspberry pi board, few dc motors (this motor works on dc using permanent magnet), LED lights and a model of 3d model of the house. The prototype model build using 3d printer and electrical bulb, doors, fan and washing machine are demonstrated. This project will be a real-world model which will have the ability to perform task of operating devices, remotely. This project will help the people to keep an eye on their house when they are not home and help to on or off some of the appliances when they are away from home.

1. Introduction

Now a day's home automation system is the growing technology. People making their lives more advanced and easier, for this fast-moving era home automation is one of the supporting hands which makes the people's life more productive. Home automation system need the power supply and maintenance in many research going on the power supply for home automation and adaptive system used in many home [1]. Home automation system enables you to perform control operation on any device in the home [2]. For collecting the data from the home and it is stored in the cloud [3]. There are many controls used in the home automation like WIFI and Bluetooth. WIFI control used to monitor in home any were from the world. Bluetooth system used only a limited range [4]. In the home for monitoring the power supply adaptive control system used for the monitoring the power [5]. Recent development of mobile application the influencing to use the wireless technology to use the mobile phone to control the home. The monitoring of the home is done by using wireless communication dives like GSM, Bluetooth and Wi-Fi dives. [6]. the network system used to connect the smart phone to the cloud and control the various home appliance using the wireless technology. The control is like remote control of lighting, live video stream, security camera, real time monitoring. Based on the application many type of microprocessor were used in the system.

2. Methodology

We can connect different devices together through the internet and actually controlling them over the internet through a web/mobile application built using Ionic Framework. Different ways and services were explored including setting-up port forwarding on the router. Setting-up port forwarding will change the configuration of the router and expose one port of the private Local Area Network (LAN) to the public [7], which is considered a security vulnerability that has to be managed carefully. Some of the services that were explored and considered to allow the communication of the web/mobile application with the Raspberry Pi connecting different devices [8]. By using some electronic switches, we can control those devices.

3. Components

Electronics and electrical there are many components to make use in this project work. One of the main components is raspberry pi it is brain of the whole system. The controller board used in this no need any extra boards. As Arduino and other controller board need some hardware boards for WIFI and Bluetooth in the raspberry pi it has the inbuilt hard ware. It acts as the minicomputer CPU. It has a provision to connect HDMI, key board, monitor, power supply cables, and SD card support. SD Card used to store the program.

3.1 Mechanical

A.C. motors are used mostly here because house hold current is alternative. The mechanism used to open the door is hold open hinges or rack and pinion or roller figure 1. Here the locks can also be controlled by the system not the man door lock for safety purposes the below given door is called swing door.

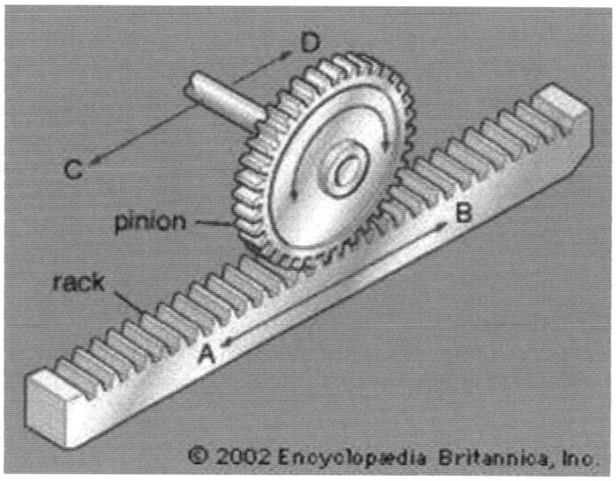

Figure 1: Rack and pinion mechanism

3.2 Raspberry pi

Using Raspberry Pi for IoT. Low priced Raspberry Pi is a credit card sized computer. It can be connected to a computer monitor or a television through HDMI cable and HDMI port found in the Raspberry Pi4. It has four USB ports, allowing the connection of different peripherals to it such as keyboard, mouse, memory stick or Wi-Fi dongle that allows its connection to wireless internet. As any computer, the Raspberry Pi needs an Operating System (OS) to run on it, to allow the interaction between it and the user. Different OS choices that can run on the Raspberry Pi are available such as Raspbian, Linux and a Windows 10 special version for the Raspberry Pi. The OS is downloaded to a micro SD card and mounted to the Raspberry Pi's SD memory reader. It also has Ethernet port that allows its connection to the internet through Ethernet cable from a router. Raspbian or some other operating system will be installed on a micro SD card that was mounted to the Raspberry Pi that allowed writing Python scripts. The needed hardware, such as sensors and LEDs, were connected, using jumper wires, to GPIO pins and the python script is written in the pi, which allowed controlling and monitoring the connected hardware for testing. The figure 2 show the image of the raspberry pi.

Figure 2. Raspberry pi 3

IOP Conf. Series: Materials Science and Engineering **623** (2019) 012002 doi:10.1088/1757-899X/623/1/012002

3.3 Android application

Using Ionic framework for web/mobile application Ionic is a free and open source SDK, built with SASS and optimized for Angular JS. It offers a library of mobile-optimized HTML, CSS and JS tools and components for developing interactive platform applications are the applications that run on different platforms such as web browsers, IOS and Android. It was used in this project to build web/mobile application to monitor and control the home model mentioned before

4 Calculations

DC motor will be used here. The simple calculation of motor is given below

4.1 DC Motor

Speed and voltage are described in the below

30 RPM, 12 VOLT, 18 WATT

MOTOR TORQUE

P = 18 kW, N = 30 rpm.

Torque = (P X 60) / (2 X 3.14 X N)

Torque = 5.72 Nm

Torque = 5.72×10^3 Nmm

4.2 Rack and Pinion Calculation

Pressure angle of rack and pinion (p) = 20° full depth system
Pitch of the rack and pinion (P) = 10 mm
Addendum (a) = 1 / P
Addendum (a) = 1 / 10 = 0.1 mm
Dedendum (D) = 1.25 / P
= 1.25 / 10
= 0.125 mm

$D_p = N_p / P_d$

No of teeth on gear (Z) $= \pi D / P$
$= \pi \times 90 / 10$
= 28 teeth
$D_p = 28 / 10$

= 2.8 mm

$\sigma = F_t / (b_a. m. Y)$

Y = Lewis form factor = 0.352

Module of gear = m = $1 / D_p$
= 1 / 2.8

= 0.357

Gear face width (b_a) = 10 mm

Gear bending stress = 110 Mpa

110 x 10^3 = F_t/ 10 x 0.357 x 0.352

F_t = 136 x $10^{\wedge 3}$N

W = Ft x v / 1000
Gear velocity = v = π x D x N / 60 x 1000
Where
D – 90mm
N = 250 rpm
= 1.178 m / s
W = F_t x v / 1000
= 136 x $10^{\wedge 3}$ x 1.178 / 1000 = 152 watts

4.3 Model

The design given below is the design of the model house which we are going to automate. There are two bedrooms, one living room the doors of the living room and other rooms can be controlled the mechanism given above the swing door. Fig 3 show the building layout

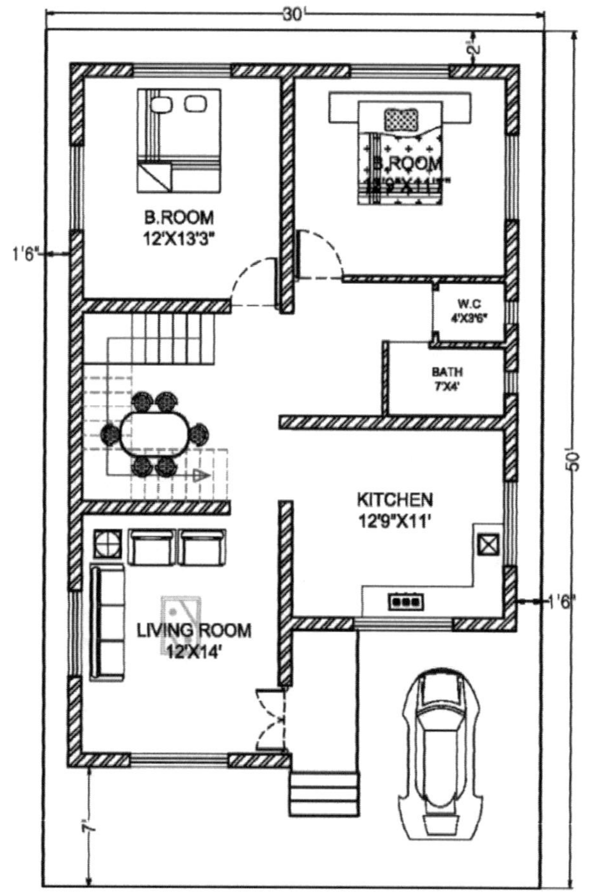

Figure .3 Building layout

The proposed system of architecture is given below in the figure 4. The controller connected to the output and the input. Based on the customer input the controller gives the output

Type of system based on network

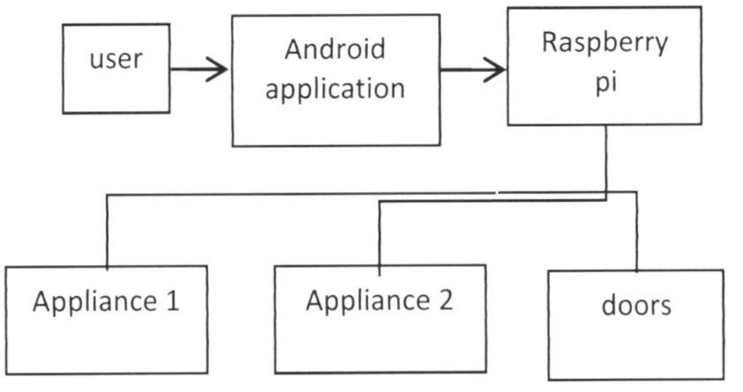

Figure4. Proposed System of Home Automation

Raspberry pi comes within built communication system which we can connect to our android or IOS phone by having an android or ios app in our we can control the appliances and the doors of our home. By using this technology can also control the temperature of our house. The temperature can be controlled using thermostat

4.4 Result and discussion

In this system we can control six devices and two doors via our smart phones. This device uses different method to control devices and this device has more spec than other system. By using raspberry pi we connect devices via internet.

Table 1: Type of connecting method

System	Cost	Speed	Real time
Bluetooth	Low	High	Yes
Voice Recognition	Low	High	Yes
ZigBee	Low	High	Yes
GSM	High	Slow	No
Internet, Wi-Fi	High	Slow	Yes
EnOcean	Low	High	Yes

The above given table refers to the many type of connection method. Bluetooth cannot be connected to the system more than 1 km. voice recognition won't work some times. ZigBee is a website-based method which is not safe. Internet is used in the system because connectivity and other

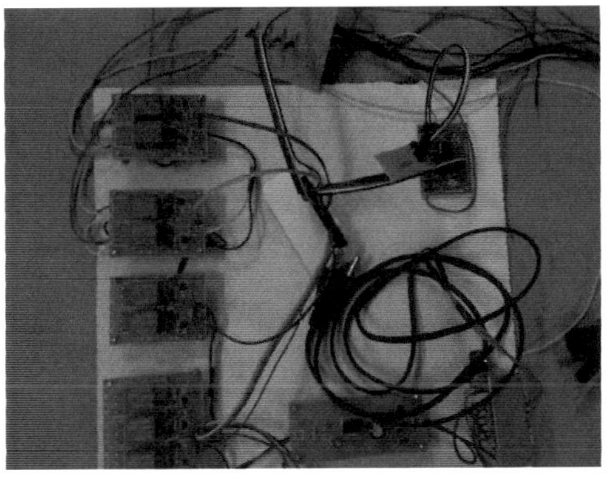

Figure 5 Electrical Components Assembly

The prototype model of home automation system is shown in figure 5. It shows the electrical part assembly. Raspberry pi used for control other relay module used for control the motor. Figure 6 shows the complete prototype model which used in the project. The model which demonstrated is only Proof of concept it is implemented in the real time without any changes

Figure 6. Prototype model of door and window system automation

In this paper discussed the home automation system. In this system the user can operate the entire home with smart phone in his hand via android or ios application the user can interface with the appliances in the home. This system can be more helpful in the modern world. The user is completely monitor their home in the mobile application. The user is controlled using the mobile app. The future work of this paper is convert into product.

5 References

[1] T. Y. Yang, C. S. Yang, and T. W. Sung, "A Dynamic Distributed Energy Management Algorithm of Home Sensor Network for Home Automation System," Proc. - 2016 3rd Int. Conf. Comput. Meas. Control Sens. Network, C. 2016, pp. 174–177, 2017.

[2] S. Majumder et al., "Smart homes for elderly healthcare—Recent advances and research challenges," Sensors (Switzerland), vol. 17, no. 11, 2017.

[3] S. Ivanović, S. Milivojša, and T. Erić, "Collection and Analysis of System Usage Data in Smart Home Automation Systems," 2017 IEEE 7th Int. Conf. Consum. Electron. - Berlin Collect., no. 689947, pp. 65–66, 2020.

[4] et al., "Smart Home Automation System using Bluetooth and Infrared," Int. J. Adv. Res. Comput. Sci. Softw. Eng., vol. 7, no. 4, pp. 269–273, 2017.

[5] T. Mehrabi, A. S. Fung, and K. Raahemifar, "Optimization of home automation systems based on human motion and behaviour," Can. Conf. Electr. Comput. Eng., pp. 15–19, 2014.

[6] U. Rawat, A. Bose, A. Vaish, and S. Gulia, "Wireless Home Automation System 1#," Int. J. Innov. Res. Electr. Electron. Instrum. Control Eng., vol. 5, no. 2, pp. 1–4, 2018.

[7] G. V. Vivek and M. P. Sunil, "Enabling IOT services using WIFI - ZigBee gateway for a home automation system," Proc. 2015 IEEE Int. Conf. Res. Comput. Intell. Commun. Networks, ICRCICN 2015, pp. 77–80, 2016.

[8] M. M. A. Jamil and M. S. Ahmad, "A pilot study: Development of home automation system via raspberry Pi," Proc. - 2015 2nd Int. Conf. Biomed. Eng. ICoBE 2015, no. March, pp. 30–31, 2015.

IOP Conf. Series: Materials Science and Engineering **623** (2019) 012003 doi:10.1088/1757-899X/623/1/012003

Dual Mode Operation of 8/6 SRM with Optimum Angle Operation in Hybrid Electric Vehicles using FLC based Controller

G.Radhakrishnan
Assistant Professor
Department of EEE
Sri Krishna College of Engineering & Technology, Coimbatore
Tamilnadu, India

Abstract. An optimum operating period of an 8/6 Switched Reluctance Motor (SRM) based on different load condition is analysed in this paper. The system is designed to analyse the speed control of machine in motoring mode and voltage control mode is carried out in this paper. Dual Mode switched reluctance motor is tested with the fuzzy logic controller, Mamdani method is used in fuzzy based system. In this paper the main objective is to concentrate on modelling the machine to maximise the average torque with minimum torque ripple in the optimised operating region with different loading conditions in dual mode region. This method also found to have increased efficiency, with the closed loop controlling action the system is robust and simple in construction. The optimised designed system has less manufacturing cost due to simple construction. The simulation is carried out using MATLAB/Simulink. '

1. Introduction

.

The development of SRM machines starts at the year of early 1920's, as the conventional system. [2]. The major drawback of the machine is due to the further advancements due to the losses faced in electromechanical energy conversion and power electronic switches. After the significant change in the field of power electronics, the SRM machines were reintroduced during the year of 1960s. The switched reluctance motor has many commercial applications in adjustable speed drives recognized to its exclusive mechanical structure and also its simple power electronic structures. The intrinsic simplicity and ruggedness of the machine contributes to its superiority over other electric machines [3].

The numerous advantages posed by SRM in its performance, low cost, high speed and acceleration capability, is shadowed by its disadvantages caused by the non-linearity such as torque ripple, acoustic noise and position sensor. SRM machine can able to operate, when the machine is integrated with power converter, controller and rotor position sensor. The performance simulation of reluctance machine for different control strategies with the experimental data's are validated. The designed SRM need to be detect the rotor position with respect to its stator using a position sensor connected to the motor shaft. A position sensor, such as an encoder, a resolver, or a Hall sensor is normally used which leads to increase in cost and size of machine with reduced reliability. As a result, several sensor-less drive techniques have been reported from the results [4]. SRM machines are

IOP Conf. Series: Materials Science and Engineering **623** (2019) 012003 doi:10.1088/1757-899X/623/1/012003

versatile with ordinary looking structure, and are hence easy to manufacture leading to their low cost [4]. This is mainly the result of the power electronic drives used here. The combination with the power electronic switches is one of the unique factors of this machine [5]. The modelling of 8/6 pole SRM machine here is done using the electro- mechanical and electromagnetic equations [6].

Trials have been done with different controllers to minimize the torque ripple, complexity in control, and non-linearity [8]. These controllers also determine the optimum turn on and turn off angle with other parameters like reference current and size of hysteresis band [9]. The optimising of the turn on and turn off angle not only reduces torque ripples but also increases average torque [10]. The turn on and turn off operation are each done by separate controllers. But, more importance is given to turn on controllers for as it minimises the torque ripple [11] and increase in the torque value is done by turn off angle control.

Sliding mode controller is also used for speed control in some cases [16] for eliminating the drive nonlinearity and it operates by setting up a state space [12]. Position sensing in SRM is a compulsion because the torque generation in SRM is subject to rotor position [13]. But, it has its own drawbacks with using external sensors like resolvers and optical sensors. Drawbacks like high cost, losses due to noise and problems with mechanical mounting usually minimises their importance in market for their accuracy and efficiency [14]. Using sensor-less control with optimised operating angle eliminates this disadvantage and increases the flexibility of SRM [12]. In this paper, the dual mode operation of this SRM eliminates the battery storage problem and increases the efficiency of the whole system with regenerative power. The closed loop controlling is done using fuzzy logic controller. The work is simulated using MATLAB/SIMULINK and results are obtained [7].

2. Modelling of SRM For Dual Mode Operation

The SRM has to be modelled to avoid discrepancies caused by non-linearity due to double saliency of the machine, leading to non-linear torque and current. Therefore, mathematical modelling is done relieving us of the tedious reference frame theory. To simplify it even more, leakage flux, magnetic saturation characteristics are neglected. The Fig 1 below shows the rotation of rotor with respect to stator for an angle of 15 degrees.

Voltage equation of motoring mode for SRM is given by

$$V = I_a R_a + e + L\frac{di}{dt} \tag{1}$$

Equating e to a function of inductance,

$$e = \omega_m . i\frac{dL}{d\theta} \tag{2}$$

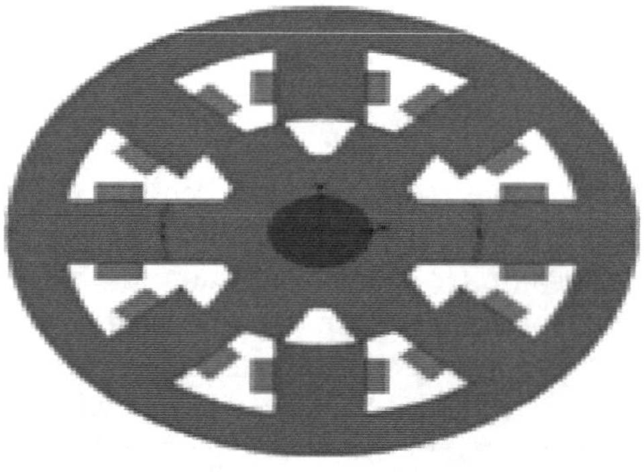

Fig 1. 8/6 Switched Reluctance Motor

$$V = i_a R_a + \omega_m i \frac{dL}{d\vartheta} + L\frac{di}{dt} \qquad (3)$$

Power absorbed from the SRM is given by

$$Vi = i_a^2 R_a + \omega_m i^2 \frac{dL}{d\vartheta} + Li\frac{di}{dt} \qquad (4)$$

$$T = \frac{P}{\omega_m} = \frac{d\left(\frac{1}{2}Li^2\right)}{dt} \qquad (5)$$

But, $\frac{d}{dt}\left(\frac{1}{2}Li^2\right) = \frac{1}{2}i^2\frac{dL}{d\vartheta} + Li\frac{di}{dt}$

Therefore, the torque equation is given by,

$$T = \frac{1}{2}i^2\frac{dL}{d\vartheta} \qquad (6)$$

Voltage equation for generating mode of the motor is given by

$$V = i_a R_a + L\frac{di}{dt} - \omega_m i\frac{dL}{d\vartheta} \qquad (7)$$

And the torque produced will be

$$T = -\left(\frac{1}{2}i^2\frac{dL}{d\vartheta}\right) \qquad (8)$$

The equations (1) to (8) give the modelling equation for the switched reluctance motor. Where, the negative sign is contributed to the falling profile of inductance (negative region) with respect to the change in rotor position

3. Dual Mode Operation of SRM

The raising region of the inductance profile of the SRM constitutes the motoring mode and the falling region constitutes the generating mode. In motoring mode, speed is the controlled parameter with the excitation given in the raising region. In generating mode, generally the output voltage is the controlled parameter with the excitation given in the falling region. If the excitation is in the optimum region, then the performance of the dual mode machine is better. Because of the non-linearity observed in the operating angle selection for different loading condition, the controller opted should be very rigid to obtain stable operation. The PID controller is replaced by fuzzy logic controller for rapid operation and faster simulation speed.

4. Optimum Angle Selection in Motoring Mode

The figure 2 shown below gives the current path in a single inductance curve showing the motoring mode operation. The optimum value of turn-on angle in the motoring mode is taken after eliminating the eliminating the resistance due to the presence of inductive coils and fringing effect

$$\vartheta_{on}^M = \vartheta_{1u} - \vartheta_{01}^M = \vartheta_{1u} - \frac{L_u . i_{ref} . \omega_r}{V_{dc}} \qquad (9)$$

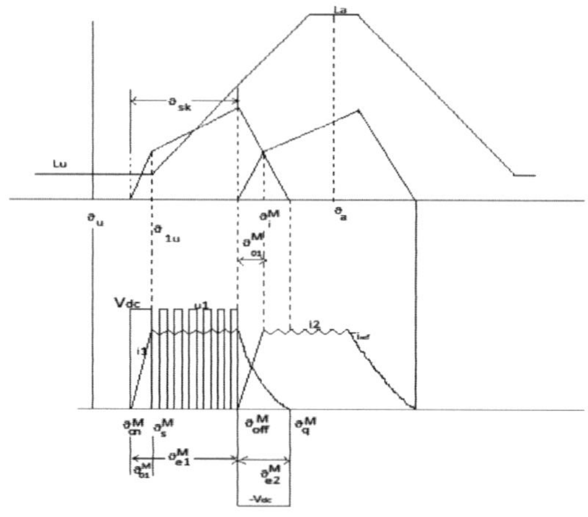

Fig 2.motoring mode operation

The optimal turn-off angle in motoring mode is given by

$$\theta_{off}^{M} = \vartheta_{1u} + (2\vartheta_{sk} - \vartheta_{e2}^{M})\left[1 - \frac{\vartheta_{01}^{M}}{\vartheta_{e2}^{M}}\right] \qquad (10)$$

Where the equation (9) and (10) gives the optimum turn on and turn off angle estimation for the motoring mode.

5. SRM Operation Dual Mode

The raising region of the inductance profile of the SRM constitutes the motoring mode and the falling region constitutes the generating mode. In motoring mode, speed is the controlled parameter with the excitation given in the raising region. In generating mode, generally the output voltage is the controlled parameter with the excitation given in the falling region. If the excitation is in the optimum region, then the performance of the dual mode machine is better. Because of the non-linearity observed in the operating angle selection for different resistance due to the presence of inductive coils and fringing effect

$$\vartheta_{on}^{M} = \vartheta_{1u} - \vartheta_{01}^{M} = \vartheta_{1u} - \frac{L_u . i_{ref} . \omega_r}{V_{dc}} \qquad (9)$$

The optimal turn-off angle in motoring mode is given by

$$\theta_{off}^{M} = \vartheta_{1u} + (2\vartheta_{sk} - \vartheta_{e2}^{M})\left[1 - \frac{\vartheta_{01}^{M}}{\vartheta_{e2}^{M}}\right] \qquad (10)$$

Where the equation (9) and (10) gives the optimum turn on and turn off angle estimation for the motoring mode.

6. Optimum Angle Selection in Generating Mode

IOP Conf. Series: Materials Science and Engineering **623** (2019) 012003 doi:10.1088/1757-899X/623/1/012003

Similarly in a switched reluctance motor, for generating mode, the optimum turn-on angle is given by

$$\vartheta_{on}^{G} = \vartheta_{q}^{G} - 2\left[\vartheta_{01}^{G} + \vartheta_{sk}\left(1 - \frac{\vartheta_{01}^{G}}{\vartheta_{e1}^{G}}\right)\right] \tag{11}$$

And, turn-off angle given by

$$\vartheta_{off}^{opt} = \vartheta_{q}^{G} - \vartheta_{01}^{G} \tag{12}$$

Where the equation (11) and (12) gives the optimum turn on and turn off angle estimation for the generating mode. The figure 3 shows the current in the generating mode operation of a switched reluctance motor.

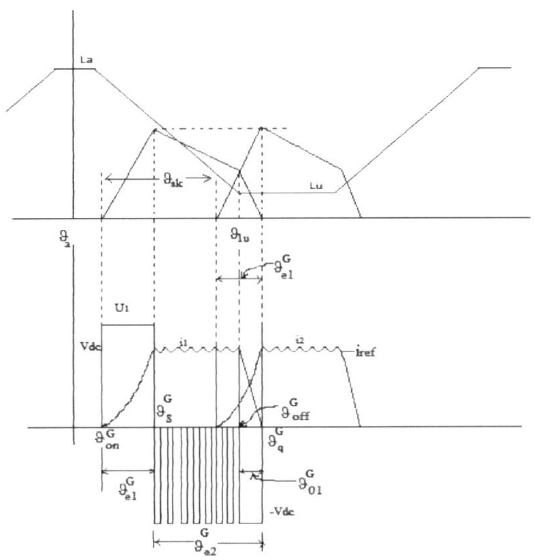

Fig 3. Generating mode operation

The operating angle of the SRM is equal to the stroke angle of the motor. This angle is always within the boundary limits. When it decreases below the given boundary limit, the torque obtained is reduced due to insufficient conduction period and when it crosses above the boundary limit, without producing torque winding losses only occurs. When it reaches the opposite slope then torque reversal occurs. It reduces the average torque and also stress occurs in the rotor shaft. To prevent this, the optimum region is always selected to be within the conduction region of 8/6 SRM.

7. Conduction Angle Estimation

From the figure 2 & 3, it is clear that the inductance changes for every change in rotor position with respect to stator. The stroke angle corresponds to the aligned position and average torque production is maximum here. When the conduction angle is less than the stroke angle, the torque production is minimised. The conduction angle decreases with increasing load. Care is taken to ensure optimum overlapping period where average torque increases and torque ripple decreases. Dual mode operation depends upon the inductance profile slope. The operating angle is advanced by an angle of

IOP Conf. Series: Materials Science and Engineering **623** (2019) 012003 doi:10.1088/1757-899X/623/1/012003

advance to obtain optimised results in the torque plot. Figure 4 shows the inductance profile of a dual mode region of switched reluctance motor.

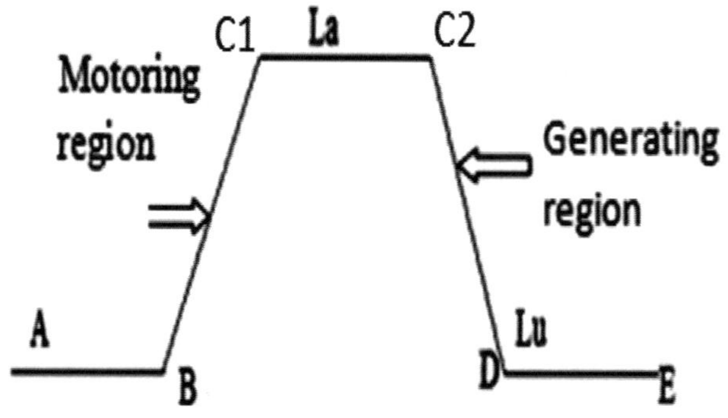

Fig 4. Inductance profile for 8/6 SRM

AB & DE - Minimum inductance region

BC1- Raising inductance region

C1C2 – Maximum inductance region

C2D- falling inductance region

$$\text{Advance angle, } \vartheta_{adv} = \frac{L_{min} I_r}{V} \quad (13)$$

The advance angle given by equation (13) is given to the turn-on angle and the turn-off angle. The turn-on angle given to energise the winding is given before the rising profile of inductance and turn-off angle to de-energise the winding is given with a period interval of fifteen degrees from the turn-on angle. Similarly, the same is done for the generating mode. This turn-on and turn-off angles are selected for obtaining better Results.

8. Main Constraints

The main constraints to be introduced are that the turn off angle should be fifteen degrees from the turn on angle.

$$\theta off - \theta on = \frac{Ns-Nr}{NrNs} * 360\ degree \quad (14)$$

Where,

Ton$\geq$ 0 degree, Toff$\leq$ 30 degree

9. Fuzzy Logic Controller

The main objective of controller is to achieve a high performance control, thereby maintain the output voltage of the system. In a fuzzy logic controller, the switching states are selected based on the predefined fuzzy variables. Here, speed control of the SR machine is done by taking into account the error in speed (input1) and the change in speed (input 2). Fuzzy logic controllers are advantageous compared to the conventional controllers when dealing with more complex processes. The difficulties in modelling non-linear models with uncertainties are generally a right choice for fuzzy controllers because of its high adaptive ness and robustness in performance under varying parameters and disturbed load.

The rules of the fuzzy system are given in the following table01.

Table.01. Fuzzy rules

Input 2 / Input 1	NL	NM	NS	Z	PS	PM	PL
NL	VB	VB	VB	B	SB	S	Z
NM	VB	VB	B	B	MB	S	VS
NS	VB	MB	B	VB	VS	S	VS
Z	S	SB	MB	Z	MB	SB	S
PS	VS	S	VS	VB	B	MB	VB
PM	VS	S	MB	B	B	VB	VB
PL	Z	S	SB	B	VB	VB	VB

The linguistic variables in table01 are

NL- Negative Large PS- Positive Small

NM- Negative Medium PM- Positive Medium

NS- Negative Small PL- Positive Large

Z- Zero VB- Very Big

MB- Medium Big SB- Small Big

VS- Very Small S- Small B- Big

IOP Conf. Series: Materials Science and Engineering **623** (2019) 012003 doi:10.1088/1757-899X/623/1/012003

Triangular membership functions are used in the fuzzy interference system (FIS) file. The output speed from the fuzzy file is given as input to control the speed of the SRM.

10. Simulation Results

Figure 5 shows the speed variation in the generating mode while changing the load torque. Figure 6 and figure 10 shows the torque variation in the generating and motoring mode respectively during load change. Figure 7 gives the gate pulse given to the motor during the motoring mode.

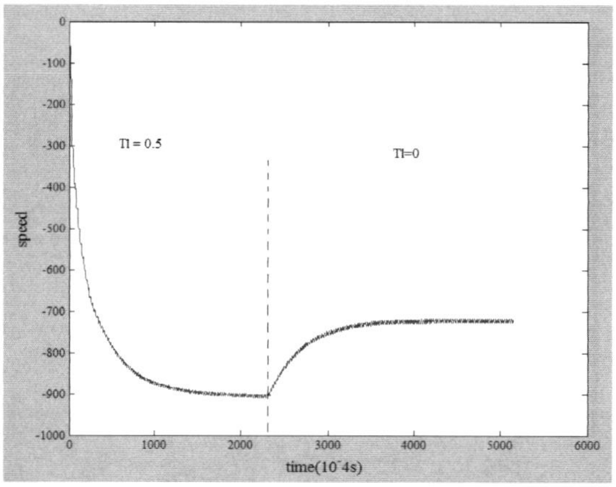

Fig 5. Speed under varying load torque for generating mode

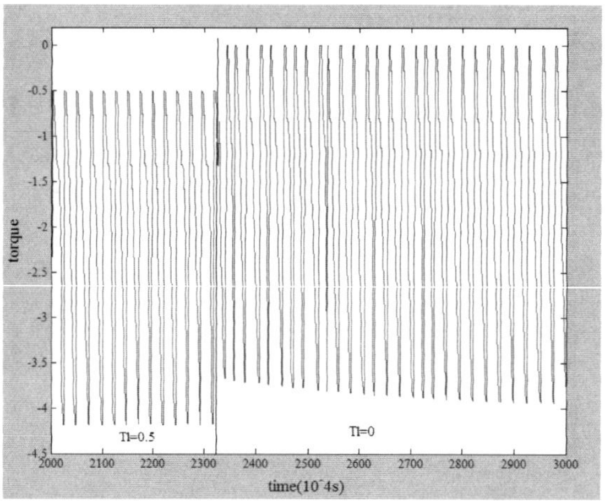

Fig 6. Generator torque under varying load torque

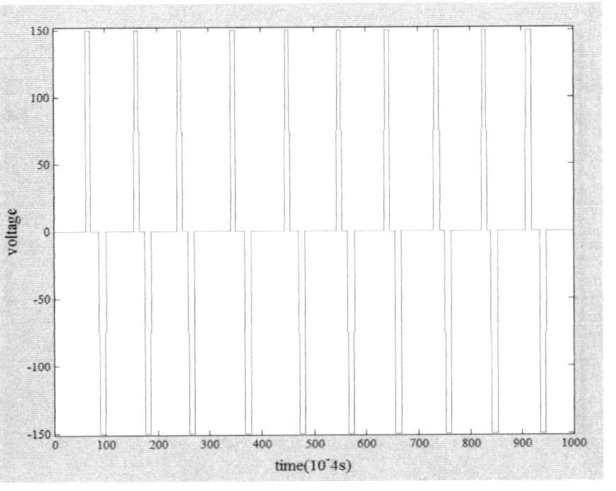

Fig 7. Gate pulse given to the SRM in motoring mode

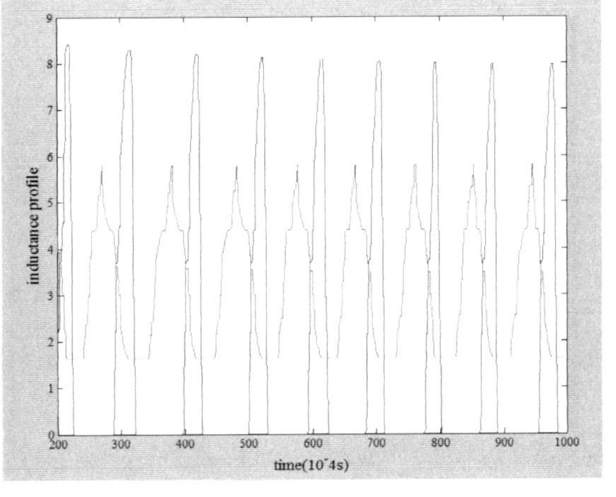

Fig 8. Inductance profile under varying load torque for generating mode

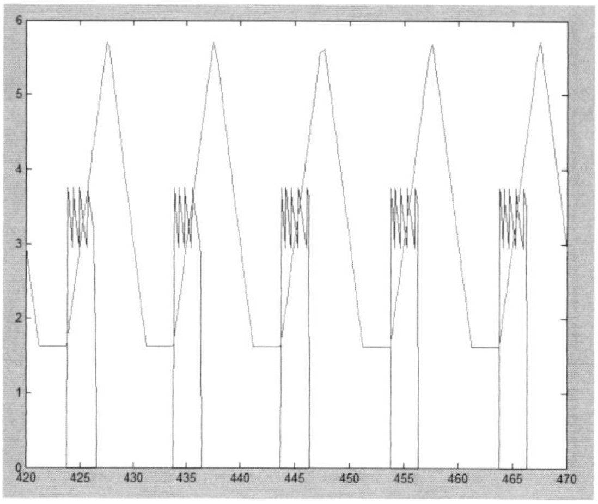

Fig 9. Inductance profile under varying load torque for motoring mode

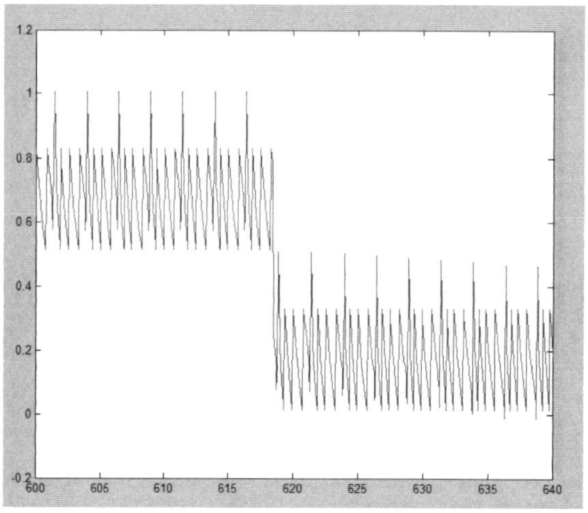

Fig 10. Motoring torque under varying load torque

Figure 8 and figure 9 shows the inductance profile of the SRM in generating and motoring mode respectively.

11. Conclusion

The dual mode operation of a Switched Reluctance Motor is implemented using MATLAB/SIMULINK platform. FLC controller is used for increased efficiency of the SRM machine. The average torque is obtained and is found to be of increased value, and the torque ripple content is obtained and found to be reduced. The results obtained from the simulation are plotted in both the mode of operations.

REFERENCE

[1] Hany M.Hasanien, S.M.Muyeen, Speed control of grid-connected switched reluctance generator driven by variable speed wind turbine using adaptive neural network controller, Electric Power Systems Research84(2012)206_213.
[2] Mukhtar Ahmad, High performance AC Drives, Springer, 2010.

[3] TJE Miller, Electronic Control of Switched Reluctance Machines, Newnes Power Engg Series, 2011.

[4]K.T.Chau and J.H.Chen, Modelling, Analysis, and Experimentation of Chaos in a Switched Reluctance Drive System, IEEE Trans on circuits and systems, VOL.50, NO.5, May 2003.

[5] Iordanis Kioskeridis and Christos Mademlis, A Unified Approach for Four-Quadrant Optimal Controlled Switched Reluctance Machine Drives With Smooth Transition Between Control Operations, IEEE Trans on Power Electronics, VOL.24,NO.1,Jan 2009.

[6] V.Vasan Prabhu, Dr.V.Rajini, Design of GT-FLC Speed Controller and Position Sensorless Control Using ANN for 8/6 SRM, Journal of Electrical Engineering.

[7] F.Soares, P.J.Costa Branco, simulation of a 6:4 switched reluctance motor based on MATLAB/Simulink environment, IEEE Trans on Aerospace and electronic systems, VOL.37, NO.#, Jul.2001.

[8] Iqbal Husain, Minimisation of torque ripple in SRM drives, IEEE Trans on Industrial Electronics, VOL.49,no.1,Feb 2002.

[9] C.Mademlis, I.Kioskeridis, optimizing Performance in current controlled switched reluctance generators, IEEE TRANS on energy Convers, VOL.20, no.3, sep 2005.

[10] C.Mademlis, I.Kioskeridis, Performance Optimization in switched reluctance motor drives with online commutation angle control, IEEE Trans on energy convers, VOL.18, no.3, Sep 2003.

[11] V.Vasan Prabhu, V.Rajini, Switched reluctance motor/generator with optimum turn-on and turn-off controller for hybrid electric vehicle application,Elixir Elec Engg.44(2012) 7527-7530., mar 2012.

[12] Syed A.Hossain, four-Quadrant and Zero-Speed Sensorless Control of a Switched Reluctance Motor, IEEE Trans on industry applications, VOL.39, NO.5, Sep/Oct 2003.

[13] R.Krishnan, Switched reluctance motor drives:modelling, simulation, analysis, design and applications, Industrial Electronics Series, CRC Press, Boco Raton:2001.

[14] Gabriel Gallegos –Lopez, A new sensorless method for Switched Reluctance Motor Drives, IEEE Trans on Industry Applications, VOL.34, NO.4, Jul/Aug 1998.

[15] Wen Ding, comparative study on dual-channel Switched Reluctance Generator Performances Under Single-Channel and Dual-Channel Operation Modes, IEEE Trans on energy conversion, Vol.27, NO.3, Sept 2012.

[16] V.Vasan Prabhu, Dr.V.Rajini, M.Balaji, V.Prabhu, comparison of PSO and GA based operating angle optimisation of 8/6 Switched Reluctance Motor with robust speed controller, J.Electr Eng Technol, Vol.10, NO.2: 742-51, 2015 .

IOP Conf. Series: Materials Science and Engineering **623** (2019) 012004 doi:10.1088/1757-899X/623/1/012004

Microinverter for wireless charging

A Maideen Abdhulkader Jeylani[1], J Kanakaraj[2], A Mahaboob Subahani[3]

[1], AssistantProfessor, Dept of EEE, Sri Krishna College of Engineering and Technology, Coimbatore
[2] Professor, Dept of EEE, PSG college of Technology, Coimbatore
[3]Assistant Professor, Dept of EEE, PSG college of Technology, Coimbatore
jeylania@skcet.ac.in,[1]jeylani7862@gmail.com

Abstract. This paper presents a multi stages grid connecting single phase micro-inverter for photo voltaic utilisation. Conventional single switch flyback inverter suffers from high voltage transients across switch, which leads to a significant reduction in efficiency, the stress on the device causes an increase in cooling arrangements and also underutilization of the device as far as the ratings are concerned. The proposed topology will have an additional switch and two clamped diodes at the primary side so as to have a reduction in the switch stress which increases the efficiency. Converter is operated in Discontinuous conduction mode to have a simple control and to avoid the stability issues. Simulation results authenticate the concert of the proposed method.

1. Introduction

With the increasing demands for the renewable energy sources, solar energy has become promising to generate electricity in a bulk amount. Most of the countries have already getting majority of their energy from the solar power plants. In the countries like India (where the sun's presence is there throughout the year) this can be more. Due to issues such as initialization costs, reliability and payback period, there was less interest for solar energy utilization. But due to advancements in the photo-voltaic technology and semiconductor devices, most of the researchers have started concentrating on converters for PV applications. So solar inverters have become an essential topic for the research. There are 3 types of solar inverters. They are 1. Central inverters 2. String inverters 3. Micro inverters. Of all, Micro-inverters have become the emerging topic for its unique qualities such as efficiency, simplicity and maintenance. Micro inverter means each solar module consists of an inverter as shown in Fig 1. Conventional micro inverters suffer from lot of problems such as high voltage transients across the switch, efficiency, complex control techniques [1]. So, there is a need for the development micro inverter which provides solution for the above mentioned issues. Fly-back converter topology has been proposed by most of the authors because of its simplicity and less component count .It was also proved that interleaved topology reduces the conduction losses of the device. Due to the occurrence of resonance among the transformer outflow inductance and MOSFET output capacitance, large voltage spikes get generated at the switch which leads to the high switching losses. And also to withstand these transients, switch should have higher voltage rating. which increases the on state resistance of the Mosfet. And also high switching frequency is also the cause of switching losses though it can reduce the size of the transformer significantly.

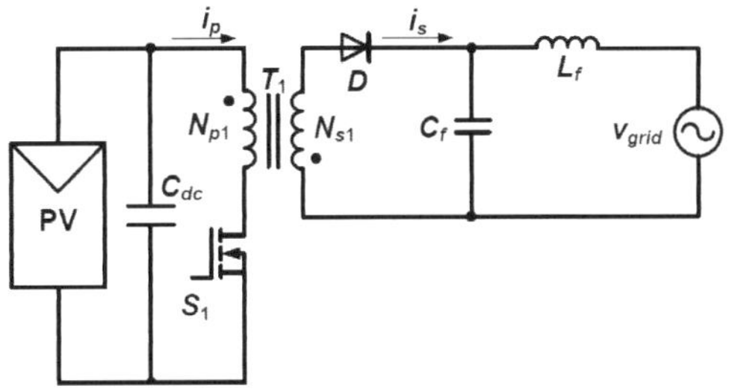

Figure 1: Conventional flyback inverter

Conventionally, Snubber circuits have been used to eliminate the problem of the voltage thorns across the flyback switch [3], nevertheless the problem is the dissipation of power in the resistor. Several other techniques have been proposed to eliminate this problem such as active clamp circuits and zero current transition methods but with these methods, the circuit complexity is getting increased. So, to overcome this problem, the two switch flyback inverter topology is much simpler .In this paper, it has been clearly given how this topology is superior compared to the conventional topology [6]. The purpose of this paper is to find a simple solution for voltage spikes across the flyback switch in a single phase grid associated flyback micro-inverters. Simulations are given to validate the theoretical analysis.

2. System Description

Fig 2 shows the circuit topology that is going to be analyzed. Single phase grid tied two stage microinverter mainly consists of phase-1 converter, phase 2 converter, single phase bridge inverter, the second order passive filer and a controller to provide pulses to the Mosfets. The S1, S2 are the main power switches and D1,D2 are the clamping diodes of the phase 1 converter and $S1^1$,$S2^1$ are the main power switches and $D1^1$,$D2^1$ are the clamping diodes for the phase 2 converter.L_{lk1},$L_{lk2}$are the leakage inductances for the converter1 and converter2 respectively.L_{m1},L_{m2} are the magnetizing inductances for the converter 1 and converter2 respectively.D3 and D4 are the flyback secondary diodes for converter 1,converter 2 respectively. Sa1-Sa4 represents the switches for the grid connected inverter .

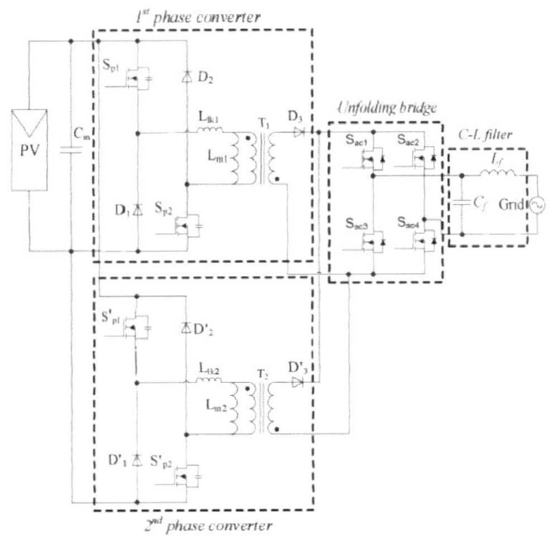

Figure 2: Interleaved microinverter

And it is also observed that as the turns ratio is smaller, the primary current stress and secondary voltage stress are more and if it is high, primary voltage stress and secondary current stress are high.

Please follow these instructions as carefully as possible so all articles within a conference have the same style to the title page. This paragraph follows a section title so it should not be indented.

3. Control Strategy

The benefits of two-phase process in interweaved method is current distribute among two phases which decreases conduction and switching losses [5]. Dipping the driving loss of the power MOSFETS and the transformer core loss are the best topographies of one phase process. So by combining these two phases by using interleaved converter, the efficiency is much more. It can be observed that the output power shape is a squared sine wave.

$$P_{out} = 2P\sin(\omega t)$$

Where P is the average output power. Fig.3 showing the working region of mono phase and poly phase operation during a bisecting line duration. As it is shown poly phase process is employed when output power is higher and mono phase processes employed when Po is under a convinced level as shown at middle line in Fig 3. It could be noticed that mono and poly phase operation is done suddenly. Phase locked loop is to detect the phase angle, magnitude and frequency of grid voltage precisely and quickly. The output of the MPPTcontrol is i_{out} which will be adjusting the reference current $i_{ref.}$ The reference signals i_{ref1} and i_{ref2} are used to have the current shape in a control technique as described by the following equations.

$$i_{ref1} = 2\sin(\omega t)\sqrt{(po)/(Lp * fs)} \quad (t<tc1, t>tc2)$$

$$i_{ref1} = \sin(\omega t)\sqrt{(po)/(Lp * fs)} \quad (tc1<t<tc2)$$

$$i_{ref2} = \sin(\omega t)\sqrt{(2po)/(Lp * fs)} \quad (tc1<t<tc2)$$

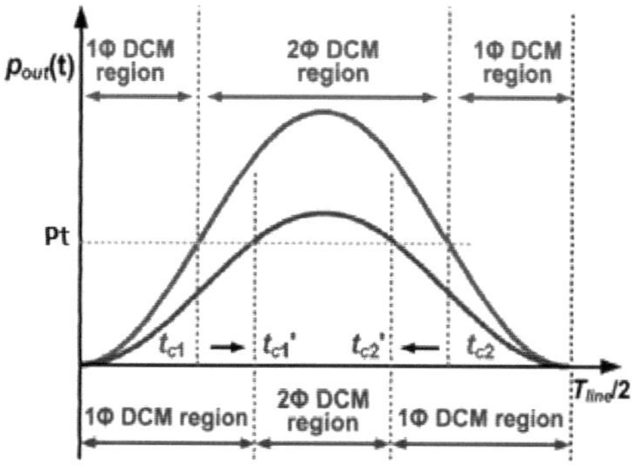

Figure 3: Operating regions of the interleaved microinverter

Best boundary condition of mono phase and poly phase operation is considered Pt=100 W. As a consequence, t_{c1} and t_{c2} can be attained as

$$t_{c1} = (1/\omega) * \sin^{-1}(\sqrt{(50/po)})$$

$$t_{c2} = (1/\omega) * (\pi - \sin^{-1}(\sqrt{(50/po)}))$$

Under low load condition, phase 1 discontinuous conduction modes of operation decreases the driving loss of the power MOSFETs and transformer core loss. As the power level rises, the conduction losses and turn-off loss of power MOSFETs, diodes and the loss of transformer become the dominant losses. Whereas 2 phase discontinuous conduction mode of operation shares the current between two interleaved phases. So the basic interest is to minimize the leading losses depends on load ailment.

4. Simulation data used in the proposed system

Data	Value	Unit
Switching frequency	50	KHz
Nominal power	200	W
Transformer turns ratio	0.5	
PV module voltage at MPPT	50	V
PV module current at MPPT	4.1	A
Leakage inductance	0.01	mH
Mutual inductance	0.980	mH
Filter inductance	1.3	mH
Filter capacitance	.33	Uf

5. Input Capacitor Selection

For the mono-stage grid-connected micro-inverter, the MPPT delivers the continual output power from the PV panel *Ppv*, whereas the power transported to the grid *pout* (*t*) is a vivacious waveform as shown in (2).Normally, the electrolytic capacitor is to measure the unhinge of the input and output power. When *Ppv* is excess to *pout* (*t*), the rest of the power is stored into the decoupling capacitor. On the contrary, when *Ppv* is smaller than *pout* (*t*), the decoupling capacitor delivers the power to the output. For *Ppv*=200 W, $\omega=2\pi f=100\pi$, *Vdc*=50 V and ΔV=2 V, the essential input capacitance is *Cdc*=6.3 mF. Four 1.7 mF electrolytic capacitors are paralleled with lower ESR.

P and O technique has been utilized to provide the maximum power point tracking.

6. Simulation Results

To support academic investigation, Matlab simulation analysis has been done for the ratings taken. Conventional flyback inverter switch voltage can be observed from Fig 4.

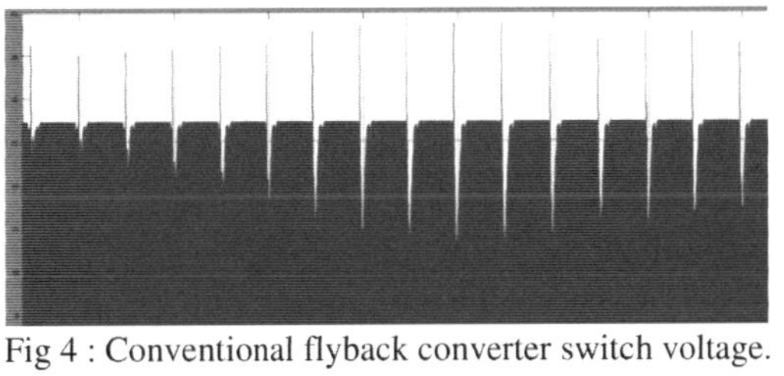

Fig 4 : Conventional flyback converter switch voltage.

IOP Conf. Series: Materials Science and Engineering **623** (2019) 012004 doi:10.1088/1757-899X/623/1/012004

It has been observed that the switch voltages have been reduced significantly as shown in Fig5.and so the switching losses which results in improved efficiency. The efficiency calculated is with this topology is 89% and is calculated under full condition by considering $R_{ds(on)}$ = 0.1 Ohm. Simulation results can be observed from the following Fig 5-10.

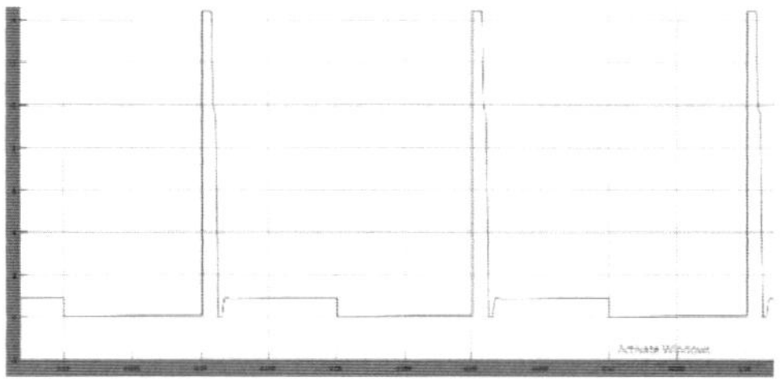

Figure 6: Flyback converter primary converter voltage

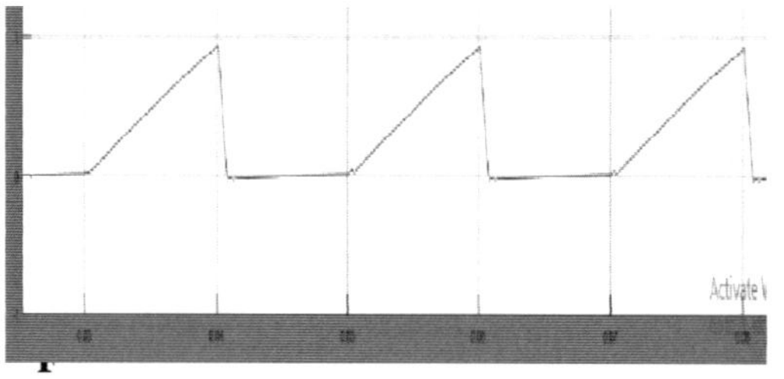

Figure 5: Interleaved converter switch

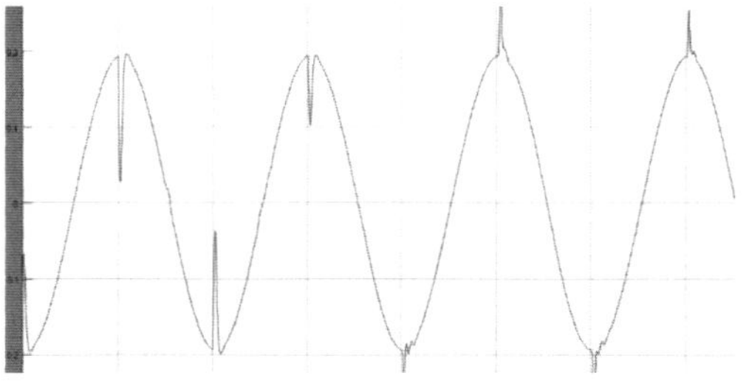

Figure 8: Dc link voltage

Figure 7: Load current for full load

Figure 9: Inverter output voltage synchronized with grid

The system is integrated to grid with the help of PLL which is useful to provide synchronization.
Each figure should have a brief caption describing it and, if necessary, a key to interpret the various lines and symbols on the figure.

IOP Conf. Series: Materials Science and Engineering **623** (2019) 012004 doi:10.1088/1757-899X/623/1/012004

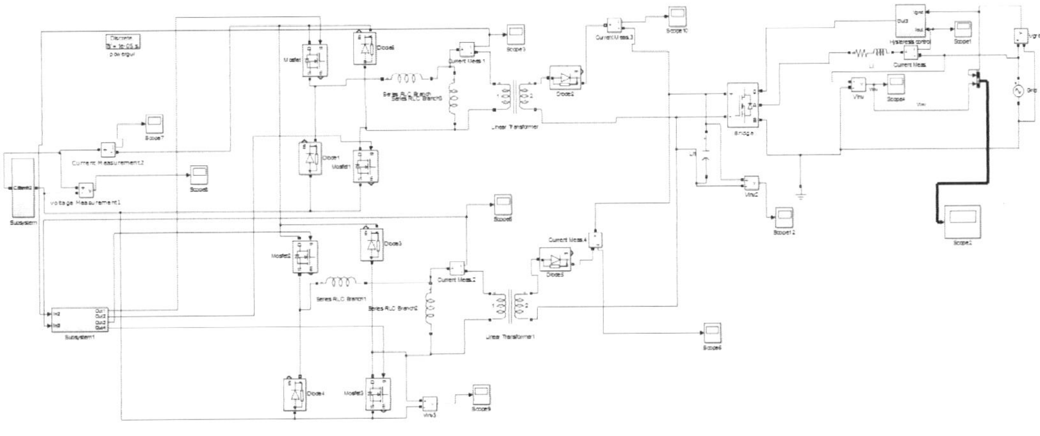

Figure 10: Matlab implementation of interleaved poly stage grid connected microinverter

7. Conclusion

This paper presents the analysis of interleaved flyback inverter which reduces the switching transients and from the simulation results ,it can be observed that switching transients have been reduced significantly compared to the conventional fly-back inverter And it was observed that the number of devices are getting increased but simple control and efficiency are the results. Still the converter can be well tuned to decrease the transformer ratings. There has to be an optimization between the current transient levels and voltage transients and the transformer ratings.

8. References

[1] S. B. Kjaer, 1. K. Pedersen, and F. Blaabjerg, "A review of single-phase grid-connected inverters for photovoltaic modules, "IEEE Trans.

[2] Young-Ho Kim, Young-Hyok Ji, Jun-Gu Kim, Yong-Chae Jung, and Chung-Yuen Won "A New Control Strategy for Improving Weighted Efficiency in Photovoltaic AC Module-Type Interleaved Flyback Inverters, "IEEE Trans Power Electron, VOL. 28, NO. 6, JUNE 2013 Ind.Appl., vol. 41, no. 5, pp. 1292-1306, Sep. 2005

[3] W. Mcmurray. Selection of snubbers and clamps to optimize the design of transistor switching converters. IEEE Trans. Ind. Appl, IA-16, pp.513-523, July-Augustl980.

[4] A. Hren, J. Korelic, and M. Milanovic, "RC-RCD clamp circuit for ringing losses reduction in a flyback converter,". IEEE Trans. Circuits and Systems-II: Express Briefs, 53, no. 5, pp. 369-373, May 2006.

[5] Abraham L Pressman, Keith Billings and Taylor Morey, Switching Power Supply Design, 3rd Ed. McGraw-Hili, 2009, pp.157-160.

[6] Zhiliang Zhang, Xiao-Fei He ,and Yan-Fei Liu, "An Optimal Control Method for Photovoltaic Grid-tied Interleaved Flyback Micro-inverters to Achieve High Efficiency in Wide Load Range, "IEEE Trans. Power Electron, VOL 28, p. 5074 - 5087, 2013.

[7] Ti .com/micro inverter design for single stage converter.

[8] Yanlin Li; Oruganti, R., "A Low Cost Flyback CCM Inverter for AC Module Application," Power Electronics, IEEE Transactions on, vol.27,no.3, pp.1295, 1303, March 2012.

IOP Conf. Series: Materials Science and Engineering **623** (2019) 012005 doi:10.1088/1757-899X/623/1/012005

Review and comparative analysis on dc-dc converters used in electric vehicle applications

Ingilala Jagadeesh[1], V. Indragandhi[2]

[1,2]School of Electrical Engineering, Vellore Institute of Technology, Vellore
Corresponding Author: indragandhi.v@vit.ac.in

Abstract: This paper is reported with different kinds of DC-DC converters such as sepic, boost and bidirectional converters. Integrating the boost, sepic, bidirectional DC-DC converters enables to identify the suitable converter for renewable energy applications with accurate power rating. The performance of non-isolated converter is assessed on the basis of these review. The solar PV conversion efficiency is low, so the converter is used to step-up/step down the voltage levels. This paper attempts to perform by examing efficiency of DC-DC converters and the voltage and current stress on the switches. A detailed review is carried out on basic PV and fuel cell based electric vehicles.

Keywords: Fuel cells, Solar PV, Boost, Sepic and Bi-directional converters, Electric vehicles.

1. INTRODUCTION

The power generated through the photo voltaic (PV) panels remains intermittent and random in nature. To maintain continuous supply from PV panels DC-DC converters are required. These converters are classified as isolated and non-isolated converters. According to the configuration of converter topologies isolated converters has two configurations full and half bridge topologies [1-5]. In addition, the half-bridge converter needs a centre-tapped transformer, which results in a complex structure, and the full-bridge converters are require a higher number of semiconductor devices and cost of the converter also increased [6-9]. In order to reduce the voltage stress caused by the leakage inductance, a DC-DC converter with an active clamp circuit is used [10]. The leakage inductance utilized in dual active bridge and the phase-shift full-bridge to achieve the soft-switching, and stored energies in leakage inductance are transmitted to the load [11-13].

IOP Conf. Series: Materials Science and Engineering **623** (2019) 012005 doi:10.1088/1757-899X/623/1/012005

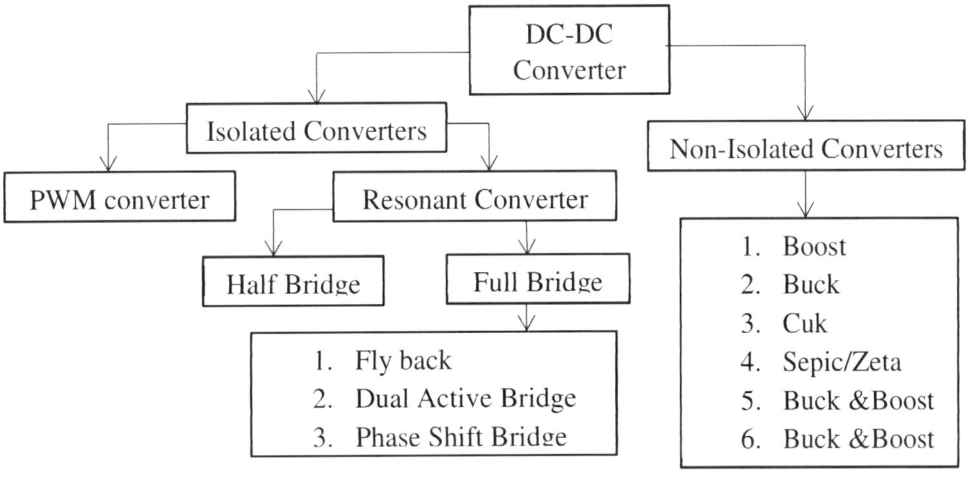

Figure. 1 Classification of DC-DC Converter

2.Non-Isolated DC-DC Converters

This topology is widely used for the application of Uninterruptible Power Supplies (UPS) and Hybrid Electric Vehicle (HEV). The design of the circuit itself helps to charging and discharging the battery bank. The above figure 1 shows the Non-isolated DC-DC converters. Bidirectional converter delivers a large voltage assortment in two operating conditions since it operates under a Zero Voltage Switching (ZVS) condition. The benefit of the BDC is as follows. 1) In both boost and buck mode the BDC attains high static voltage gain. 2) If we want to oprate the converter as a bidirectional condition three active switches are required. 3) To design these converters less number of passive elements are required. 4) The switching and conduction losses is reduced by using voltage clamping circuit, ZVS and synchronous rectification.

2.1 BOOST CONVERTER

The conventional boost converters used for medium and low power applications. The input current of the converter is non-pulsating and structure is simple. The simple boost converters are used for low and medium power levels. Electromagnetic interface is the major drawback of the basic converter. These converters have non-pulsating input current and simple structure. The conventional switched inductor boost converter (SIBC) having more switching losses at the main switches and the switching diodes having more conduction losses. The converter having the advantages are soft switching, low duty loss and high voltage gain. The converter conduction losses reduced by adding the auxiliary switches [1]. The interleaved coupled inductor boost converter needs a single active soft switching module to achieve the soft switching property. These converters share the input and output currents and also reduce the ripple currents [2]. For extensive range of input voltage, load current variations and switching frequency the ZVZCS provides soft commutation to the main switches [3]. If the number of capacitors and diodes increases higher voltage gain is attained but the cost of the converter becomes increases. By increasing the number of diodes and capacitors a high gain can be attained [4]. The boost converter can't track maximum power point at low irradiation condition. High voltage-gain and low-cost boost converter topologies are

- Coupled inductor with boost converter
- Switched capacitor with boost converter
- Coupled inductor with boost converter and switched capacitor
- Switched capacitor with inductor and boost converter [5].

IOP Conf. Series: Materials Science and Engineering **623** (2019) 012005 doi:10.1088/1757-899X/623/1/012005

TABLE-1

Reference	Switching frequency In kHz	Input Voltage In volts	Number of switches	Number of capacitors	Number of inductors	Output Voltage In volts	Efficiency In (%)
[1]	100	30	4	2	3	200	94.76
[2]	50	100	3	1	3	250	95.12
[3]	100	50	4	1	4	120	98.2
[4]	24	30	1	3	2	300	92.9

2.2 Coupled inductor with boost converter

To obtain higher gain of the voltage and realizing continuous input current the coupled boost converter having extra diodes. The coupled inductor with boost converter as shown in fig.2 reversal coupled inductor T-source boost converter, coupled inductor Y- source boost converter, coupled inductor T- source boost converter topologies are analyzed. Based on the high efficiency and voltage gain the coupled inductor Y-source boost converter is better than the quasi-Y-source boost converter [6]. The circuit cost and volume is reduced when the two input inductors are coupling each other. Yie-Tone Chen proposed a converter with two main switches and one active soft switching module to attain soft switching property [7]. Because of coupled inductor leakage inductace, high voltage spike occurs at the main switch. The coupled inductor with cascaded boost converter gives higher voltage gain and efficiency [8].

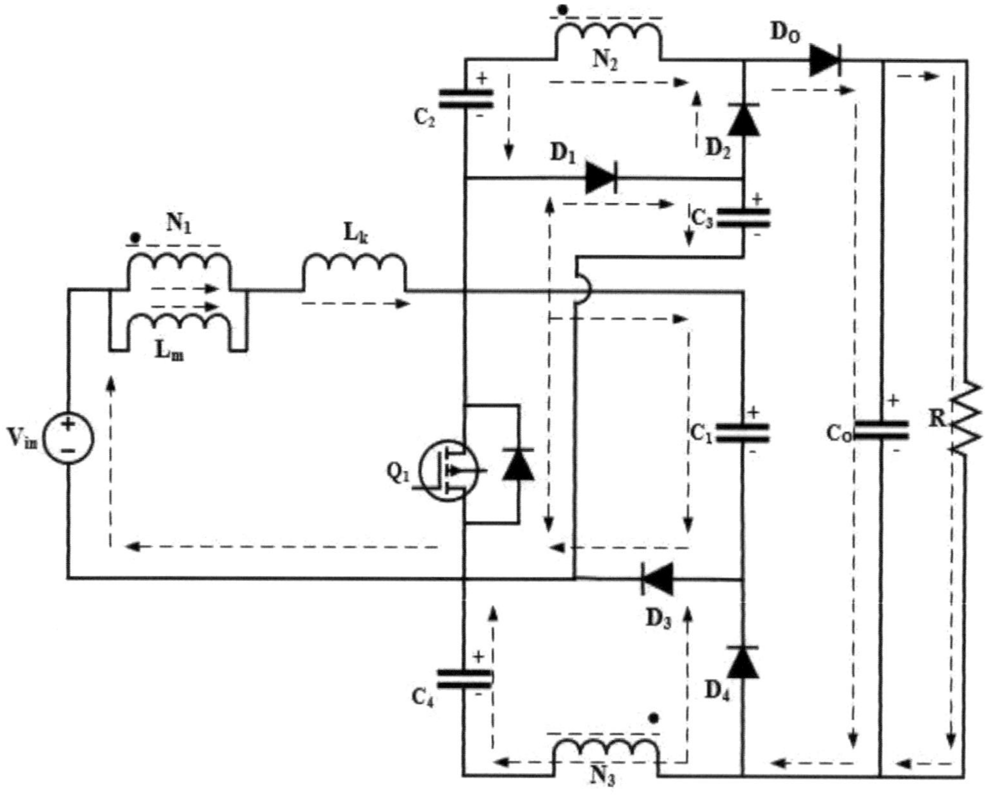

Figure. 2 Coupled inductor with boost converter [9].

To form a hybrid multiplier cell in coupled inductor the secondary windings are used. In addition to maximizing the voltage gain, and minimizes the voltage stress on the converter the coupled inductor type converter is preferred [9].

2.3 Switched capacitor with boost converter

To enhance the regulation of output voltage Robert stala proposed a converter. Fig. 3 shows the switched capacitor type boost converter. By increasing the voltage ratio, the inrush current problem occurs in the circuit. Based on the inrush current the circuit parameters needs to be designed [10].

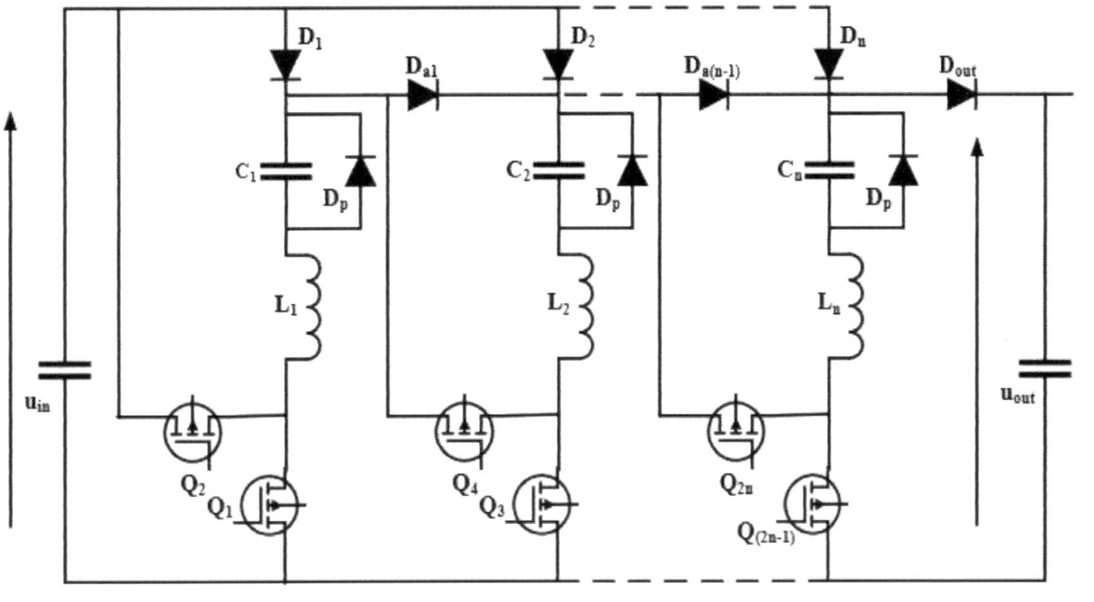

Figure.3 Switched capacitor type boost converter [10]

The two capacitors are connected in parallel to the single dc source to form a switched capacitor cell. The advantages are high efficiency less complexity and appropriate boosting property. To reach a greater number of output voltage levels with extended configuration two switched capacitor cells are connected. [11]. Minh-khai Nguyen proposed a converter with small duty cycle with high voltage gain. The converter decreases the conduction loss on the switches and decrease the voltage stress. The converter gives 200V output voltage with the 25-50V input voltage and operating power 200W [12]

2.4 Coupled inductor with boost converter and switched capacitor

The voltage stress on the main switches reduced by using low voltage rated transistors. The structure of the converter easily extended for high voltage gain. Fig. 4 shows the coupled inductor with boost converter and switched capacitor. In the converter soft switching condition achieved by utilizing leakage inductance with the help of diodes. To improve the efficiency small on state resistance with low voltage rated MOSFETs can be chosen. With the same duty ratio and the same input voltage the voltage stress on the main switches are less [13].

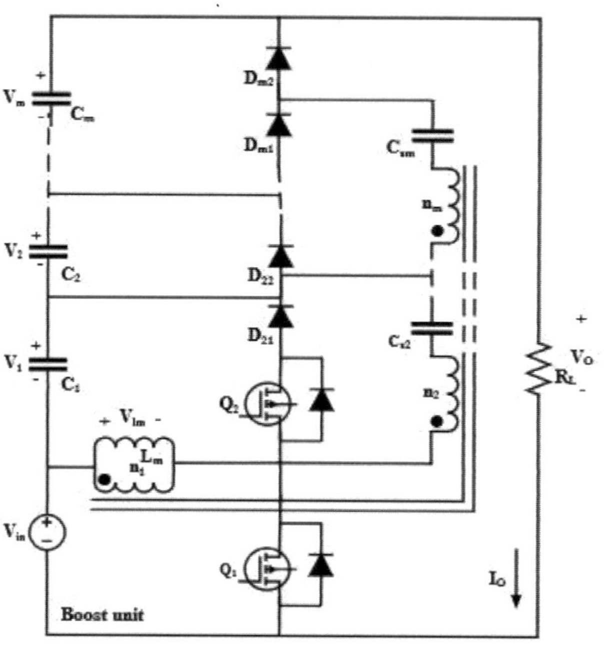

Figure. 4 Coupled inductor with boost converter and switched capacitor

To eliminate the hard-switching phenomenon diodes are ZCS turned on/off and transistors are ZVS turned on. The voltage stress on the converter diodes and transistors equal to the input voltage, remains low. The resonance tank consists of two small capacitor and one inductor. The converter output characteristics, voltage-gain curves are analyzed [14].

2.5 Switched capacitor with inductor and boost converter

The combination of conventional switched boost network through the switched capacitor or switched inductor cell have maximum voltage gain, The voltage stress on the switches and diodes. Fig. 5 shows the switched capacitor with inductor and boost converter circuit. The switched capacitor with inductor having the advantage of expandability. By adding three diodes and one inductor the extra cell can be easily cascaded. Compared to other boost converters switched capacitor converter gives low duty ratio with more voltage gain [15].

IOP Conf. Series: Materials Science and Engineering **623** (2019) 012005 doi:10.1088/1757-899X/623/1/012005

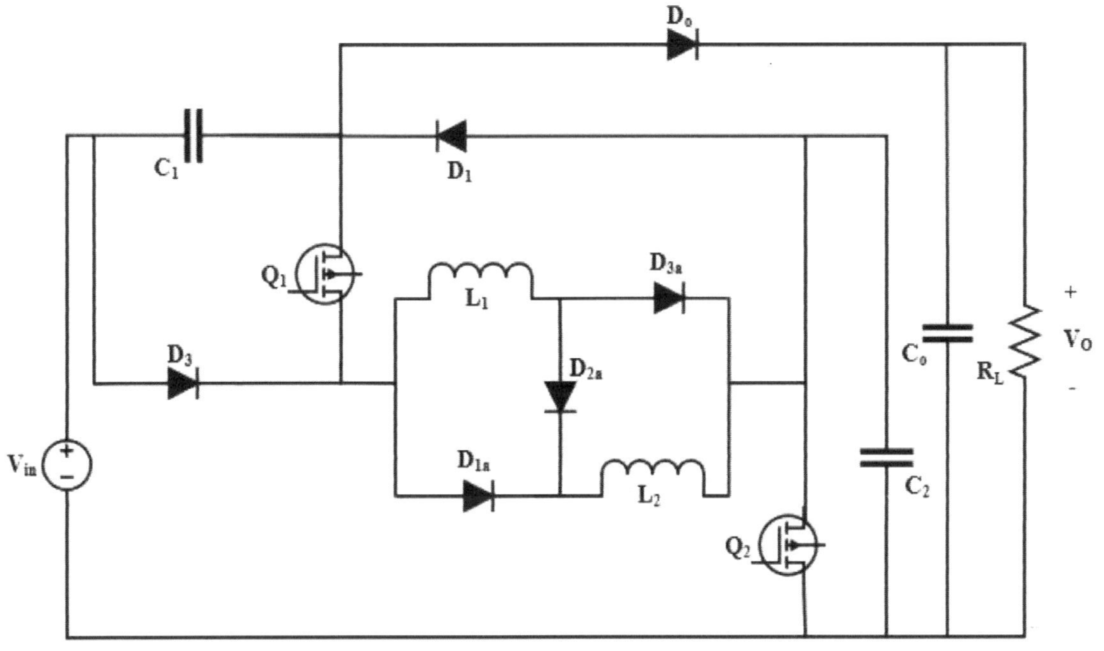

Figure. 5 switched capacitor with inductor and boost converter

The coupled inductor leakage inductance is used to the load. The switched capacitor having inrush current problem due to the leak inductance of the coupled inductor restrained the inrush current [16].

PV system performance is also influenced by selection of the converter. If the panel resistance is equal or below the load resistance then boost converter track maximum power point. By reducing the stress on switches, the converter efficiency increases. The input ripple current of buck-boost, buck and zeta converters are greater than that of cuk, zeta and SEPIC converters [17].

The leakage inductance reduces the voltage gain of the coupled inductor DC-DC converter. Coupled inductor with voltage multiplier cell is used to increase the voltage gain of the converter. The voltage stress on the switch is reduced by using high gain cell. If high gain cell increases the voltage conversion ratio increases [18]. The combination of self-lift Cuk and boost converter reduces the voltage stress of the diodes and switches [19].

3. Buck converter
Although the modified series-capacitor (SC) high conversion ratio (HCR) DC-DC buck converter having extra capacitor to eliminate the start-up voltage stress on the switches. Compared with conventional SC buck converter the input capacitance of these converter is 0.5 V_{in} [20]. Based on the size of MOSFETs the parasitic capacitance and on resistances are changed. The efficiency of the buck converter reduced by switching and conduction losses. Gate split technique is used to adjusts the switching and conduction losses [21]. To minimize the output voltage ripples of the buck converter the resolution of the analog to digital converter have to be increased [22]. Yi Liu et all proposed a dc-dc buck converter to limit the current and voltage ripples [23].

4. Sepic Converter
The maximum power point tracking of the PV system is achieved by the single ended primary inductor (SEPIC) converter with fuzzy logic controller. Fig. 6 shows the SEPIC converter circuit. Even though the input signal is not smooth the output signal is smooth [24].

IOP Conf. Series: Materials Science and Engineering **623** (2019) 012005 doi:10.1088/1757-899X/623/1/012005

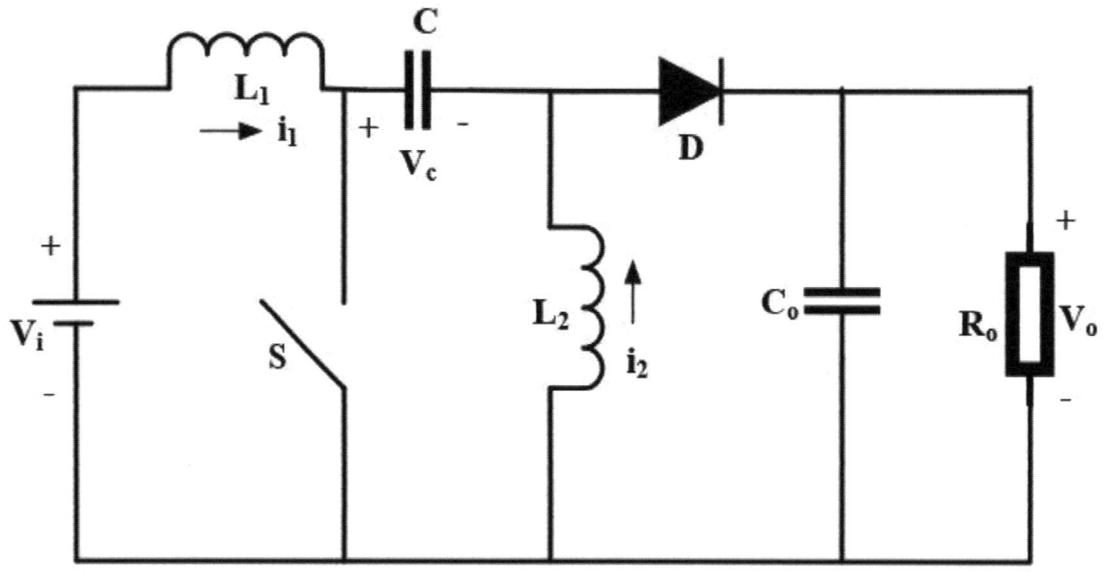

Figure. 6 SEPIC converter

Emilio Mamarelis proposed a converter, designing a sliding mode controller for PV applications with maximum power point technique [25]

5. NON-ISOLATED BI-DIRECTIONAL CONVERTER

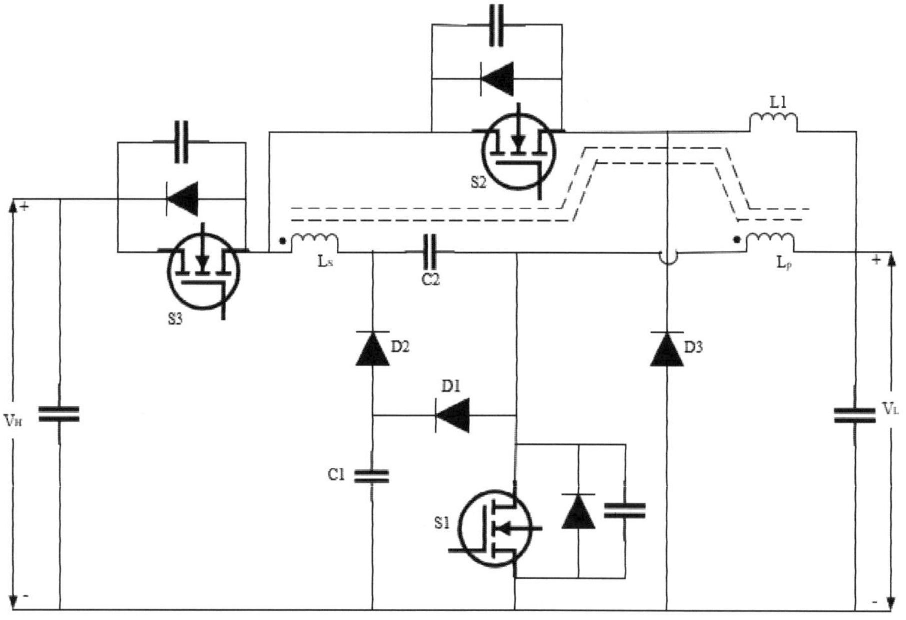

Figure. 7 ZVS Non-Isolated Bi-directional Converter [26]

IOP Conf. Series: Materials Science and Engineering **623** (2019) 012005 doi:10.1088/1757-899X/623/1/012005

Table 2. Efficiency of the Non-isolated bidirectional converter with different output power

Output Power in Watts	Efficiency in %	
	Buck Mode	Boost Mode
50	87.5	92.5
100	90.5	95.5
150	92	96
200	93	96.5

Fig. 7 shows the ZVS non-isolated Bi-directional converter. A Switched-Capacitor (SC) BDC with a high step-up/step-down static voltage gain is conferred for EV with a hybrid energy source scheme. Also the presented converter has the benefits of being a simple circuit, a lesser number of components, an extensive voltage-gain limit, a squat voltage stress, and a common ground. Furthermore, the efficiency of the converter is raised since the synchronous rectifiers permit ZVS turn-on and turn-off deprived of needing any additional hardware [26].

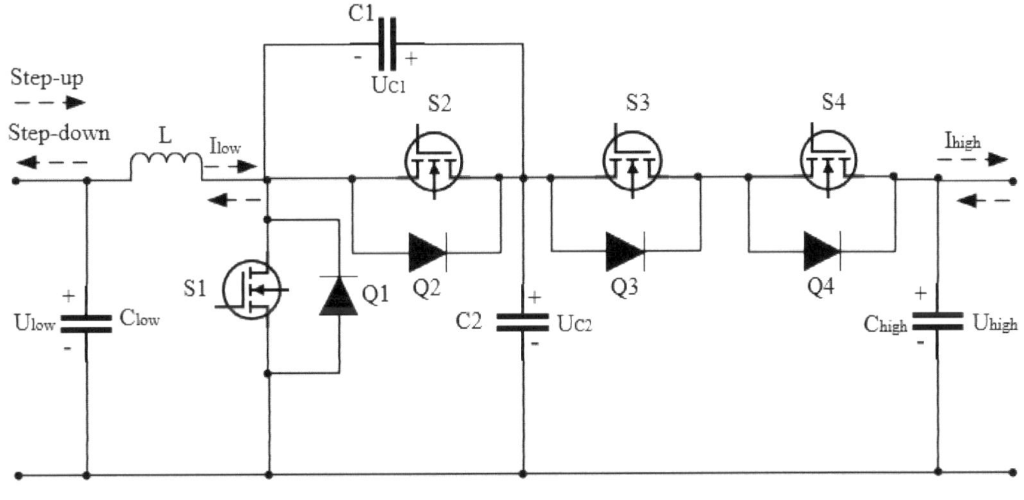

Figure. 8 Switched – Capacitor BDC [27]

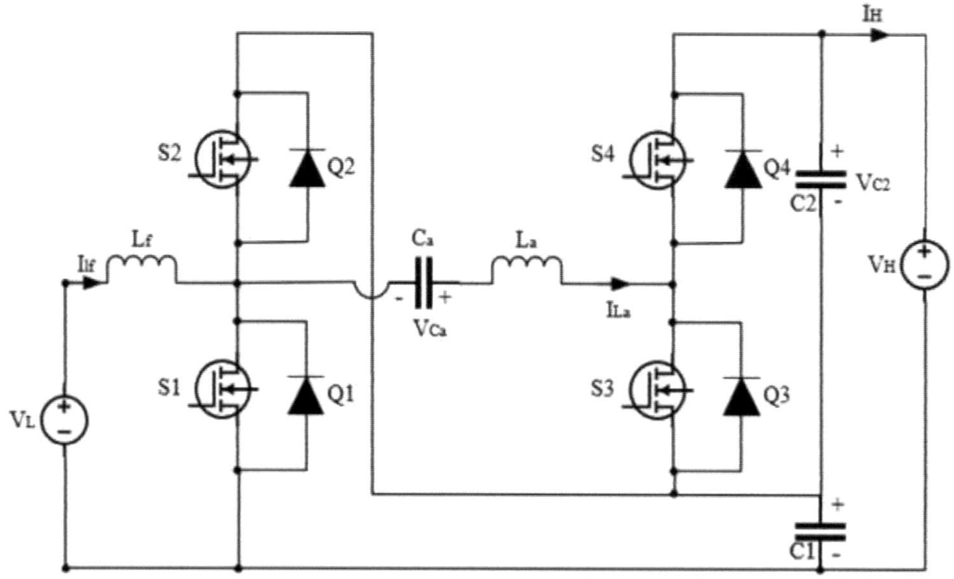

Figure. 9 Proposes high step-up soft switched bidirectional dc-dc converter [28]

Fig. 8 shows the switched capacitor BDC. The passive component size and stress current is decreased by applying interleaving scheme. A two-phase interleaved type of the high step up converter is depicted in Fig. 9.

TABLE 3. ANALYSIS OF THE DC-DC CONVERTERS

Parameter	**Converter in [29]**	**Converter in [30]**	**Converter in [31]**	**Converter in [32]**	**Converter in [33]**
Voltage stress on MOSFETs	$\frac{V_0}{1+2N-ND}$	$\frac{V_0}{1+ND}$	$\frac{V_0}{2(1+N)}$	$\frac{V_0}{1+N}$	$\frac{V_0}{2}$
Soft switching of MOSFETs	ZVS	Hard switching	ZVS	Hard switching	ZVS
Number of MOSFETs	2	1	2	1	2
Voltage stress on output diode	$\frac{NV_0}{1+2N-ND}$	$\frac{NV_0}{1+ND}$	$\frac{V_0}{2}$	$\frac{NV_0}{1+N}$	$\frac{V_0}{2}$
Soft switching of diodes	ZCS	Hard switching	Hard switching	Hard switching	ZCS
Diodes	3	2	4	3	2
Number of magnetic components	1	1	1	1	2
Voltage gain	$\frac{1+2N-ND}{1-D}$	$\frac{1+ND}{1-D}$	$\frac{2(1+N)}{1-D}$	$\frac{1+N}{1-D}$	$\frac{2}{1-D}$

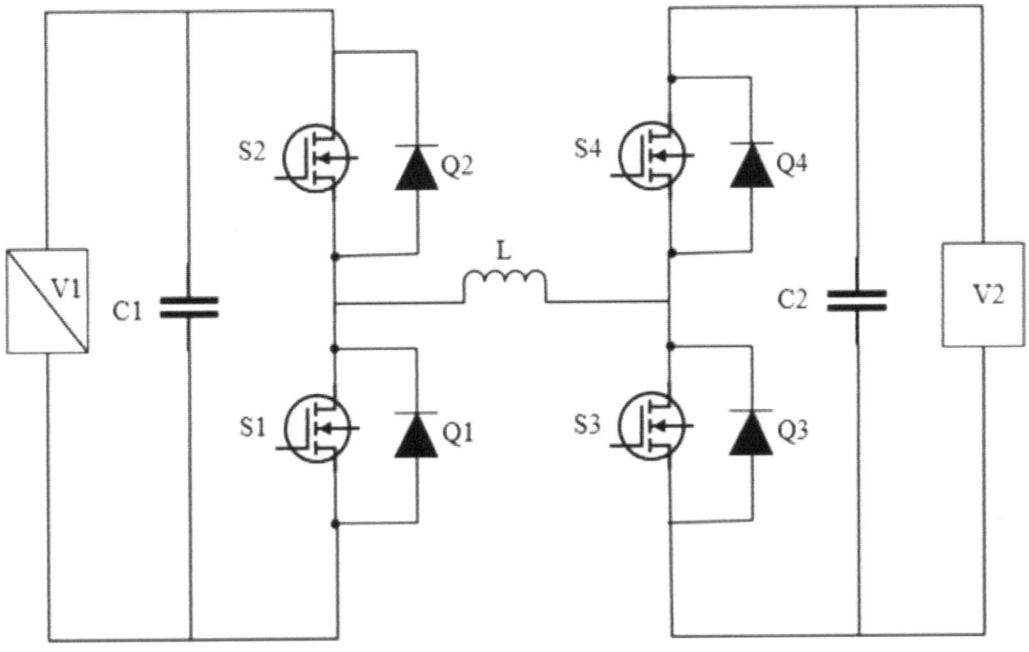

Figure. 10 CBB-IIM [34]

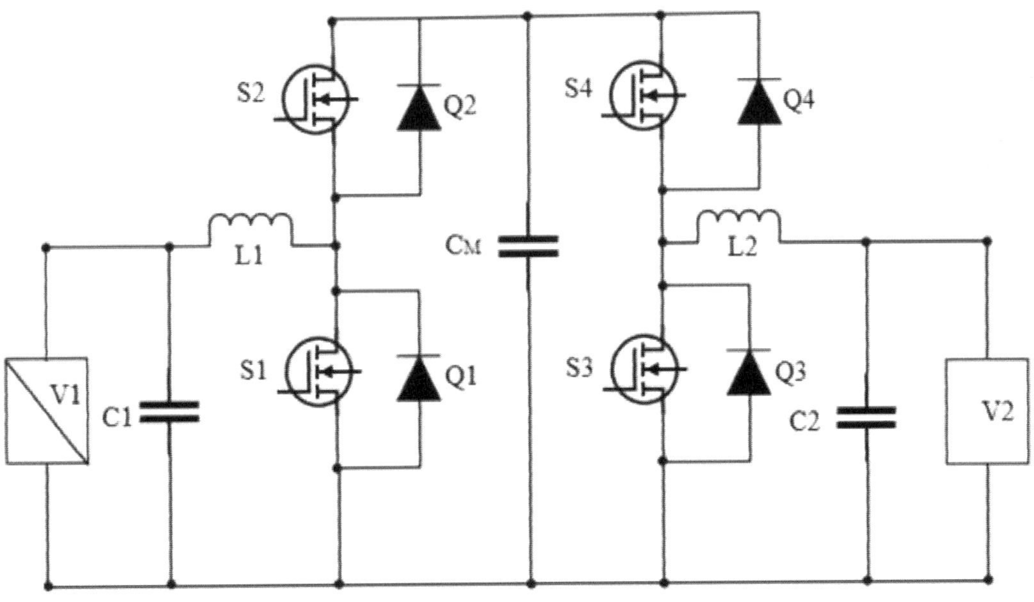

Figure. 11 CBB-CIM [34]

A dc–dc converter with Multi-Input and Multi-Output (MIMO) competence is valuable for EV/HEVs which uses numerous input sources or needs numerous auxiliary outputs. Multiple input decisions are desirable to

IOP Conf. Series: Materials Science and Engineering **623** (2019) 012005 doi:10.1088/1757-899X/623/1/012005

unite two sources such as an ultra-capacitor and a battery pack consolidation [35]–[37]. In heavy HV such as fleet trucks, several auxiliary outputs will need dissimilar output voltage range [38], [39]. Nowadays, the mechanical engine belt compressor for Air Conditioning (AC) structures is swapped by an electrically compelled compressor to decrease emissions and improve gas mileage. The compressor motor is compelled by a high voltage (220–400 V) at 3–5-kW power necessities [40], [41]. The circuit of CBB-IIM as shown in Fig. 10. The usage of CBB-CIM and CBB-IIM topologies with several inputs and/or several outputs is evaluated in the subsequent. The CBB-CIM topology with auxiliary outputs associated to *V*out1 and *V*out2 is exposed in Fig. 11.

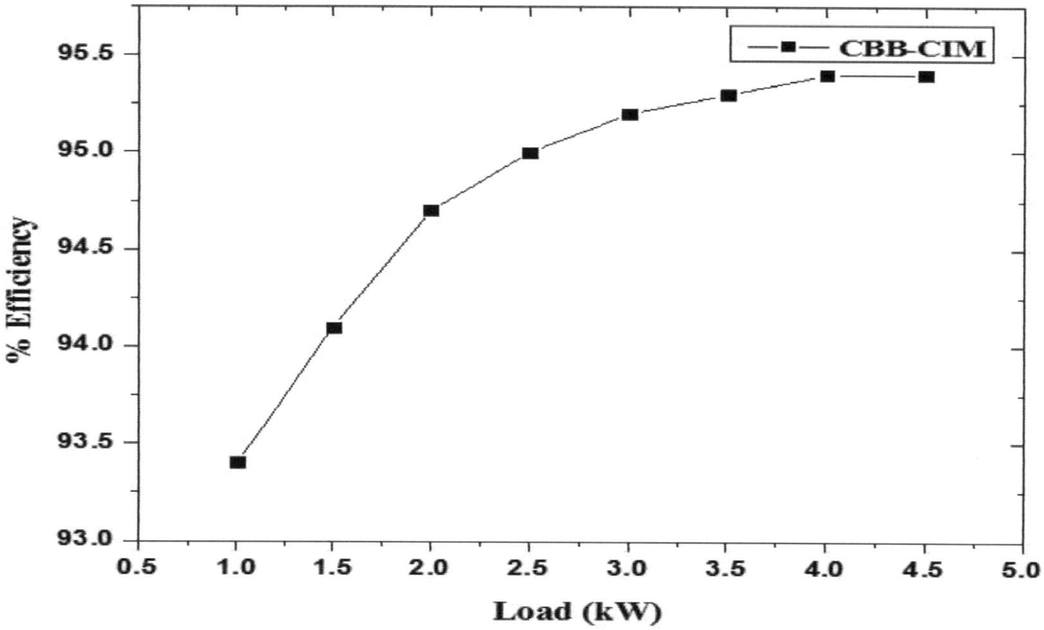

Figure. 12 Efficiency with load variation of the CBB-CIM topology for the multiple input case

For an EV charging station, multi-input and multi-output instance, CBB-CIM provides better enactment as the input-side and output-side reins are sovereign. Fig. 12 shows the efficiency with load variation of the CBB-CIM topology. System controller elasticity and consistency are enhanced with CBB-CIM.

6. CONCLUSION

This paper proposes different kinds of FC based non-isolated DC-DC converters. Each converter has its own importance for the required applications. By using the coupled inductor with ZVS operation, the stress on the semiconductor devices reduces. By using the suitable DC-DC converter between the source and dc bus the efficiency of the system increases. The voltage stress on the capacitors and power semiconductor devices are reduces with the help of bidirectional interleaved switched capacitor type DC-DC converter. By using the interleaved converter, current ripples at the low voltage side is less. The voltage stress on the diodes, power switches and conduction losses are reduced in the coupled inductors and switched capacitor technique. The life time of the fuel cell increases by using the parallel input, series output DC-DC boost converter because the ripple current at the input is low and the voltage stress on the switches are 50% of the output voltage.

References

[1] Yao, Tong, Chenhao Nan, and Rajapandian Ayyanar. "A new soft-switching topology for switched inductor high gain boost." *IEEE Transactions on Industry Applications* 54, no. 3 (2018): 2449-2458.

[2] Chen, Yie-Tone, Zih-Ming Li, and Ruey-Hsun Liang. "A novel soft-switching interleaved coupled-inductor boost converter with only single auxiliary circuit." *IEEE Transactions on Power Electronics* 33, no. 3 (2018): 2267-2281.

[3] Ashique, Ratil H., and Zainal Salam. "A Family of True Zero Voltage Zero Current Switching (ZVZCS) Nonisolated Bidirectional DC–DC Converter With Wide Soft Switching Range." *IEEE Transactions on Industrial Electronics* 64, no. 7 (2017): 5416-5427.

[4] Saravanan, S., and N. Ramesh Babu. "Analysis and implementation of high step-up DC-DC converter for PV based grid application." *Applied Energy* 190 (2017): 64-72.

[5] Taghvaee, M. H., M. A. M. Radzi, S. M. Moosavain, Hashim Hizam, and M. Hamiruce Marhaban. "A current and future study on non-isolated DC–DC converters for photovoltaic applications." *Renewable and sustainable energy reviews* 17 (2013): 216-227.

[6] Liu, Hongchen, Yuliang Ji, Liuchao Wang, and Pat Wheeler. "A family of improved magnetically coupled impedance network boost DC–DC converters." *IEEE Transactions on Power Electronics* 33, no. 5 (2018): 3697-3702.

[7] Chen, Yie-Tone, Zih-Ming Li, and Ruey-Hsun Liang. "A novel soft-switching interleaved coupled-inductor boost converter with only single auxiliary circuit." *IEEE Transactions on Power Electronics* 33, no. 3 (2018): 2267-2281.

[8] Lee, Sin-Woo, and Hyun-Lark Do. "High step-up coupled-inductor cascade boost DC–DC converter with lossless passive snubber." *IEEE Transactions on Industrial Electronics*65, no. 10 (2018): 7753-7761.

[9] Hu, Xuefeng, Jianzhang Wang, Linpeng Li, and Yongchao Li. "A three-winding coupled-inductor DC–DC converter topology with high voltage gain and reduced switch stress." *IEEE Transactions on Power Electronics* 33, no. 2 (2018): 1453-1462.

[10] Stala, Robert, Zbigniew Waradzyn, Adam Penczek, Andrzej Mondzik, and Aleksander Skała. "A Switched-Capacitor DC–DC Converter With Variable Number of Voltage Gains and Fault-Tolerant Operation." *IEEE Transactions on Industrial Electronics* 66, no. 5 (2019): 3435-3445.

[11] Barzegarkhoo, Reza, Majid Moradzadeh, Elyas Zamiri, Hossein Madadi Kojabadi, and Frede Blaabjerg. "A new boost switched-capacitor multilevel converter with reduced circuit devices." *IEEE Transactions on Power Electronics* 33, no. 8 (2018): 6738-6754.

[12] Nguyen, Minh-Khai, Truong-Duy Duong, and Young-Cheol Lim. "Switched-capacitor-based dual-switch high-boost DC–DC converter." *IEEE Transactions on Power Electronics* 33, no. 5 (2018): 4181-4189.

[13] Ye, Yuanmao, K. W. E. Cheng, and Sizhe Chen. "A high step-up PWM DC-DC converter with coupled-inductor and resonant switched-capacitor." *IEEE Transactions on Power Electronics*32, no. 10 (2017): 7739-7749.

[14] Li, Shouxiang, Yifei Zheng, Bin Wu, and Keyue Ma Smedley. "A family of resonant two-switch boosting switched-capacitor converter with ZVS operation and a wide line regulation range." *IEEE Transactions on Power Electronics* 33, no. 1 (2018): 448-459.

[15] Zhu, Xiaoquan, Bo Zhang, Zhong Li, Hong Li, and Li Ran. "Extended switched-boost dc-dc converters adopting switched-capacitor/switched-inductor cells for high step-up conversion." *IEEE Journal of Emerging and Selected Topics in Power Electronics* 5, no. 3 (2017): 1020-1030.

[16] Liang, Tsorng-Juu, Shih-Ming Chen, Lung-Sheng Yang, Jiann-Fuh Chen, and Adrian Ioinovici. "Ultra-large gain step-up switched-capacitor DC-DC converter with coupled inductor for alternative sources of energy." *IEEE Transactions on Circuits and Systems I: Regular Papers* 59, no. 4 (2012): 864-874.

[17] Reddy, K. Jyotheeswara, and N. Sudhakar. "High voltage gain interleaved boost converter with neural network based MPPT controller for fuel cell based electric vehicle applications." *IEEE Access* 6 (2018): 3899-3908.

[18] Zhang, Yun, Chuanzhi Fu, Mark Sumner, and Ping Wang. "A wide input-voltage range quasi-z-source boost dc–dc converter with high-voltage gain for fuel cell vehicles." *IEEE Transactions on Industrial Electronics* 65, no. 6 (2018): 5201-5212.

[19] Hu, Xuefeng, Jianzhang Wang, Linpeng Li, and Yongchao Li. "A three-winding coupled-inductor DC–DC converter topology with high voltage gain and reduced switch stress." *IEEE Transactions on Power Electronics* 33, no. 2 (2018): 1453-1462.

[20] Kim, Kisu, Honnyong Cha, Siho Park, and Il-Oun Lee. "A modified series-capacitor high conversion ratio DC–DC converter eliminating start-up voltage stress problem." *IEEE Transactions on Power Electronics* 33, no. 1 (2018): 8-12.

[21] Park, Young-Jun, Ju-Hyun Park, Hong-Jin Kim, Hocheol Ryu, SangYun Kim, YoungGun Pu, Keum Cheol Hwang, Youngoo Yang, Minjae Lee, and Kang-Yoon Lee. "A design of a 92.4% efficiency triple mode control DC–DC buck converter with low power retention mode and adaptive zero current detector for IoT/wearable applications." *IEEE Transactions on Power Electronics* 32, no. 9 (2017): 6946-6960.

[22] Kim, Sang-Yun, Young-Jun Park, Imran Ali, Truong Thi Kim Nga, Ho-Cheol Ryu, Zaffar Hayat Nawaz Khan, Seong-Mun Park et al. "Design of a high efficiency dc–dc buck converter with two-step digital pwm and low power self-tracking zero current detector for iot applications." *IEEE Transactions on Power Electronics* 33, no. 2 (2018): 1428-1439.

[23] Liu, Yi, Meng Huang, Huai Wang, Xiaoming Zha, Jinwu Gong, and Jianjun Sun. "Reliability-oriented optimization of the LC filter in a buck DC-DC converter." *IEEE Transactions on Power Electronics* 32, no. 8 (2017): 6323-6337.

[24] Mohanraj, K., and S. Bharathnarayanan. "Three Level SEPIC For Hybrid Wind-Solar Energy Systems." *Energy Procedia* 117 (2017): 120-127.

[25] Gopi, R. Reshma, and S. Sreejith. "Converter topologies in photovoltaic applications–A review." *Renewable and Sustainable Energy Reviews* 94 (2018): 1-14.

[26] Aamir, Muhammad, Saad Mekhilef, and Hee-Jun Kim. "High-gain zero-voltage switching bidirectional converter with a reduced number of switches." *IEEE Transactions on Circuits and Systems II: Express Briefs* 62, no. 8 (2015): 816-820.

[27] Zhang, Yun, Yongping Gao, Lei Zhou, and Mark Sumner. "A Switched-Capacitor Bidirectional DC–DC Converter With Wide Voltage Gain Range for Electric Vehicles With Hybrid Energy Sources." *IEEE Transactions on Power Electronics* 33, no. 11 (2018): 9459-9469.

[28] Kwon, Minho, Secheol Oh, and Sewan Choi. "High gain soft-switching bidirectional DC–DC converter for eco-friendly vehicles." *IEEE Transactions on Power Electronics* 29, no. 4 (2013): 1659-1666.

[29] Shitole, Amardeep B., Shelas Sathyan, H. M. Suryawanshi, Girish G. Talapur, and Pradyumn Chaturvedi. "Soft-switched high voltage gain boost-integrated flyback converter interfaced single-phase grid-tied inverter for SPV integration." *IEEE Transactions on Industry Applications* 54, no. 1 (2017): 482-493.

[30] Liang, Tsorng-Juu, and K. C. Tseng. "Analysis of integrated boost-flyback step-up converter." *IEE Proceedings-Electric Power Applications* 152, no. 2 (2005): 217-225.

[31] Molavi, Navid, Ehsan Adib, and Hosein Farzanehfard. "Soft-switched non-isolated high step-up DC–DC converter with reduced voltage stress." *IET Power Electronics* 9, no. 8 (2016): 1711-1718.

[32] Das, Moumita, and Vivek Agarwal. "Design and analysis of a high-efficiency DC–DC converter with soft switching capability for renewable energy applications requiring high voltage gain." *IEEE Transactions on Industrial Electronics* 63, no. 5 (2016): 2936-2944.

[33] Park, Yohan, Byoungkil Jung, and Sewan Choi. "Nonisolated ZVZCS resonant PWM DC–DC converter for high step-up and high-power applications." *IEEE Transactions on Power Electronics* 27, no. 8 (2012): 3568-3575.

[34] Ahmed, Adeeb, Mehnaz Akhter Khan, Mohamed Badawy, Yilmaz Sozer, and Iqbal Husain. "Performance analysis of bi-directional dc-dc converters for electric vehicles and charging infrastructure." In *2013 IEEE Energy Conversion Congress and Exposition*, pp. 1401-1408. IEEE, 2013.

[35] Lee, Il-Oun, and Gun-Woo Moon. "Half-bridge integrated ZVS full-bridge converter with reduced conduction loss for electric vehicle battery chargers." *IEEE Transactions on Industrial Electronics* 61, no. 8 (2013): 3978-3988.

[36] Monteiro, Vítor, J. G. Pinto, and João Luiz Afonso. "Operation modes for the electric vehicle in smart grids and smart homes: Present and proposed modes." *IEEE Transactions on Vehicular Technology* 65, no. 3 (2016): 1007-1020.

[37] Lee, Young-Joo, Alireza Khaligh, and Ali Emadi. "Advanced integrated bidirectional AC/DC and DC/DC converter for plug-in hybrid electric vehicles." *IEEE Transactions on vehicular technology* 58, no. 8 (2009): 3970-3980.

[38] Quraan, Mahran, Pietro Tricoli, Salvatore D'Arco, and Luigi Piegari. "Efficiency assessment of modular multilevel converters for battery electric vehicles." *IEEE Transactions on Power Electronics* 32, no. 3 (2016): 2041-2051.

[39] Lee, Il-Oun. "Hybrid DC–DC converter with phase-shift or frequency modulation for NEV battery charger." *IEEE Transactions on Industrial Electronics* 63, no. 2 (2016): 884-893.

[40] Abdelrahman, Ahmed S., Zekiye Erdem, Yosra Attia, and Mohamed Z. Youssef. "Wide Bandgap Devices in Electric Vehicle Converters: A Performance Survey." *Canadian Journal of Electrical and Computer Engineering* 41, no. 1 (2018): 45-54.

[41] Beraki, Mebrahtom Woldelibanos, João Pedro F. Trovão, Marina S. Perdigão, and Maxime R. Dubois. "Variable Inductor Based Bidirectional DC–DC Converter for Electric Vehicles." *IEEE Transactions on Vehicular Technology* 66, no. 10 (2017): 8764-8772.

PV based high level hybrid multilevel inverter

M.Senthilkumar[1], N.Loganathan[2]

[1]Assistant Professor /Department of EEE,Sri Krishna College of Engineering & Technology,Coimbatore, Tamil nadu, India, *senthilkumarm@skcet.ac.in*
[2]Assistant Professor /Department of EEE, Sri Krishna College of Engineering& Technology, Coimbatore, Tamilnadu, India, *loganathann@skcet.ac.in*

Abstract-This paper presents a PV based high level hybrid multilevel inverter. SEPIC converter is the technologically advanced converter derived from the buck-boost converter. The proposed scheme is used in the SEPIC-converter to eradicate the output ripples and well enhance the output voltage level. Cascaded H Bridge and Neutral point clamped (NPC) inverter topologies are combined to obtain 127 level of output voltage using reduced number of switches and DC sources. To increasing the voltage level, Total Harmonic Distortion (THD) can be reduced and hence reduction in size of the filter. Simulation work is done using MATLAB and then to verify the performance of suggested scheme.

Keywords- SEPIC converter, Hybrid multilevel inverter, Cascaded H Bridge & NPC inverters, Total Harmonic Distortion.

1. INTRODUCTION

The constraint of conservative energy system and environmental problems equivalent to heating and pollution is prompting the planet toward the improvement of renewable energy sources. The electrical occurrence of Photovoltaic (PV) plays a very significant part in power generation among more outdated earliest energy sources corresponding to oil, coal, nuclear, hydro and wind [1]. PV array power directly feed into small loads equal to lighting systems and DC motors. Further sophisticated application required electronic converters to electricity from the PV array. These converters are used to normalize the output voltage and current in the load, to control the power flow in grid-connected systems and maximum power point tracking (MPPT) is used to obtain maximum power in the PV array panel. Converters with maximum power point tracking technique are used to detect the continuous maximum instantaneous power of the PV array. The results of suggested scheme is operating in different environment conditions of the PV array with slight modification. The entire operation of the system an MPPT procedure is required to obtain the maximum instantaneous power can be extracted and delivered to the load.

Over the few past decades, completely different MPPT techniques are implemented and proposed [2] PV system is used to track the maximum instantaneous power from arrays. The systems vary in sensors needed, complexity, cost, convergence speed, vary of effectiveness, implementation of hardware, popularity, and in other respects. The best MPPT technique is analyzed in [2], [3] and [4] in Perturb and Observe (P&O) methodology. The output of PV panel obtained from DC voltage with contains high voltage ripples and it is not constant. DC-DC converters are engaged to reduce the

ripples not withstanding alteration in load voltage. The traditional device doesn't meet the load demand containing a lot of ripples in the output voltage and parasitic effects. To beat this drawback in advanced developed DC-DC SEPIC device was used [5].
SEPIC device is the established from buck-boost converter, that performs buck or boost operation [6], [7]. Sepic converters have some important characteristics such as high power density, high voltage transfer gain, and reduced ripple in output voltage and current in the PV array. Its further filter parts may be accustomed to eradicate the ripples in output voltage and effectually increase the output voltage level. It also overcomes the parasitic effects in the classical dc-dc device. The various electrical converter topologies for interfacing electrical phenomenon modules to the grid is bestowed in [8] that involves two major tasks. First one is used to insert a curving current into the grid and second one is used to see the PV module(s) operation at the MPP, is used to detect the operating condition and anywhere most of the energy is captured. This is often circulated by mistreatment MPP tracker. So as to attain a utilization magnitude ratio of 98% and the amplitude of the ripple voltage should be less than the 8.5% [8]. To interface the PV module(s) with the grid connected systems should maintain some standards in the electrical converter topologies. Nowadays Multi level inverter (MLI) technologies has been achieved its extensive recognition within the space of high power medium voltage energy management. It includes associate with nurturing an array of power semiconductors and DC voltage sources, the output of that generate voltages with stepped waveforms. Compared with a two-level voltage source inverter (VSI), the structure of VSI will turn out a lot of ranges in the voltage levels within the output voltage [9]. Hence, by increasing the quantity of levels in the MLI, the output voltage has more number of steps required in generating a wave with shape that includes and reduced total harmonic distortion (THD). The Diode clamped, Flying capacitor and Cascaded H-bridge inverter are the three main multilevel inverter structures that area unit employed in industrial applications with separate dc sources. In flying capacitor and diode-clamped inverter there is a problem of capacitor voltage balancing and this problem is overcome in cascaded H-bridge inverter. But in the cascaded H-bridge inverter, separate dc sources area unit required for every H-bridge. It will lead to higher range of dc input sources [10]-[12]. So as to boost power quality and scale back EMI issue, it is needed to extend the number of levels [13]. The range of the amount the quantity of switches becomes too high with higher number of voltage levels if all the DC busses area unit at same voltage; this conjointly will increases the number of gate drivers and better physical phenomenon of conduction loss of devices. All on top of issue can have negative impact on price, reliability (reliability is inversely proportional to number of components) and efficiency. The asymmetric structure of multilevel inverter can increase the number of levels without dramatic increase in number of switches [14]. So, to beat the top of mentioned the issue two structure of multilevel inverter topologies are mixed to get higher range of output voltage level by mistreatment lesser number of switches and sources [15], [16]. During this planed system in Cascaded H-Bridge & Neutral Point Clamp topologies are hybridized to form this topology. Increasing the number of voltage levels provides a lot of flexibility to scale back Total Harmonic Distortion (THD) of the system and successively reduction in cost and size of the power filter.

2. PROPOSED TOPOLOGY DESCRIPTION

In this suggested scheme the electrical phenomenon of photovoltaic (PV) is used as a DC supply. The DC-DC Sepic converter provides a constant DC source and additionally went to eliminate or reduce the output voltage and current ripples. Here, to induce a most power from the panel, to come up with a change pulse for the converter switches the MPPT with management unit is employed. A usual electrical converters only having 30 to 40 percentage of occurrence in irradiation into voltage.
The most of the power from PV panels and they have to operate at their maximum power point (MPP). That's why the controllers of all solar energy in electronic converters use some technique for maximum power point tracking (MPPT). There was a different MPPT techniques [1] are available. The Figure.1 shows the block diagram of suggested scheme.

IOP Conf. Series: Materials Science and Engineering **623** (2019) 012006 doi:10.1088/1757-899X/623/1/012006

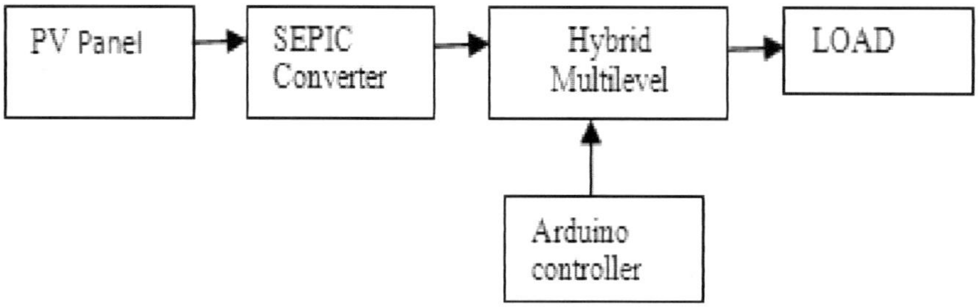

Figure.1. Block diagram of suggested system

Among the various techniques in maximum power point tracking (MPPT). In our suggested system use the process is Perturb & Observe (P&O) method. It's the simplest and most commonly used method. In this process, the sign of the last perturbation and increment in the power are used to decide what the next perturbation should be. If there is an increment in the power, the perturbation operate should be in the same direction and if the power decreases, then the next perturbation should be in the opposite direction. Based on these facts, the algorithm is implemented. The process is repeated until the MPP is reached [4].

21. Converter circuit operation

The DC-DC change in technology is mainly used in the field of power electronics and drives. These converters are mainly used in industrial applications and computer hardware circuits. The proposed DC-DC SEPIC topology provides stable and ripple free output voltage and current [8]. DC voltage obtained from the PV panel contains high voltage ripples and it's not constant. The traditional convertor doesn't meet the load demand containing additional ripples on the output voltage and parasitic effects.

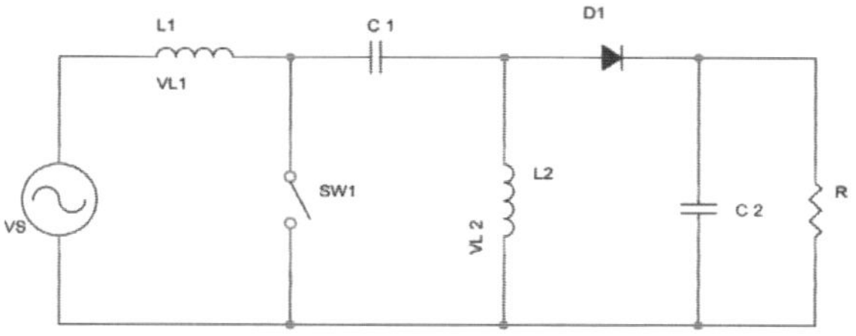

(a)

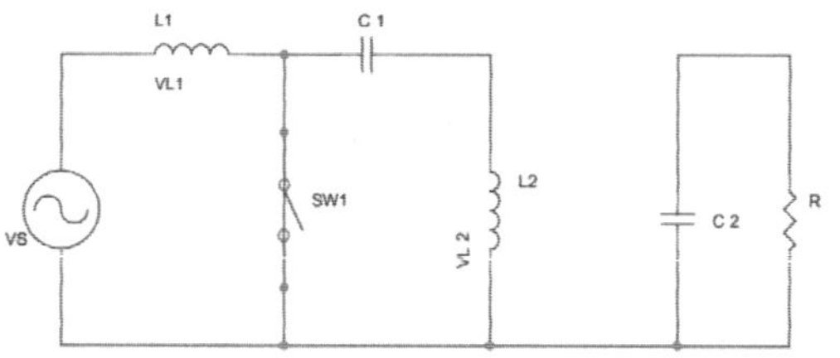

(b)

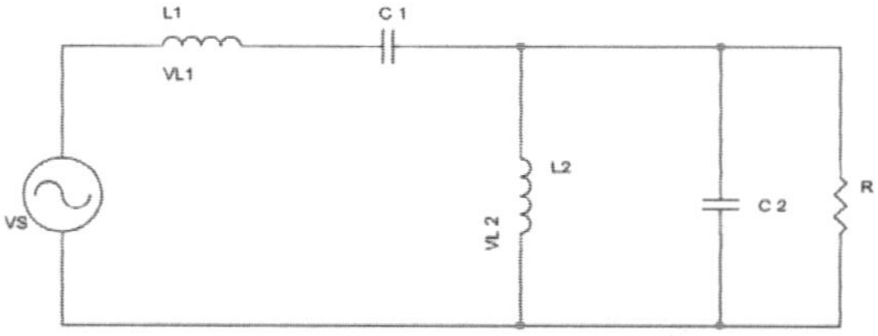

(c)

Figure.2. SEPIC converter circuit operations

In Figure. 2(b) when the switch is ON the inductor L1and L2 gets charging in reverse flow, in that time there is no output across the load R. In Figure. 2(c) when the switch is in OFF position the same current of L1.the L2 will continuous in negative direction. However the L1 being charged during the OFF cycle and L2 being discharged during the ON cycle.

2.2 Multi-level Inverter circuit operations

It has a more number of advantages it has drawbacks in the manner of higher levels because of using more number of semiconductor switches and isolated DC sources. This may leads to vast size and price of the inverter is very high. In order to overcome this problem the new multilevel inverter topology is proposed with reduced number of switches and sources. As a result it is possible to reduce or even to eliminate output ripples. The unified analysis and design consideration for hybrid multilevel inverter has been proposed in [15]. In this proposed system the Cascaded H-Bridge & Neutral Point Clamp topologies are hybridized.

In Figure. 3(a) when the switch SW6 is turned ON the current flow will be in the direction of VS6 (+) – SW6-D5- D4- D3- D2- D1- SW7- RL Load (+)- RL Load (-)- SW10- VS6 (-). During this time the diode D6 will not conduct. The output voltage and current will be positive. The switches in the H-Bridge SW7 and SW10 are turned ON for the operation of inverter in the positive half cycle. The inverter operation for the first level in negative cycle is shown in fig.3 (b), when the switch SW5 is turned ON the current flow will be in the direction of VS5 (+)-SW5-D4-D3-D2-D1-SW8-RL Load (-)-RL Load (+)-SW7-VS5 (-). The output voltage and current will be negative. The switches SW8 and

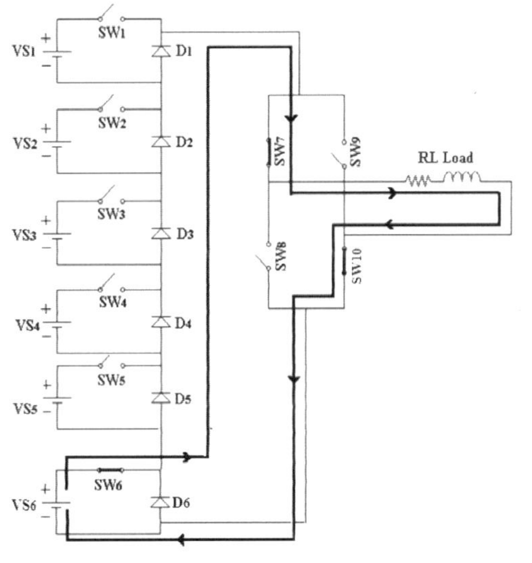

(A)

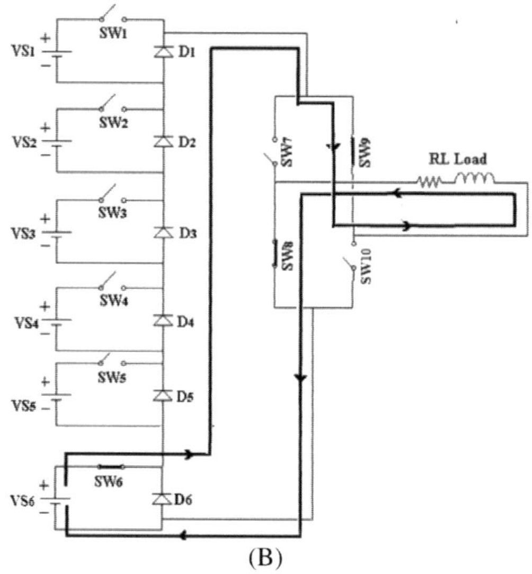

(B)

Figure.3. (a) Inverter operation for the first level in positive cycle
(b) Inverter operation for the first level in negative cycle

SW9 in the H Bridge are turned ON for the operation of inverter in the negative half cycle. Similarly the remaining level is obtained by alternatively ON and OFF of the switches as per the switching states given in the following table. The switching state is given for the 63 level of positive half cycle only. It will be same for the negative cycle but the difference is, when the switches SW7 and SW10 are conducting the inverter output voltage and current will be in positive half cycle, when the switches SW8 and SW9 are conducting the inverter output voltage and current will be in the negative half cycle.

Table.1 Switching States

Level	Neutral point clamped inverter(NPC)						Cascaded H-bridge inverter(CHB)			
	S1	S2	S3	S4	S5	S6	S7	S8	S9	S10
1	0	0	0	0	0	1	1	0	0	1
2	0	0	0	0	0	0	1	0	0	1
3	0	0	0	0	1	1	1	0	0	1
4	0	0	0	1	0	0	1	0	0	1
5	0	0	0	1	0	1	1	0	0	1
6	0	0	0	1	1	0	1	0	0	1
7	0	0	0	1	1	1	1	0	0	1
8	0	0	1	0	0	0	1	0	0	1
9	0	0	1	0	0	1	1	0	0	1
10	0	0	1	0	1	0	1	0	0	1
11	0	0	1	0	1	1	1	0	0	1
12	0	0	1	1	0	0	1	0	0	1
13	0	0	1	1	0	1	1	0	0	1
14	0	0	1	1	1	0	1	0	0	1
15	0	0	0	1	1	1	1	0	0	1
16	0	1	0	0	0	0	1	0	0	1
17	0	1	0	0	0	1	1	0	0	1
18	0	1	0	0	1	0	1	0	0	1
19	0	1	0	0	1	1	1	0	0	1

20	0	1	0	1	0	0	1	0	0	1
21	0	1	0	1	0	1	1	0	0	1
22	0	1	1	1	1	0	1	0	0	1
23	0	1	1	1	1	1	1	0	0	1
24	0	1	1	0	0	0	1	0	0	1
25	0	1	1	0	0	1	1	0	0	1
26	0	1	1	0	1	0	1	0	0	1
27	0	1	1	0	1	1	1	0	0	1
28	0	1	1	1	0	0	1	0	0	1
29	0	1	0	1	0	1	1	0	0	1
30	0	1	0	1	1	0	1	0	0	1
31	0	1	0	1	1	1	1	0	0	1
32	1	0	0	0	0	0	1	0	0	1
33	1	0	0	0	0	1	1	0	0	1
34	1	0	0	0	1	0	1	0	0	1
35	1	0	0	0	1	1	1	0	0	1
36	1	0	1	1	0	0	1	0	0	1
37	1	0	1	1	0	1	1	0	0	1
38	1	0	1	1	1	0	1	0	0	1
39	1	0	1	1	1	1	1	0	0	1
40	1	0	1	0	0	0	1	0	0	1
41	1	0	1	0	0	1	1	0	0	1
42	1	0	1	0	1	0	1	0	0	1
43	1	0	0	0	1	1	1	0	0	1
44	1	0	0	1	0	0	1	0	0	1
45	1	0	0	1	0	1	1	0	0	1
46	1	0	0	1	1	0	1	0	0	1
47	1	1	0	1	1	1	1	0	0	1
48	1	1	0	0	0	0	1	0	0	1
49	1	1	0	0	0	1	1	0	0	1
50	1	1	1	0	1	0	1	0	0	1
51	1	1	1	0	1	1	1	0	0	1
52	1	1	1	1	0	0	1	0	0	1
53	1	1	1	1	0	1	1	0	0	1
55	1	1	1	1	1	0	1	0	0	1
55	1	1	1	1	1	1	1	0	0	1
56	1	1	1	0	0	0	1	0	0	1
57	1	1	0	0	0	1	1	0	0	1
58	1	1	0	0	1	0	1	0	0	1
59	1	1	0	0	1	1	1	0	0	1
60	1	1	0	1	0	0	1	0	0	1
61	1	1	0	1	0	1	1	0	0	1
62	1	1	0	1	1	0	1	0	0	1
63	1	1	0	1	1	1	1	0	0	1

The parameter and its values which is used in simulation is given below in table 2.

Table.2 Values used in Simulation

S. No	**Parameters**	**Values used in simulation**
1	DC input voltage	19.1 V
2	Inverter output voltage	230 V
3	Total number of switches	11
4	Inductance (L1, L2)	45μH, 1.5mH

IOP Conf. Series: Materials Science and Engineering **623** (2019) 012006 doi:10.1088/1757-899X/623/1/012006

5	Capacitance (C1, C2)	20μF, 100μF
6	Number of level	127

In this proposed system the SEPIC converter boost up the input DC voltage to 230V and it can be splitted into six input sources for the inverter switches by using multiple output DC/DC converter. The ripples in the output can be eliminated by using filter elements (L2 & C2). By alternatively turning ON and OFF of the inverter switches (SW_1-SW_6) the 127 level of output voltage is obtained and it is converter into AC by means of CHB switches (SW_6-SW_8) which perform the operation of conventional inverter.

3. SIMULATION RESULTS

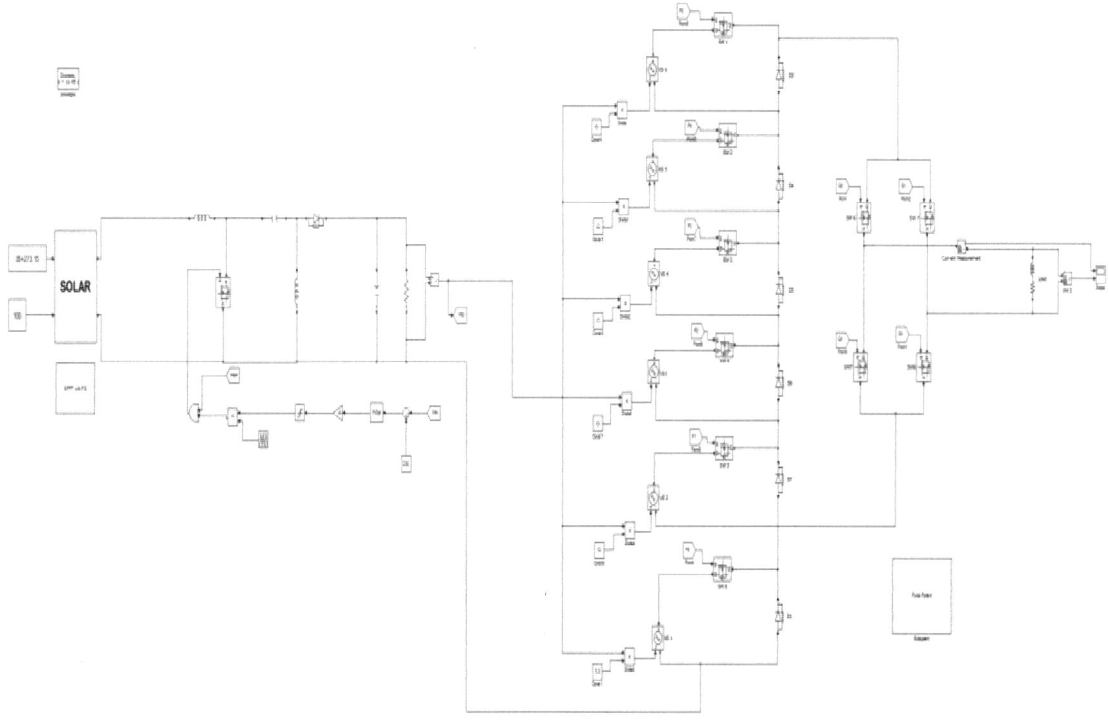

Figure 4. Simulation circuit for proposed system

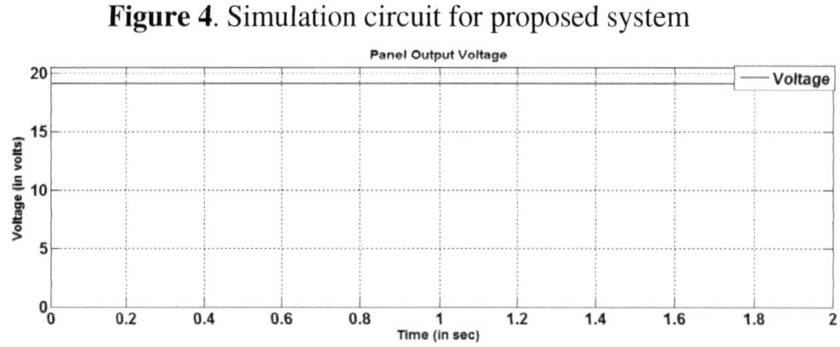

(a)

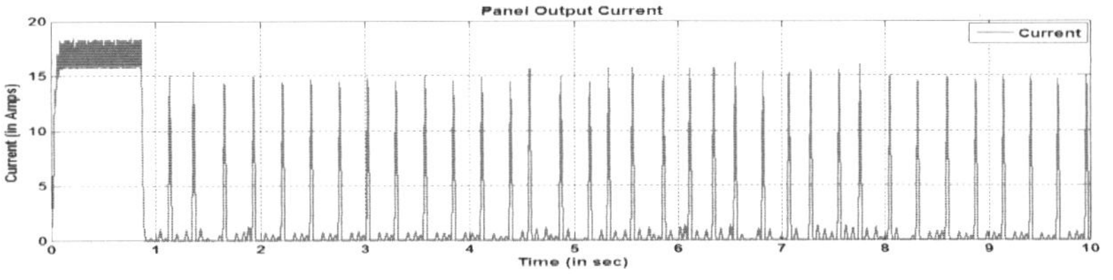

(b)

Figure.5. (a) PV panel output voltage
(b) PV panel output current

The Figure 5(a) shows the output voltage waveform for solar panel. The voltage magnitude is about 19.1 V. This voltage is not constant enough. To converter this voltage into constant DC this will be send into DC-DC SEPIC converter. The fig 5(b) shows the output current waveform for solar panel. The current magnitude is around 15 A.

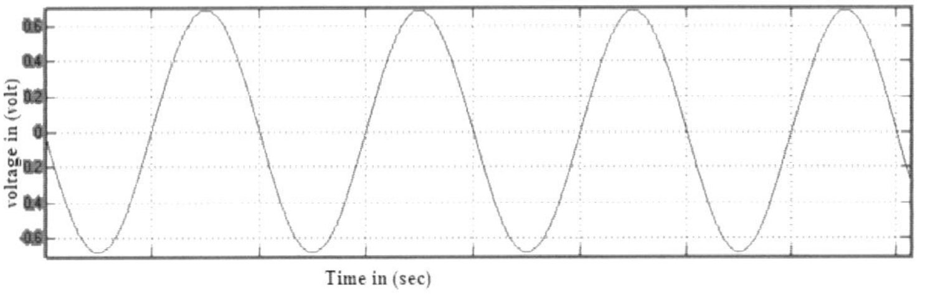

Figure 6. Inverter output127 level

The Figure 6(a) shows the output voltage waveform for proposed Hybrid multi-level inverter. The output voltage having 127 level. Hence THD can be reduced by using proposed inverter topology.

4. COCLUSION

This paper presents a PV based hybrid multilevel inverter. During this planned DC-DC sepic converter the extra filter parts eliminate the output ripples and effectively enhance the output voltage level. It overcomes the parasitic problems present in the classical dc-dc device. The 127 level of output voltage is obtained by using proposed Hybrid multilevel inverter topology. Hence the THD is reduced. The presented concepts have been verified in simulations and validated experimentally.

REFERENCES

[1] A. J. Basanth, S. P. Natarajan, T. S. Sivakumaran, "Performance Analysis of Positive Output Super-Lift Re-Lift Luo Converter With PI and Neuro Controllers,"*IOSR Journal of Electrical and Electronics Engineering (IOSR-JEEE)* e-ISSN: 2278-1676, p-ISSN: 2320-3331, Volume 6, Issue 3, pp 21-27, May. - Jun. 2013.

[2] J. P. Benner and L. Kazmerski, "Photovoltaics gaining greater visibility," *IEEE* Spectr., Vol. 29, No. 9, pp. 34–42, Sep. 1999..

[3] G. Buticchi, D. Barater, E. Lorenzani, C. Concari, and G. Franceschini, "A Nine-Level Grid-Connected Converter Topology for Single-Phase Transformerless PV Systems," *IEEE Transactions on industrial electronics,* Vol.61, No.8, pp 3915-3960, Aug. 2014.

[4] S. K. Chattopadhyay, C. Chakrabort and B. C. Pae, "Cascaded H-Bridge & Neutral Point Clamped Hybrid Asymmetric Multilevel Inverter Topology for Grid Interactive Transformerless Photovoltaic Power Plant," *IEEE,* 2012.

[5] T. Esramand P. L. Chapman, "Comparison of Photovoltaic Array Maximum Power Point Tracking Techniques," *IEEE Transactions on energy conversion,*Vol. 22, No. 2, pp. 439-449, June 2007.

[6 J. Huang and K. Corzine, "Extended Operation of Flying Capacitor Multilevel Inverters,".

[7] Y. Jiao and F.L. Luo, "An Improved Sliding Mode Controller for Positive Output Luo Converter," *IEEE,* 2009.

[8] S. B. Kjaer, J. K. Pedersen, and F. Blaabjerg , "A Review of Single-Phase Grid-Connected Inverters for Photovoltaic Modules," *IEEE transactions on industry applications,* Vol. 41, No. 5, september/october 2005.

[9] K. L. Lian, J. H. Jhang, and I. S. Tian, "A Maximum Power Point Tracking Method Based on Perturb-and-Observe Combined With Particle Swarm Optimization," *IEEE Journal of photovoltaic's,* Vol. 4, No. 2, pp. 626-633, March 2014..

[10] F. L. Luo and Hang ye, Advanced DC/AC inverters: Applications in renewable energy.

[11] A. Manikandan , N.Vadivel, "Design and Implementation of Luo Converter For Electric Vehicle Applications," *International Journal of Engineering Trends and Technology (IJETT),* Vol. 4, Oct. 2013..

[12] M. Malinowski, K. Gopakumar, J. Rodriguez, and M. A. Pérez, "A Survey on Cascaded Multilevel Inverters," *IEEE Transactions on industrial electronics,* Vol. 57, No. 7, pp 2197-2206, July 2010.

[13] A. Nami, F. Zare, A. Ghosh, and F. Blaabjerg, "A Hybrid Cascade Converter Topology with Series-Connected Symmetrical and Asymmetrical Diode-Clamped H-Bridge Cells," *IEEE Transactions on power electronics,* Vol. 26, No.1, pp. 51- 65, Jan 2011.

[14 M. H. Rashid, Power electronics handbook devices, circuits and applications, Third edition 2011..

[15] C. Rech, and J. R. Pinheiro, "Hybrid Multilevel Converters: Unified Analysis and Design Considerations," *IEEE Transactions on industrial electronics,* Vol. 54, No. 2, pp. 1092-1104, Apr. 2007.

[16] J. Rodríguez, and F. Z. Peng "Multilevel Inverters: A Survey of Topologies, Controls, and Applications," *IEEE Transactions on industrial electronics,* Vol. 49, No. 4, pp. 724-738, Aug. 2002.

.[17] J. Rodriguez, S. Bernet, P. K. Steimer, and I. E. Lizama, "A Survey on Neutral-Point-Clamped Inverters," *IEEE Transactions on industrial electronics,* Vol. 57, No. 7, pp. 2219-2230, July 2010.

.[18] M. G. Villalva and E. Ruppert, "Analysis and Simulation of the P&O MPPT Algorithm Using a Linearized PV Array Model," *IEEE,* 2009.

IOP Conf. Series: Materials Science and Engineering **623** (2019) 012007 doi:10.1088/1757-899X/623/1/012007

Current control techniques of single phase shunt active power filter- a review

[1]Rameshkumar Kanagavel, [2]Indragandhi.V

[1]Research Scholar, VIT University, Vellore, Tamilnadu, India
[2]Associate Professor, VIT University, Vellore, Tamilnadu, India

Abstract. This manuscript gives a complete survey on current control for single phase Shunt Active Power Filter (SAPF). Various current control techniques like linear current control, Hysteresis, Model Predictive Current Control (MPCC) and other techniques are discussed. The primary advantages and disadvantages of current controllers in SAPF have been presented in a table. This manuscript offers a suitable reference for forthcoming work in power quality applications and its associated research.

1. Introduction

In distribution network the power quality problem is mainly due to the nonlinear loads.The NLLs like uninterruptible power supplies, Grid-connected inverters, light controllers, electric furnaces, temperature controllers, solid-state ac voltage regulators, and arc furnaces cause power quality problems at the consumer ends. Because of the nonlinear loads the major problems in the power sector applications like overheating of the transformer, machine vibrations, inaccurate power metering and disturbance to nearby communication networks [1-4].
In order to improve the power quality conventional solution passive filters used which is of low cost, simple complication and improved efficiency, but still has major drawbacks like fixed compensation, huge size and resonance problems.So, to overcome the above mentioned cons the most effective techniquefor improving power quality the use of Shunt Active Power Filter (SAPF) is preferred.
The SAPF concept was firstly proposed in 1971 by Sasaki and Machida for removing current harmonics[5]. The SAPF operate in different operating conditions and higher frequencies effectively, which leads to giving enhanced performances. The general aim of a SAPF is to reduce harmonic distortions caused by non-linear loads and improve power factors for reactive loads. The single-phase active power filter is typically used in lower power ratings. It is suitable for educational buildings and commercial with computers load [6, 7]. The efficient performance of a SAPFmostlybased upon three different areas such as estimate the reference source current, current controller, and selection of parameters in SAPF.
To increase the performance of the SAPF researchers have been proposed various extraction technique with numerous current control methods were developed such as Hysteresis, Linear current control, Repetitive, parabolic, Ramp time, current mode, Dead Beat, One cycle, Predictive, Model predictive current control and Artificial intelligence method. The literature explored in this paper involves publications from IEEE, IET, and Elsevier, etc. from 1991 to 2016. This manuscript aims to review and discuss various current control methodologies of single phase SAPF.

2. Review on Current Control Techniques in SAPF

The SAPF system performance mainly depends on the excellence of the applied current control method. The hysteresis and linear current control techniques are mainly used in SAPF filter. The advanced other current control approaches, such as Parabolic Current control, current mode control, Ramp time current control, Delta modulation control, MPCC and artificial intelligence based current control approaches are used to control the single phase SAPF switches

2.1 Hysteresis current controller

From the literature review, it has been noticed that numerous researchers give more importance to the hysteresis controller for SAPF application [8-18]. This method is used to controls inverter switches, it can force the actual current to go up and down to track the input reference current of SAPF. The current error signal e (t), which is the difference between the filter reference current and actual filter current. The Hysteresis current controller is shown in Fig.1a, 1b and Fig. 1c.

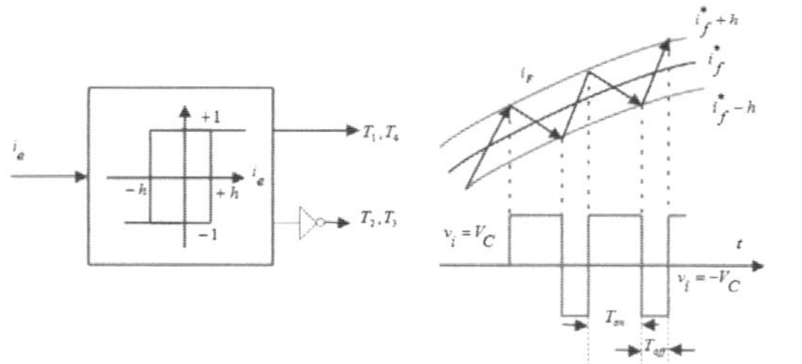

Figure. 1. a. Single band hysteresis current control

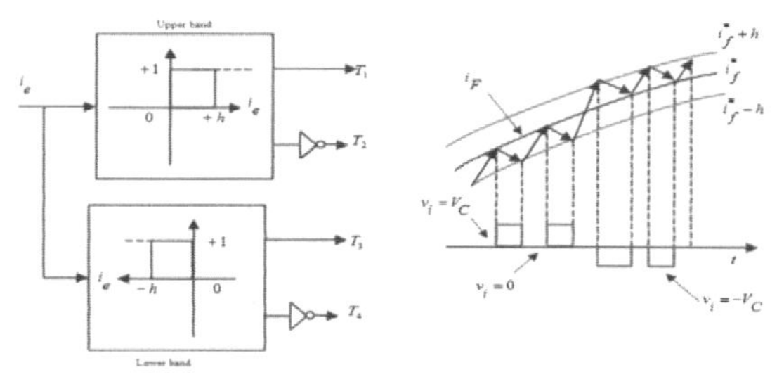

Figure.1. b. Double band hysteresis current control

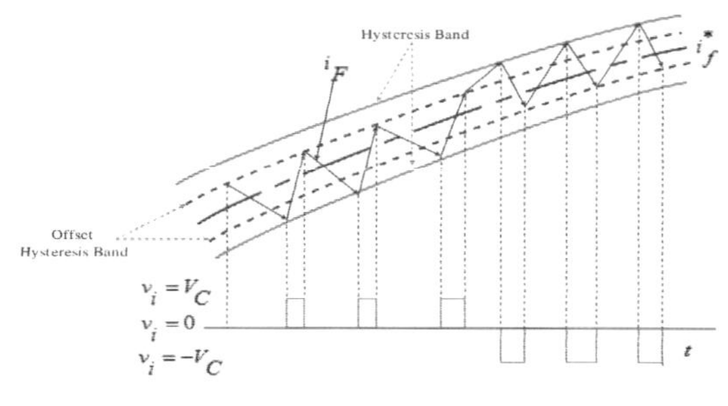

Figure. 1. c three level Hysteresis band

The benefit of the hysteresis control techniques is simple to implement, maximum current limit, high accuracy and fast response in transient condition. The drawbacks of this method are dependent on load

parameters, varying switching frequency with system voltage, generation of sub harmonics which affect the quality power system.
In [29] author proposed the concept of limiting the power devices maximum switching frequency. The calculation of current ripple and the switching frequency by using simple algebraic equation as a function of the hysteresis band, dc bus voltage and inductive output filter value in [30]. The controller has benefits such as simple to implement and fast transient response. Also, it has some problems like mostly operation is dependent on load parameters and switching frequency variations with system voltage.
In [14] the author proposes a new generalized frequency domain formulation based method. It was established based on the Tsypkin's method, the relation between hysteresis bandwidth and the maximum allowable switching frequency of controller were found. The converter switching frequency is mostly based on the converter's parameter, and the load and line of the distribution system.
In [16] the author proposed that a three-level Hysteresis current controller be developed by taking two HCC. By using an extra offset hysteresis band over the exiting two-level based Hysteresis band which have a slight difference in switching frequency as illustrated in Fig 1 (c).In [34] the author introduce a double band based HCC method for SAPF as shown in Fig 1. (b). In this method is to permit access to the zero level of the SAPF input supply voltage due to that an inverter switch is only operated through a half cycle, whereas during another cycle, it remains either on or off. This method gives much lower average switching frequency and low total harmonic distortion when compared with conventional hysteresis based current control for lower hysteresis band values. This method offers better dynamic performance under bridge rectifier with R and RL load conditions.

2.2 Linear Current controller
The active current wave shaping controller to produce the gating signals to the power converter is proposed in [22] [19]. The active current wave shaping controller consisting of a subtraction circuit, an error amplifier, and a PWM modulator. The source current of SAPF is subtracted from the reference source current of SAPF and then sent to an error amplifier if the circuit. The output signal from the error amplifier is fed to the PWM modulator for generation of gating signals. It can be illustrated as

$$V_m(t) = K(i_{S,ref}(t) - i_S(t)) \quad (1)$$

where K is the gain of the error amplifier. In [21] the author proposed modified PWM technique for controlling power converter. The controlling signal is made from the comparison between the source current and its reference. The actual supply current is subtracted from the reference supply current, and then the error current signal is sent to PI controller, it has a limiter circuit at its output side. Subsequently, the controlling signal from the controller is equalled with PWM triangular carrier signal as shown in Fig .2. If the controlling signal output is more than the triangularpwm wave, then the converter upper switching devices are switched on, and also the lower switching devices are switched off, and vice versa. The control signal can be represented in the mathematical model:

$$V_m(t) = K_P\left(i_{S,ref}(t) - i_S(t)\right) + \frac{K_I}{s}\left(i_{S,ref}(t) - i_S(t)\right) \quad (2)$$

The dynamic response is enhanced by changing the values of $K_P$and K_I. The PI controller is mainly used for its high DC gain, and to remove the steady state error of the system. The main drawback of this method is bandwidth, which remains unchanged gives delays and substantial error in the following of the high value of harmonics components of the supply reference current. The gain values of the PI controller were found by trial and error method or by optimization method. The modified PWM technique method has the advantage of less THD value when compared to standard PWM.

IOP Conf. Series: Materials Science and Engineering **623** (2019) 012007 doi:10.1088/1757-899X/623/1/012007

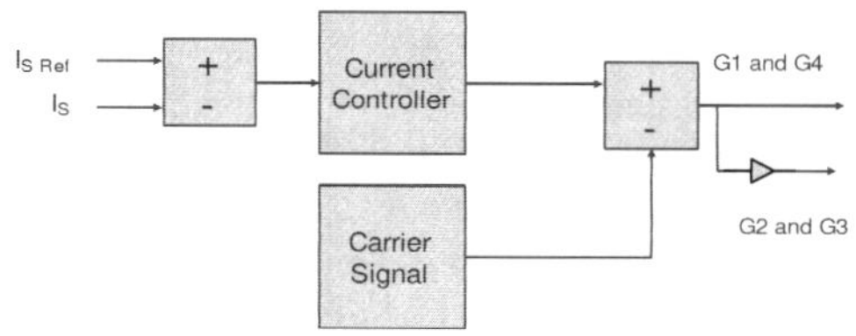

Figure. 2 a PI-PWM Current controller

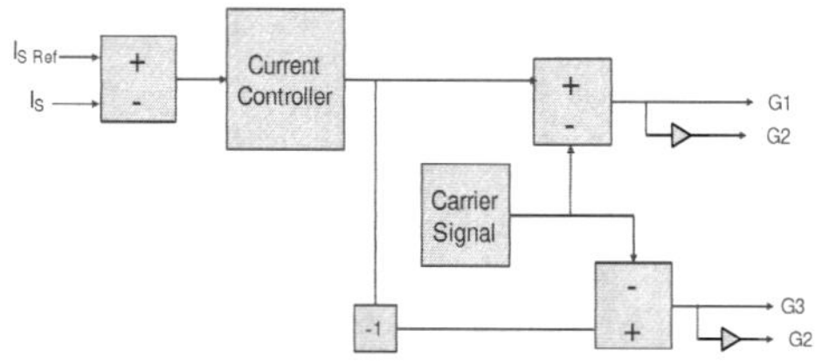

Figure. 2 b PI-M-PWM Current controller

Fig 2 (a) and 2 (b) explains the two different switching techniques for controlling single phase SAPF, which is discussed by the author in [20]. This periodic sampling and Pulse Width Modulation (PWM) for two different switching frequencies are used. This method is simple to implement, but it does not work with a constant switching frequency and PWM method operating with avariable duty cycle and aconstant predetermined switching frequency. In this method different types of nonlinear loads were consider to test and analyse the single phase SAPF such as a linear RL load, a controlled bridge rectifier with RL type load, and a same rectifier with RC type load. In rectifier with RL load case source current THD is minimum by using Periodic Sampling method when compared with PWM method. In single-phase, SAPF prototype model with different switching methods was discussed [21].

2.3 Parabolic Current control

The parabolic current control comes under the category of variable hysteresis band current control technique. By using this method to generate the variable hysteresis bands according to the periodic parabolic function to achieve the desired control characteristics with fixed switching frequency and better transient response. The parabolic function is reset when the error signal hits it [22].

2.4 RAMP time current controller

The current controller of the ramp type works so that the current of the inductor accurately tracks the desired reference current with a fixed switching frequency. This technique to generate the gating signal of inverter is based on the instant at which the zero crossing occurs. This control technique is almost continuous and has a excellent dynamic efficiency like a standard hysteresis controller. The input error signal to the PI controller is a binary signal (ε), which can point out. The error polarity between the actual and reference compensating reference. And the value of next switching instant is obtained by making use of polarity of error signal and switching of controllerThis control technique is designed to generate an accurate control at a moderately small switching frequency band with slight low order harmonics. The ramp type controller is classified into three types

1. Ramp-time current controller

2. Digital-ramp time current controller

3. Polarized Ramp-time current controller

The working concept of Digital Ramp time current controller is discussed in [25]. It is similar to the normal ramp time controller. The main difference between two ramp time control methods is the generation of error polarity signal. Digital Ramp time controller has advantages without using analogue circuits like ADC, comparator, and DAC to produce ε signal digitally by using internal digital source current by the FPGA controller. By comparing two ramp-time controllers, the normal ramp-time controller is not substantially affected by sampling frequency and it provides better output than the Digital ramp-time controller because of the accuracy of zero crossing of present errors in the analog domain. At the same time, for highly demanding power electronics applications like APF, the digital ramp time controller can be used due to the option of choosing multisampling strategy for its operation. The SAPF can be controlled by the Polarized Ramp-time Current Control (PRCC). It works with a set switching frequency based on the notion of Zero Average Current Error (ZACE), and it does not produce any extra low-order harmonics to the scheme. The PRCC technique has a large bandwidth and a quick transient reaction [26].

2.5 Current-Mode Controller

With a fixed switching frequency, the current mode control technique is used in SAPF. It can be classified as

1. Peak current mode control (Peak CMC)
2. Valley current mode control (Valley CMC).
3. Average current mode control (Average CMC)

In the peak CMC technique, when switching on condition, it senses the inductor maximum current and information is used to turn off the switch. On the other side, it senses the inductor peak current when switching off condition in the Valley CMC technique, and that information is used to turn on the switch. This switching mechanism is called the trailing edge and the leading edge modulation. The Peak CMC method is mostly used in various power converters like rectifiers and converter applications due to it naturally give over current protection. Whereas the valley CMC is used in output current regulators for DC/AC converter and APF, because of its non-average ripple nature, it is unable to deliver the regulated current with zero DC component.

The Peak CMC scheme has drawbacks of a low gain vale of current control loop and poor noise immunity. To overcome the above problem, the Average CMC method is used it has high gain value in current control loop, excellent noise immunity and it is tracking the current program with a high degree of accuracy [45].

2.6 Delta Modulation based current control

The delta modulation system is the variation of the standard hysteresis controller, which provides constant voltage during the converter switching period. It operates based on a comparison between fixed tolerance band and current error. When the positive current error occurs during the operation, the inverter output voltage must be positive and the inverter output voltage must be negative if the negative current error occurs. When compared to traditional hysteresis current controller a small phase lag addition in the tracking process is minimized. The main advantage of this method is an excellent dynamic response and simple implementation [24].

2.7 Dead-beat current controller

Since the early 1990s, the dead-beat controller has been used for present inverter control and shunting active filters. This control technique requires detail knowledge of parameters like filter inductance value, line voltage, present active filter current or supply current and the value of reverence current. The primary disadvantage of this technique is the delay in the control phase. In order to minimize the harmful delay, a unique technique is implemented to compensate for the mistake of nth and (n-1)th instantly by identifying the error at nth instant with the output at (n-1)th instant. [33]. A modified dead beat controller is used in [34]. By the end of the modulation period , it can estimate the period directly when a switching device is switched on to generate the actual phase current track to its reference. The average value of inverter output voltage can be represented as

IOP Conf. Series: Materials Science and Engineering **623** (2019) 012007 doi:10.1088/1757-899X/623/1/012007

$$u_{inv} = \frac{L}{T_S}\left(i_{ref} - i_{act}\right) + u_{sys} \quad (3)$$

where u_{inv} is the average value of inverter output, T_S is the sampling time, and u_{sys} is the system voltage. i_{ref}, i_{act} is the reference and actual phase current respectively. It has a simple control algorithm, and it minimizes the number of current sensors and reduces the cost and the size of the system [35-37].

2.8 Model Predictive current controller

Next to previously mentioned current control strategies, there are a few current control techniques are utilized to improve the SAPF execution. A standout amongst the most prominent strategies for current control in SAPF is MPC control method. MPC control technique was developed in the process control sector in 1970, and its interest has only reached its peak in latest days. Its gaining popularity is due to the accessibility of quick processor that in a fraction of seconds can solve many complicated issues. The main benefits of the MPC based control technique are the utilization of variable switching frequency,absence of modulator, lower complexity and the possibility of online optimization for inverters and the addition of constraints. The MPC has recently developed an approach to control of various power converters.

The MPCC of three-Level NPC converter based single phase SAPF is used for compensation of load harmonics [38]. MPCC of single phase SAPF is presented [39]. Photovoltaic integrated SAPF to provide harmonic reactive power compensation using MPC control was given in [40]. This method having better current tracking capability in dynamic conditions. The discrete mathematical of the SAPF, DC Link voltage and equivalent power system impedance are used to design the MPC controller to find out the future reference current. Most of the control methods involve complex calculations, need of modulation stage to produce the switching signals and thus its real-time implementation is time-consuming and challenging. MPC scheme does not require internal current control loops and modulation stage.

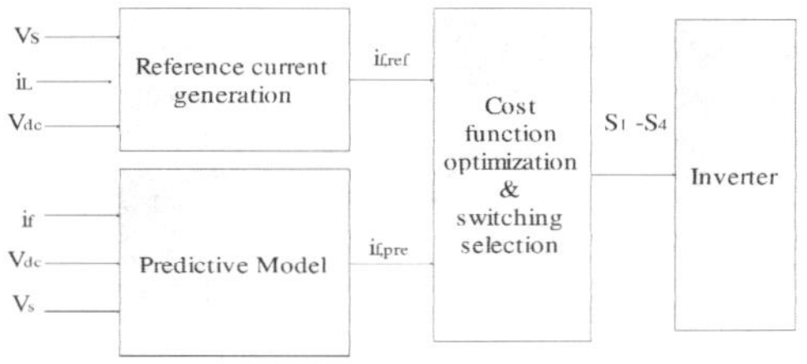

Figure.3. Model Predictive current control schematic diagram

Figure. 3 demonstrates the schematic diagram of a MPC based SAPF. The essential control of MPCC is to limit the error between the references and estimated load current which is characterized as cost function.

To achieve this goal, the SAPF switching state that will reduce ' g ' is selected and given at the time of the next sampling instant. The main problem of MPC strategy is variable frequency and involve more design computation. This technique, however, is based on a simple approach and can be easily implemented in real time owing to the rapid growth of controllers such as microprocessors, digital signal processors and field-programmable gate arrays.

In addition, the MPC strategy is flexible ; the controller design includes additional limitations such as minimizing switching frequency and minimizing switching loss. In [42], author mixed predictive control with the traditional space vector modulation system to obtain steady switching frequency operation, but this technique is complicated to implement in real time.

2.9 Intelligent Current Control Techniques

The neural networks belong to the artificial intelligence techniques family, fuzzy logic, genetic algorithms. Artificial Neural network (ANN) is a connection of some artificial neurons which simulate a biological brain system. NN can train either offline or online when it is used in a system control [41]. The ANN based current control of SAPF. The error signal is provided as an ANN input and ANN produces an appropriate switching signal for harmonic and reactive power compensation to the SAPF. In this method, a fixed switching frequency operation is achieved. In this control approach, the conventional controllers are replaced by the fuzzy controller, which possess two input and one output signal. The error (e) and its derivative (de) are used as the input to the fuzzy controller while the output is the command (Cde). The Intelligent Current Control methods do not require a model of SAPF; they need precise knowledge about operation and behavior of SAPF systems.

After reviewing the different current control techniques, we summarized the control methods as shown in Table 1. In the previous sections, current control techniques are studied, and significant advantages and disadvantages were discussed in Table 1. This table gives a better idea to the researcher to select the optimal controller for designing a SAPF. By exploring different current controllers due to their dynamic performance, tracking behavior and switching frequency, it is concluded that MPC control is a better choice for APFapplications. The drawback of MPC controller is that switching frequency variations. To solve this problem the combined predictive control with traditional space vector modulation scheme to prefer or to achieve constant switching frequency operation optimized MPC controller is used.

Table 1. Comparison of current controller

Type of Current Controller	Advantages	Disadvantages
Hysteresis Controller	• Simple and Robust • Fast transient response • Doesn't need complex circuits for real-time implementation • High robustness	• Variable switching frequency • High current ripple • Not preferable in lower power level applications due to switching loss.
PI-PWM Controller	• Fixed switching frequency • Small calculation time • Simple to implement	• Accurate tuning is needed • Non zero steady-state error • Poor dynamic response
Parabolic Current control	• Fast transient response • High accuracy control with constant frequency switching	• Due effect of dead time the stable duty cycle is reduced, switching frequency deviation occur and current tracking error increased. • Inadequate resolution of parabolic carrier generators could influence the current control performance. • Noise from the control board influences the control accuracy

Ramp time current controller	• Fixed switching frequency • Good dynamic performance	• Relatively complex
Current-Mode Controller	• Fixed switching frequency	• It is mostly not preferable in VSI application due presence of significant DC component in the current
Delta Modulation based current control	• Easy implementation • Simple control process due to inherent self-carrier generating feature	• Asymmetrical operation
Dead-beat current control	• Fixed switching frequency • High dynamic performance • Digital in nature	• Low robustness • High computational power • Relatively complex
Model Predictive Control	• Possibility to include nonlinearities • High dynamic response • Allow attaining more accurate current control with less harmonic distortion • System perturbations and dead times can be compensated	• Variable switching frequency • Needs many calculations • Selection of weighting factor is not analytical or numerical
Intelligent Current Control	• Fixed switching frequency operation	• Need precise knowledge about operation and behavior of SAPF systems

3. Conclusion

In this manuscript, a classification and comparison of different SAPF control techniques were evaluated.In addition, the present control technique proposed and used by different researchers is up to date and has been recognized and summarized with its pros and cons. The comparisons table is showed to discuss the various current controllers in detail. From the review, the authors discovered that predictive current controllers have a wide range of developments in recent days, which may give better and more efficient results than the existing controllers. It is envisaged that this manuscript will be a suitable one-stop reference source for engineers, manufacturers, and researchers involved in the topic discussed.

4. References

[1]. Ataide, M. V., and J. A. Pomilio. "Single-phase shunt active filter: a design procedure considering harmonics and EMI standards." Industrial Electronics, 1997. ISIE'97. *Proceedings of the IEEE International Symposium* on. Vol. 2. IEEE, 1997.

[2]. Qazi, SajidHussain, and MohdWazir Mustafa. "Review on active filters and its performance with grid connected fixed and variable speed wind turbine generator." *Renewable and Sustainable Energy Reviews* 57 (2016): 420-438.

[3]. Chatterjee, Kishore, B. G. Fernandes, and Gopal K. Dubey. "An instantaneous reactive volt-ampere compensator and harmonic suppressor system." *IEEE Transactions on Power Electronics* 14.2 (1999): 381-392.

[4]. Pottker, Fabiana, and Ivo Barbi. "Power factor correction of non-linear loads employing a single phase active power filter: control strategy, design methodology and experimentation." Power

Electronics Specialists Conference, 1997. *PESC'97 Record., 28th Annual IEEE*. Vol. 1. IEEE, 1997.

[5]. Sasaki, H., and T. Machida. "A new method to eliminate AC harmonic currents by magnetic flux compensation-considerations on basic design." *IEEE Transactions on Power Apparatus and Systems* 5 (1971): 2009-2019.

[6]. Mahela, Om Prakash, and Abdul GafoorShaik. "Topological aspects of power quality improvement techniques: A comprehensive overview." *Renewable and Sustainable Energy Reviews* 58 (2016): 1129-1142.

[7]. Singh, Bhim, Ambrish Chandra, and Kamal Al-Haddad. Power quality: problems and mitigation techniques. *John Wiley & Sons*, 2014.

[8]. Mahanty, R. "Indirect current controlled shunt active power filter for power quality improvement." *International Journal of Electrical Power & Energy Systems* 62 (2014): 441-449.

[9]. Afghoul, Hamza, et al. "Design and real time implementation of fuzzy switched controller for single phase active power filter." *ISA transactions* 58 (2015): 614-621.

[10]. Antchev, M., M. Petkova, and M. Petkov. "Single-phase shunt active power filter using frequency limitation and hysteresis current control." *Power Conversion Conference-Nagoya,* 2007. PCC'07. IEEE, 2007.

[11]. Zabihi, Sasan, and FiruzZare. "A new adaptive hysteresis current control with unipolar PWM used in active power filters." *Australian Journal of Electrical and Electronics Engineering* 4.1 (2008): 9-16.

[12]. Rafael, Ordonez, and Sadarnac Daniel. "Hysteresis describing method for control of an active filter." *Proc. IEEE-IECON*. Vol. 8. 2008.

[13]. Mishra, Mahesh K., and K. Karthikeyan. "An investigation on design and switching dynamics of a voltage source inverter to compensate unbalanced and nonlinear loads." *IEEE Transactions on Industrial Electronics* 56.8 (2009): 2802-2810.

[14]. Gupta, Rajesh. "Generalized frequency domain formulation of the switching frequency for hysteresis current controlled VSI used for load compensation." *IEEE Transactions on Power Electronics* 27.5 (2012): 2526-2535.

[15]. Komurcugil, Hasan. "Double-band hysteresis current-controlled single-phase shunt active filter for switching frequency mitigation." *International Journal of Electrical Power & Energy Systems* 69 (2015): 131-140.

[16]. Varaprasad, O. V. S. R., and DVSS Siva Sarma. "An improved three level Hysteresis Current Controller for single phase shunt active power filter." *Power Electronics (IICPE),* 2014 IEEE 6th India International Conference on. IEEE, 2014.

[17]. Rukonuzzaman, M., and M. Nakaoka. "Single-phase shunt active power filter with harmonic detection." *IEE Proceedings-Electric Power Applications* 149.5 (2002): 343-350.

[18]. Khadem, Shafiuzzaman, MalabikaBasu, and Michael Conlon. "Harmonic power compensation capacity of shunt active power filter and its relationship with design parameters." *IET Power Electronics* 7.2 (2014): 418-430.

[19]. Wu, J-C., and H-L. Jou. "Simplified control method for the single-phase active power filter." *IEE Proceedings-Electric Power Applications* 143.3 (1996): 219-224.

[20]. Bojoi, RaduIustin, et al. "Enhanced power quality control strategy for single-phase inverters in distributed generation systems." *IEEE Transactions on Power Electronics* 26.3 (2011): 798-806.

[21]. Neves, Pedro, et al. "Experimental results of a single-phase shunt active filter prototype with different switching techniques." Industrial Electronics, 2007. ISIE 2007. *IEEE International Symposium* on. IEEE, 2007.

[22]. Daniyal, H., et al. "Comparing current control methods using an active power filter application as the benchmark." *Power Engineering Conference, 2008. AUPEC'08.* Australasian Universities. IEEE, 2008.

[23]. Rahmani, Salem, et al. "Modified PWM with indirect current control technique applied to a single-phase shunt active power filter topology." *Industrial Electronics Society, 2003. IECON'03.* The 29th Annual Conference of the IEEE. Vol. 2. IEEE, 2003.

[24]. Petit, Johann F., HortensiaAmarís, and Guillermo Robles. "Control schemes for shunt active filters to mitigate harmonics injected by inverted-fed motors." *15th PSCC,* Liege (2005): 22-26.
[25]. Daniyal, Hamdan, et al. "Design and development of digital ramptime current control technique." *IEEE Transactions on Industrial Informatics* 9.2 (2013): 992-1002.
[26]. Tumbelaka, Hanny H., and Lawrence J. Borle. "Harmonic Mitigation Using a Polarized Ramp-time Current-Controlled Inverter." *TELKOMNIKA* (Telecommunication Computing Electronics and Control) 8.3 (2010): 235-244.
[27]. Hamad, M. S., A. M. Fahmy, and M. Abdel-Geliel. "Power quality improvement of a single-phase grid-connected PV system with fuzzy MPPT controller." *Industrial Electronics Society, IECON* 2013-39th Annual Conference of the IEEE. IEEE, 2013.
[28]. Johann Petit Suárez, HortensiaAmarís and Guillermo Robles "Current control schemes for three-phase four wire shunt active power filters: a comparative study" *Rev. Fac. Ing. Univ. Antioquia* N. 52 pp. 206-214, 2010.
[29]. Petit Suárez, Johann, HortensiaAmarís, and Guillermo Robles. "Current control schemes for three-phase fourwire shunt active power filters: a comparative study." *RevistaFacultad de Ingeniería Universidad de Antioquia* 52 (2010): 206-214.
[30]. Petit, Johann F., HortensiaAmarís, and Guillermo Robles. "Control schemes for shunt active filters to mitigate harmonics injected by inverted-fed motors." *15th PSCC, Liege (2005):* 22-26.
[31]. Tumbelaka, Hanny H., and Lawrence J. Borle. "Harmonic Mitigation Using a Polarized Ramp-time Current-Controlled Inverter." *TELKOMNIKA* (Telecommunication Computing Electronics and Control) 8.3 (2010): 235-244.
[32]. Daniyal, Hamdan. Investigation of digital current control based on zero average current error concept. University of Western Australia, 2011.
[33]. Petit, Johann F., HortensiaAmarís, and Guillermo Robles. "Control schemes for shunt active filters to mitigate harmonics injected by inverted-fed motors." *15th PSCC, Liege (2005)*: 22-26.
[34]. Han, Byung-Moon, Byong-YeulBae, and Seppo J. Ovaska. "Reference signal generator for active power filters using improved adaptive predictive filter." *IEEE Transactions on Industrial Electronics* 52.2 (2005): 576-584.
[35]. Fahmy, A., et al. "Power quality improvement in three-phase four-wire system using a shunt APF with predictive current control*." IECON 2012-38th Annual Conference on IEEE Industrial Electronics Society*. IEEE, 2012.
[36]. Hamad, M. S., A. M. Fahmy, and M. Abdel-Geliel. "Power quality improvement of a single-phase grid-connected PV system with fuzzy MPPT controller*." Industrial Electronics Society, IECON* 2013-39th Annual Conference of the IEEE. IEEE, 2013.
[37]. Johann Petit Suárez, HortensiaAmarís and Guillermo Robles "Current control schemes for three-phase four wire shunt active power filters: a comparative study" *Rev. Fac. Ing.* Univ. Antioquia N. 52 pp. 206-214, 2010.
[38]. Petit Suárez, Johann, HortensiaAmarís, and Guillermo Robles. "Current control schemes for three-phase fourwire shunt active power filters: a comparative study." *RevistaFacultad de Ingeniería Universidad de Antioquia* 52 (2010): 206-214.
[39]. Acuna, Pablo, et al. "A single-objective predictive control method for a multivariable single-phase three-level NPC converter-based active power filter." *IEEE Transactions on Industrial Electronics* 62.7 (2015): 4598-4607.
[40]. Rameshkumar, K., Indragandhi, V., Palanisamy, K., &Arunkumari, T. Model Predictive Current Control of Single Phase Shunt Active Power Filter. *Energy Procedia*, (2017) 117, 658-665.
[41]. Kanagavel, R., Indragandhi, V. R., Palanisamy, K., &Kannan, R. "A Novel Current Control Technique for Photo Voltaic Integrated Single Phase Shunt Active Power Filter" *International Journal of Renewable Energy Research (IJRER),* 7(4), 1709-1722.
[42]. Lamchich, MoulayTahar. "ANN current controller based on PI-Fuzzy adaptive system for shunt power active filter." *Advances in Power Electronics* 2012 (2012).

Disturbance Analysis and Implementation of High Voltage Gain Non-Isolated DC-DC Converter for Renewable Applications

Arunkumari T[1], Indragandhi V[2*], Vikramjeet Singh Sidhu[3], K T M Udayanga Hemapala[4]

[1,2,3]School of Electrical Engineering, Vellore Institute of Technology, Vellore, India
[4]Department of Electrical Engineering, University of Moratuwa, Sri Lanka
Corresponding Author: indragandhi.v@vit.ac.in

Abstract. A modified high gain single switch high-voltage converter is presented in this paper. The proposed converter is a blend of a voltage tripler and a SEPIC-boost mode converter. Its operational modes are presented in detail. This converter has advantages that include a single switch, reduced component stress and size, and reduced ripple output voltage and input ripple current. The simulation results of the suggested converter and a conventional multilevel SEPIC converter are presented. The prototype of a 30 V to 400 V input, 400 W output is developed and the performance of the designed topology is evaluated. The theoretical and experimental results conclude that this converter is suitable for high-voltage applications.

Index Terms—High voltage gain, SEPIC- boost mode converter, Ripple, Voltage Tripler.

1. Introduction

Converters and inverters play a vital role in converting renewable energy to electrical energy [1-4].The PV energy output has low voltage, hence to step up the voltage, high ratio converters are used. High voltage step-up PV energy is utilized for industrial, residential, and grid connection purposes. DC-DC converters are of two types, with and without Transformer. As transformer introduces loss, non-isolated converters are often found to be better suited for PV energy conversion [5]. Hence, non-isolated boost, buck, and buck-boost converters are preferred for conversion. The boost converter is utilized for high voltage conversion with large duty cycles but is highly stressful to the switch affecting the conversion efficiency [6].

To improve the conversion ratio, various high gain topologies such as dual boost converters, Single-Ended Primary Inductor Converter (SEPIC), Cuk converters, etc. [7-12].Different techniques have been used to improve the voltage conversion ratio, such as switched capacitor technique, coupled inductive method, voltage lift technique, etc., but are replete with drawbacks such as high voltage stress on the switch, pulsating input current, complexity of the circuit, and high cost of the components [13-14].

These techniques have been combined with normal boost converters and quadratic converters. However, the focus is on SEPIC converters as their performance is better than other converters. SEPIC converter has been reported to be better than conventional boost converts in terms of efficiency and transient response. The SEPIC converter is utilized to bring down reverse recovery loss and to enhance the power factor correction [15]. In a study, the SEPIC converter was implemented with voltage lift technique and obtained a positive–positive DC voltage with high conversion gain [16]. In [17-18] the SEPIC converter was combined with high frequency transformer to sustain the inductor current during the conduction period.
The SEPIC-Integrated Boost converter (SIB), a combination of SEPIC and boost, was designed to improve the PV voltage ratio. SIB was found capable to reduce the diode loss and reverse recovery current [19-20]. The SEPIC converter using voltage multiplier technique was suggested in some studies [21-24]. The modified sepic converter has been proposed in [25] and here in this manuscript the extension along with load and line disturbance analysis is proposed.

To overcome these drawbacks, a non-isolated modified high gain SEPIC converter with a single switch design is proposed here. The benefits of the proposed converter are a high voltage conversion ratio, low switching voltage, reduced output ripple current, low duty control, lesser number of components, easy control, and cost effectiveness compared to other converters. The proposed modified high gain SEPIC converter is depicted in Fig. 1. The presented converter provides 400 W from input of 30 V.
This paper is structured as follows: Section II details the operation of the proposed converter with Continuous Conduction Mode (CCM), in section III the simulation results are explained .Section V discusses the disturbance analysis results, and Section VI concludes the work.

2. Operating principle

The high gain SEPIC converter proposed is shown in Fig.1.Following are some of the assumptions made to simplify the circuit design and operation of the proposed converter:
1) The converter operates under Continuous Conduction Mode at the steady state condition.
2) Semiconductor devices are considered to be ideal in nature.
3) The capacitor utilized has high storage voltage, therefore it is assumed to be constant.

The converter design consists of main switch S_1, inductors L_1, L_2, diodes D_1 to D_6, capacitors C_1 to C_5, and output capacitor C_6. The Voltage Tripler (VT) circuit is combined with the SEPIC converter to increase the voltage gain of the converter. The switching voltage in the semiconductor device is kept reduced. The proposed converter operation in CCM mode is explained below.

IOP Conf. Series: Materials Science and Engineering **623** (2019) 012008 doi:10.1088/1757-899X/623/1/012008

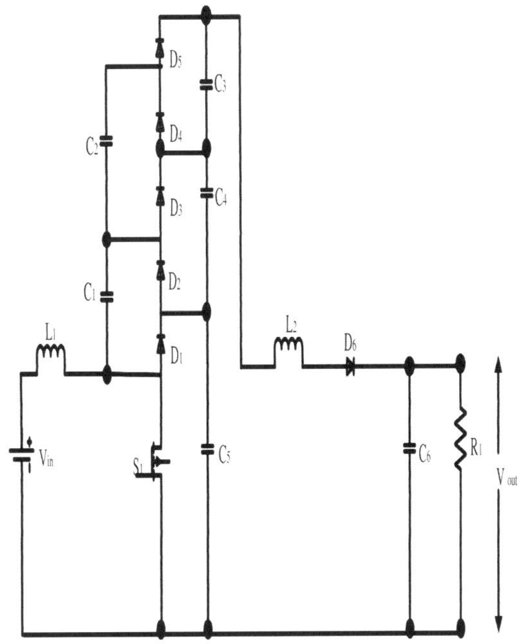

Figure.1 Proposed single switch DC-DC Converter

2.1. Continuous conduction mode operation

The proposed converter functions in two modes as shown in Figs. 2 a) and b)

Mode I [t0–t1]: As the switch S_1 is switched ON, diodes D_2,D_4, and D_6 are turned ON. Diode D_1, D_3 and D_5 are reverse biased. The voltage V_{in} is delivered to L_1 and $V_{C3} - V_{C2} - V_{C1}$ is delivered to L_2. These inductors help in storing the energy. The output capacitor C_6 discharges the energy required to the load for its operation. With the switch OFF, this mode ends. Also, the current in diodeD_1 and D_3 attain zero at $t=t_1$.

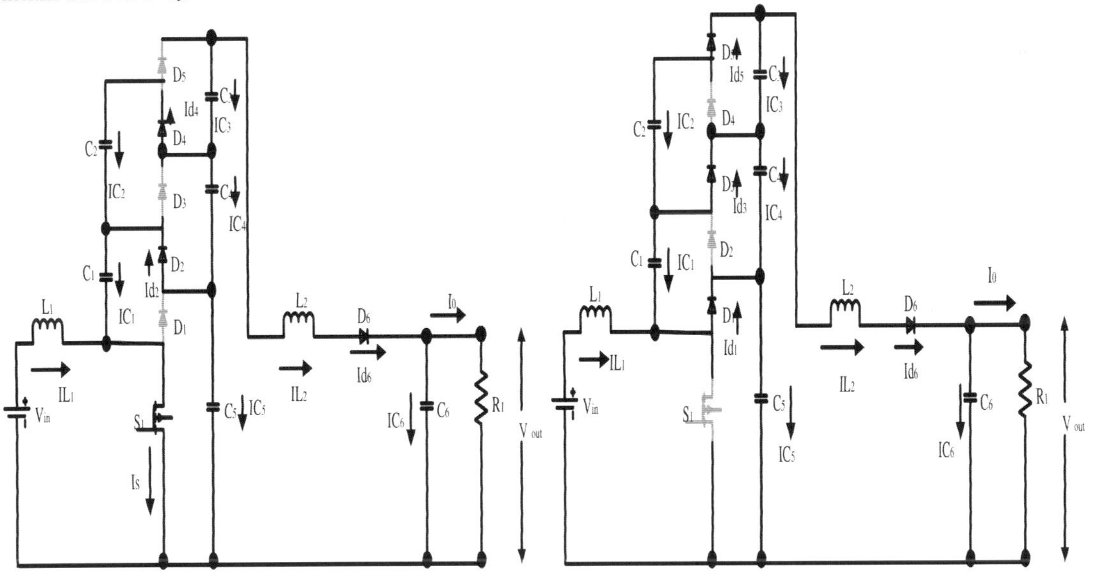

a) Turn ON mode of proposed converter converter

b) Turn OFF mode of proposed

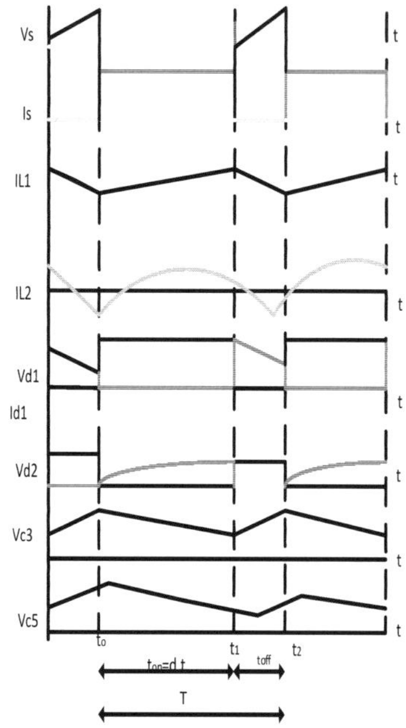

b) CCM operation waveforms

Figure.2 a,b,c Modes Of Operation

Mode II [t1–t2]: With the switch S_1 turned OFF, the diodes D_2 and D_4 are in OFF condition. The diodes D_1, D_3 and D_5 are in forward condition. The inductors L_1 and L_2 charge the capacitors. The load receives the energy by discharging mode of the capacitor. This operation proceeds with the switch turned ON. This continues to the next cycle. The main operational waveform is represented in Fig. 2 c). The total capacitive voltage equals the output voltage of the converter.

$$V_{C6} = V_{C3} + V_{C4} + V_{C5} \quad (1)$$

3. Simulation results and analysis

In this section, the simulation results of the proposed converter are explained. The input of 30 V is simulated to obtain an output of 400 V with 400 W. Simulation Parameters of the proposed converter are depicted in Table 1.

Table 1 Proposed converter parameters

Components	Parameter
Input voltage	30 V
Output voltage	400 V
Input current	15.2 A
Output current	1 A
Inductor L_1, L_2	205, 180 μH
Capacitor C_1, C_2, C_4	2.2μF
Capacitor C_3	2.2μF
Capacitor C_0	40 μF

IOP Conf. Series: Materials Science and Engineering **623** (2019) 012008 doi:10.1088/1757-899X/623/1/012008

Resistor	400 Ω

A comparison with the conventional converters shows that the efficiency of the proposed converter is good and it is well suited for a high-voltage gain conversion scheme. The principal advantage of the converter is that it has lesser conduction and switching loss, reduced number of switches, and reduced size of the converter. The ripple voltage and current in proposed converter is less compared to other conventional converters. The proposed converter is well suited for inverter processes for converting DC sources to AC.

3.1. Simulative analysis of the proposed converter

The proposed converter is simulated in MATLAB and results are depicted as below.

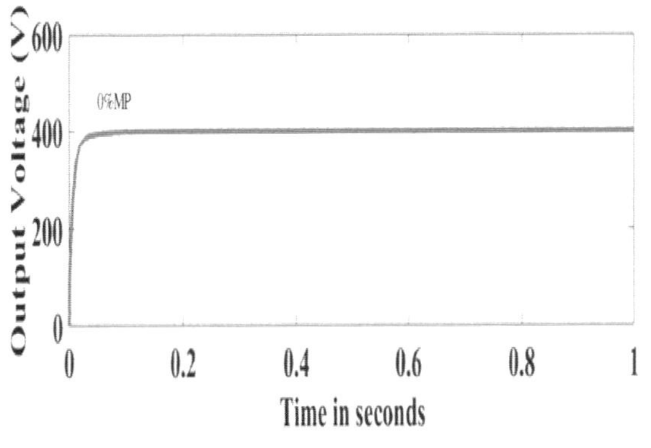

Fig. 3 Output voltage (V_0) of proposed converter

As seen in Fig. 3, the output voltage attained from the proposed converter is lower in ripple, the maximum over shoot is 0.1%, the settling time is 0.5ms, and the steady state error is 0.6%. This output suggests that the converter is operating in a stable state. The output current is depicted in Fig. 4. It can be noted that the output current is disturbance-free and it is constant, which helps the converter to be in a steady state. In Fig.5 the inductor current, capacitor voltage and diode voltage is depicted.

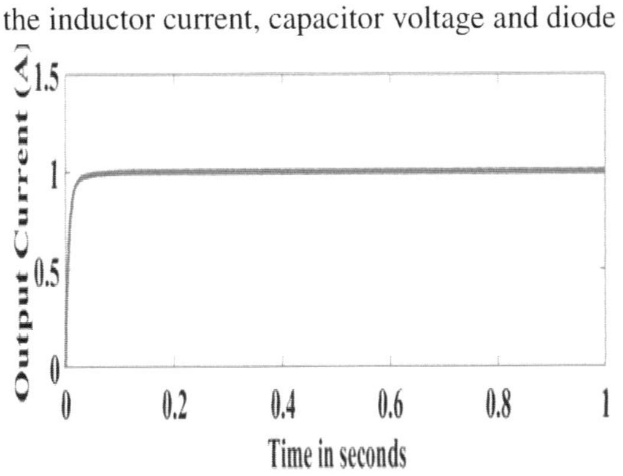

Figure. 4 Output current (I_0) of proposed converter

IOP Conf. Series: Materials Science and Engineering **623** (2019) 012008 doi:10.1088/1757-899X/623/1/012008

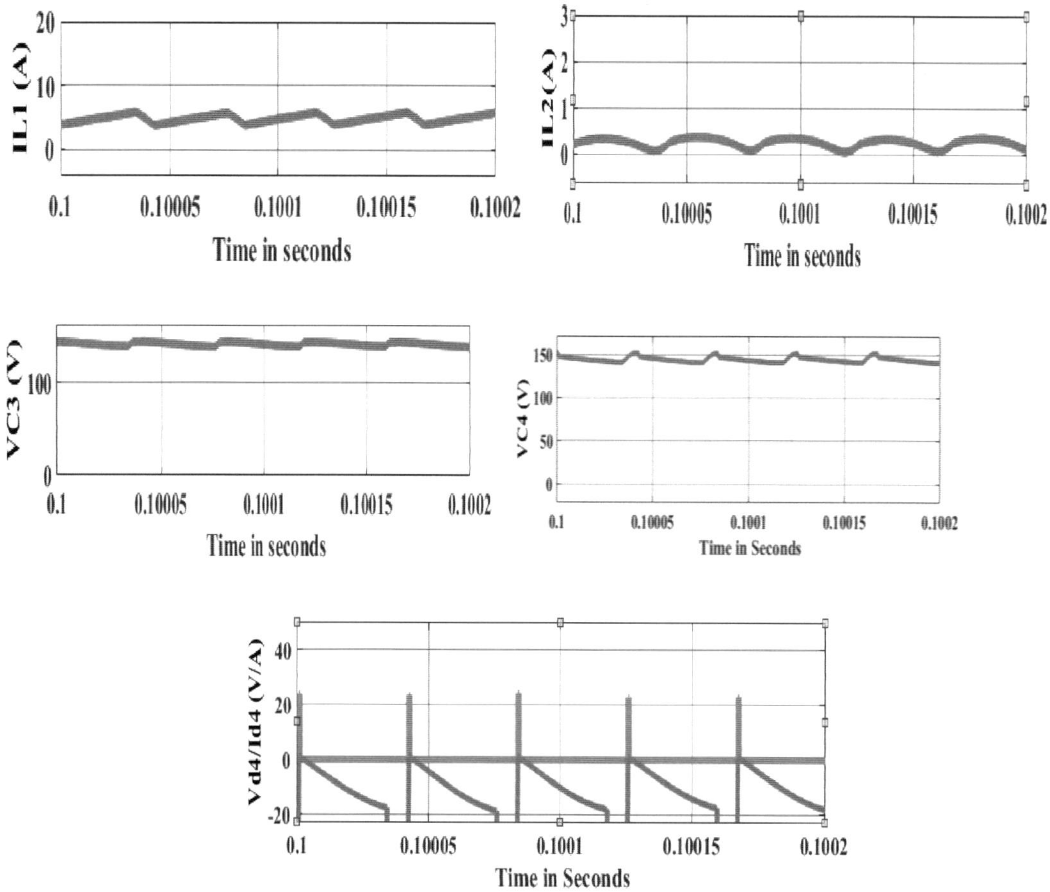

Figure .5a) Inductor current, b) capacitor voltage and c) diode voltage of the proposed converter

4. Disturbance analysis

4.1. Load Disturbance

The Proposed converter is tested under load disturbances such as at 10W, 20W to 100W .As the load is disturbed the converter is stable and provides the stable output even under disturbances. The fig 6 shows the stability of the converter even under different load disturbance. When the load is disturbed, the converter currents and output voltage slightly varies and attains stability at different load conditions. The hardware set up is depicted in fig.10

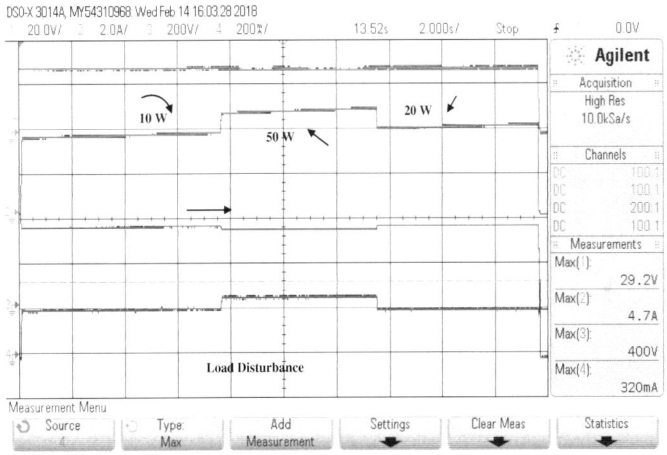

Figure.6 Load disturbance

4.2. Line Disturbance

Under Line disturbance, the input voltage is disturbed and it's not uniform as 30 V. The input voltage was disturbed as 28V, 32V etc. Even under different line disturbance condition the output voltage is maintained stable and constant. The fig 7 shows the line disturbance and stability of the system.

4.3. Line and Load Disturbance

As of next step, the proposed converter is tested under both line and load disturbance together. When load and line are disturbed together, the converter maintains stability with stable output. Hence under Dynamic conditions the converter is stable and it is verified experimentally. The fig.8 represents the converter stability under different load and line disturbances.

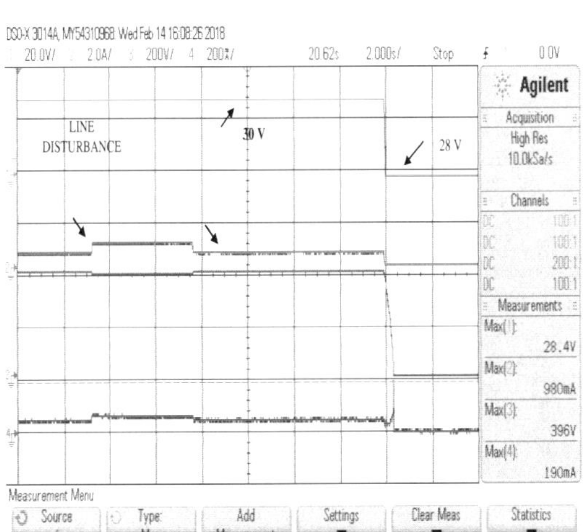

Figure.7 Line Disturbance

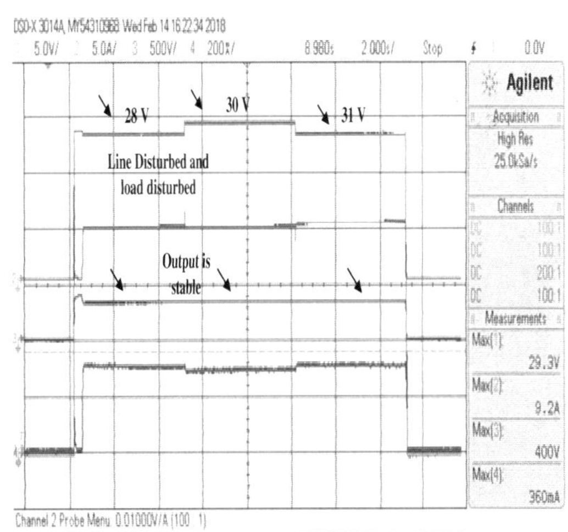

Figure.8 Line and Load disturbance

4.4. Efficiency of the proposed converter

The proposed converter attains an efficiency of 97.3% and it has been tested experimentally. The fig.9 shows the experimental analysis of efficiency of the proposed converter. Hence the proposed converter is highly stable and it suits for PV application. The proposed converter has fulfilled the needs such as high efficiency with reduced duty cycle and reduced ripple levels. All these needs are fulfilled by the proposed converter and it's proved and tested experimentally.

IOP Conf. Series: Materials Science and Engineering **623** (2019) 012008 doi:10.1088/1757-899X/623/1/012008

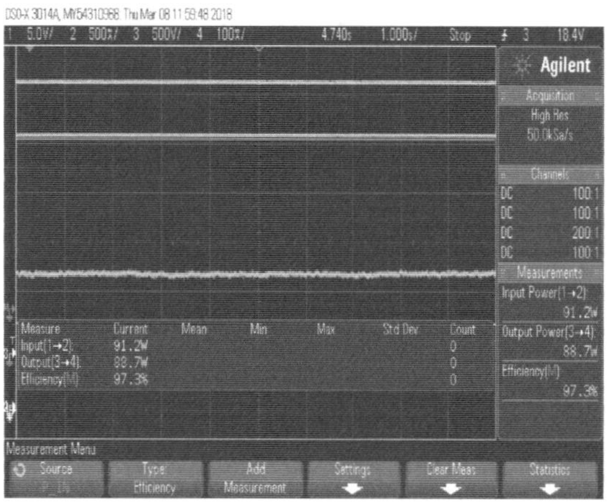

Figure.9 Efficiency of the proposed converter

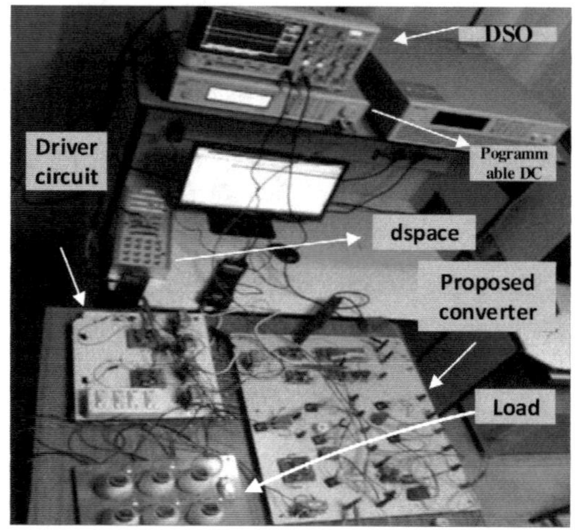

Figuer.10 Hardware Prototype

5. Conclusion

A high step-up voltage gain modified SEPIC converter is proposed in this paper. The working and operation of the proposed converter in CCM mode is explained. The simulation analysis of the proposed converter is presented. A prototype was developed in the laboratory and its performance verified. The experimental results prove that the proposed converter attains a high stepping up of voltage (up to three times) with an efficiency of 97% and reduced switching stress. Also the disturbance analysis of the proposed converter is detailed.

6. References

[1] Arunkumari, T. and Indragandhi, V.,. An overview of high voltage conversion ratio DC-DC converter configurations used in DC micro-grid architectures. *Renewable and Sustainable Energy Reviews*, *77*, pp.670-687,Sep 2017.

[2] Blaabjerg, F. and Ma, K.,. Future on power electronics for wind turbine systems. *IEEE Journal of Emerging and Selected Topics in Power Electronics*, *1*(3), pp.139-152, Sep 2013.

[3] Saravanan, S. and Babu, N.R., 2017. Analysis and implementation of high step-up DC-DC converter for PV based grid application. *Applied Energy*, *190*, pp.64-72.

[4] Kazimierczuk, M.K.,. *Pulse-width modulated DC-DC power converters*. John Wiley & Sons.

[5] Lai, J.S., 2009. Power conditioning circuit topologies. *IEEE Industrial Electronics Magazine*, *3*(2),Aug 2015.

[6] Calais, M., Myrzik, J., Spooner, T. and Agelidis, V.G.,. Inverters for single-phase grid connected photovoltaic systems-an overview. In *Power Electronics Specialists Conference, 2002. pesc 02. 2002 IEEE 33rd Annual*(Vol. 4, pp. 1995-2000).,2002 IEEE.

[7] Liang, T.J., Chen, S.M., Yang, L.S., Chen, J.F. and Ioinovici, A.,. Ultra-large gain step-up switched-capacitor DC-DC converter with coupled inductor for alternative sources of energy. *IEEE Transactions on Circuits and Systems I: Regular Papers*, *59*(4), pp.864-874, April 2012

[8] Gules, R., Dos Santos, W.M., Dos Reis, F.A., Romaneli, E.F.R. and Badin, A.A.,. A modified SEPIC converter with high static gain for renewable applications. *IEEE transactions on power electronics*, *29*(11), pp.5860-5871, Nov 2014.

[9] Hu, X. and Gong, C.,. A high voltage gain DC–DC converter integrating coupled-inductor and diode–capacitor techniques. *IEEE transactions on power electronics*, *29*(2), pp.789-800,Feb 2014.

[10] Changchien, S.K., Liang, T.J., Chen, J.F. and Yang, L.S.,. Step-up DC–DC converter by coupled inductor and voltage-lift technique. *IET power electronics*, *3*(3), pp.369-378 May 2010.

[11] Freitas, A.A.A., Tofoli, F.L., Júnior, E.M.S., Daher, S. and Antunes, F.L.M.,. High-voltage gain dc–dc boost converter with coupled inductors for photovoltaic systems. *IET Power Electronics*, *8*(10), pp.1885-1892, June 2015.

[12] Khalilzadeh, M. and Abbaszadeh, K.,. Non-isolated high step-up DC–DC converter based on coupled inductor with reduced voltage stress. *IET Power Electronics*, *8*(11), pp.2184-2194,Nov 2015.

[13] Wu, G., Ruan, X. and Ye, Z.,.Nonisolated high step-up DC–DC converters adopting switched-capacitor cell. *IEEE Transactions on Industrial Electronics*, *62*(1), pp.383-393 Jan 2015.

[14] Poorali, B., Torkan, A. and Adib, E.,. High step-up Z-source DC–DC converter with coupled inductors and switched capacitor cell. *IET Power Electronics*, *8*(8), pp.1394-1402, Apr 2015.

[15] Kwon, J.M., Choi, W.Y., Lee, J.J., Kim, E.H. and Kwon, B.H.,. Continuous-conduction-mode SEPIC converter with low reverse-recovery loss for power factor correction. *IEE Proceedings-Electric Power Applications*, *153*(5), pp.673-681.Sep 2006.

[16] Zhu, M. and Luo, F.L.,. Series SEPIC implementing voltage-lift technique for DC–DC power conversion. *IET Power Electronics*, *1*(1), pp.109-121,Mar 2008.

[17] Al-Saffar, M.A., Ismail, E.H., Sabzali, A.J. and Fardoun, A.A.,.An improved topology of SEPIC converter with reduced output voltage ripple. *IEEE Transactions on Power Electronics*, *23*(5), pp.2377-2386, Sep 2008.

[18] deMelo, P.F., Gules, R., Romaneli, E.F.R. and Annunziato, R.C.,. A modified SEPIC converter for high-power-factor rectifier and universal input voltage applications. *IEEE Transactions on Power Electronics*, *25*(2), pp.310-321, Feb 2010.

[19] Park, K.B., Moon, G.W. and Youn, M.J.,.Nonisolated high step-up boost converter integrated with sepic converter. *IEEE Transactions on Power Electronics*, *25*(9), pp.2266-2275,Sep 2010.

[20] Rosas-Caro, J.C., Mayo-Maldonado, J.C., Valdez-Resendiz, J.E., Salas-Cabrera, R., González-Rodríguez, A., Salas-Cabrera, E.N., Cisneros-Villegas, H. and Gonzalez-Hernandez, J.G.,. Multiplier SEPIC converter. In *Electrical Communications and Computers (CONIELECOMP), 2011 21st International Conference on* (pp. 232-238). IEEE. Feb 2011.

[21] Middlebrook, R.D., California Institute of Technology,. *Transformerless dc-to-dc converters with large conversion ratios*. U.S. Patent 4,654,769 Nov 1984.

[22] Hu, J., Sagneri, A.D., Rivas, J.M., Han, Y., Davis, S.M. and Perreault, D.J.,. High-frequency resonant SEPIC converter with wide input and output voltage ranges. *IEEE Transactions on Power Electronics*, *27*(1), pp.189-200, Jan 2012.

[23] Zhu, M. and Luo, F.L., 2007, May. Implementing of developed voltage lift technique on SEPIC, Cuk and double-output DC-DC converters. In *Industrial Electronics and Applications,. ICIEA 2007. 2nd IEEE Conference on*(pp. 674-681). IEEE, ay 2007.

[24] Han, Y. and Perreault, D.J.,. Analysis and design of high efficiency matching networks. *IEEE Transactions on Power Electronics*, *21*(5), pp.1484-1491 Sep 2006.

[25] Arunkumari, T. and Indragandhi, V., 2018. A NOVEL SINGLE SWITCH HIGH STEP UP DC-DC CONVERTER FOR PV BASED APPLICATION. *International Journal of Renewable Energy Research (IJRER)*, *8*(2), pp.1085-1097.

[26] Rosas-Caro, J.C., Sanchez, V.M., Vazquez-Bautista, R.F., Morales-Mendoza, L.J., Mayo-Maldonado, J.C., Garcia-Vite, P.M. and Barbosa, R., 2016. A novel DC-DC multilevel SEPIC converter for PEMFC systems. *International Journal of Hydrogen Energy*, *41*(48), pp.23401-23408.

Extraction and conversion of exhaust heat from automobile engine in to electrical energy

A Venkatesh[1], **A Nazar Ali**[2], **R Jaiganesh**[3] and **V Indragandhi**[4]

[1] Assistant Professor, Department of EEE, Rajalakshmi Engineering College, Chennai, Tamilnadu, India.

[2]Associate Professor, Department of EEE, Rajalakshmi Engineering College, Chennai, Tamilnadu, India.

[3] Assistant Professor, Department of EEE, K.Ramakrishnan College of Technology, Trichy, Tamilnadu, India.

[4]Associate Professor, Department of EEE, VIT University, Vellore, Tamilnadu, India.

Abstract. The ever increasing demand for energy is leading to rapid depletion of our resources. Inspite of the advancement in renewable technology .It is therefore important that the available energy conserved and utilized efficiently. In automobiles about 60-70% of energy out of the total heat supplied by fuel is dissipated as waste heat and just about 30 to 40 percent is transformed into useful mechanical work [1-3].This paper focus on the recovery and utilization of waste heat into useful electrical energy by using Seebeck effect where in Thermoelectric Generators (TEG) are placed on the exhaust surface for converting heat directly into electrical energy. Thermoelectric generators are highly doped semiconductor solid state devices. The voltage throughput of the thermoelectric generator is given to a dc super lift converter circuit. The output of the super lift converter is used for battery charging.

1. Introduction

1.1. General

The thermoelectric impact is an immediate transformation of temperature contrasts to voltage and vice versa. An output is produced by the thermoelectric device only in the presence of a temperature gradient. This can be explained due to the fact that , the temperature gradient causes electrons to gain more kinetic energy, leading to its diffusion from the hotter junction to the cooler junction. This impact can be utilized to create power. It depends on Thomson effect Seebeck effect, and Peltier effect. Large amount of hot flue gases is produced from boilers, kilns, ovens, IC engines and furnaces. The availability of waste heat from various sources is tabulated below:

Table I Availability of Waste Heat

Types of Device	Temperature (^{0}C)
Steam boiler exhausts	230-480
Nickel refining furnace	1370-1650
IC engine exhausts	120-250
Gas turbine exhausts	370-540
Heat treating furnaces	425 – 650
Steel heating furnaces	925-1050
Air conditioner condenser	32–43

Although it is not possible to completely recover the energy lost in waste gases, efficient revival and utilization of this waste heat could save a significant amount of primary fuel.
The advantages of waste heat recovery are:
1. Improved combustion process efficiency:
2. Most vehicles are still powered by either Compression Ignition (CI) or Spark Ignition (SI) engines. The waste heat usage is reflected in reduced primary fuel consumption, reduced process cost.
3. Reduction in Air pollution
4. Combustion of fuel releases toxic gases like Carbon Monoxide, Nitrogen Oxide and Hydrocarbons into the atmosphere. Since heat recovery reduces fuel consumption, it consequently reduces the amount of toxic gases that pollutes the atmosphere.
5. Increased battery efficiency:
6. The recovery and smart usage of waste heat can be used for energizing some of the circuits involved in the additional features of automobiles, such as rear view camera, and reduces the load on battery.

2. Proposed System

In this section, the block diagram of a thermo electric converter to convert waste heat obtained on exhaust surface is discussed and the working of various blocks in block diagram is explained in detail. Thermoelectric Generators are devices that work on the underlying principle behind the phenomenon called the Seebeck effect, to convert heat into electrical energy. These devices can be useful in multiple low power remote applications [5].

2.1 Existing System

Fuel cells and solar panels are used to generate the electricity to the vehicles. The Cost of this system is high. Regenerative braking system is applied to generate the power. The output voltage is very low in regenerative braking system.

2.2 Proposed system

The proposed system is used to generate the power from heat energy using Peltier device. Proposed system is interfaced with super capacitor and super lift converter. By using this system, it will distribute the power to the rechargeable battery. The charged Power is distributed to the vehicle power supply such as lamps and etc.

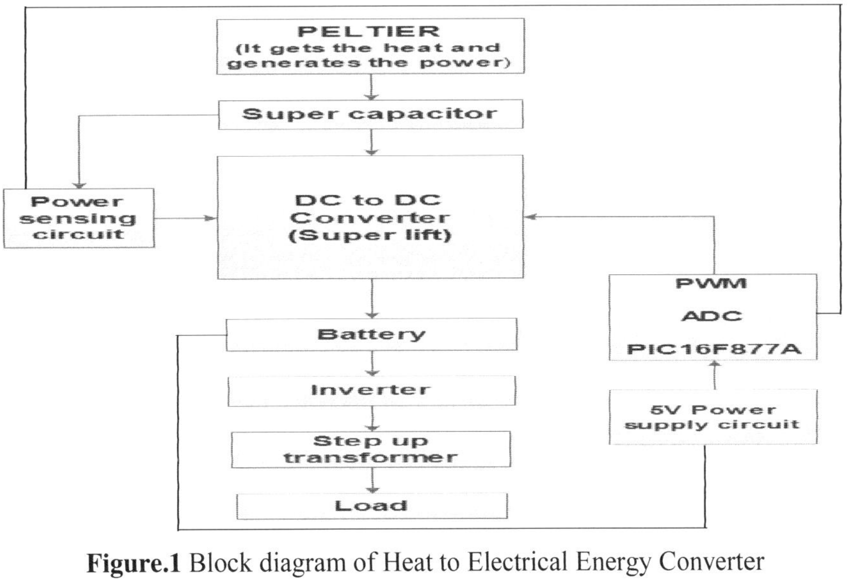

Figure.1 Block diagram of Heat to Electrical Energy Converter

IOP Conf. Series: Materials Science and Engineering **623** (2019) 012009 doi:10.1088/1757-899X/623/1/012009

The thermo electric energy harvester device used is Peltier. Based on Peltier Effect is heat energy is converted into electrical energy and the piezoelectric converts vibration energy into electrical energy. These electrical energies have been stored in super capacitor. The super capacitor works like a battery. The super capacitor is a fast charging and slower discharging capacitor than other types of capacitor. The super capacitor output is given to the Boost converter. The function of Boost converter is to convert the output voltage greater than the input voltage. The Boost converter output is given to the battery for battery charging purpose. The battery power is given to the Inverter. The Inverter output power is distribute to the load [4].

2.3 Components Description

The System consists of Thermo Electric Convertor (TEC 1-12706), Power Mosfet-IRFP460, Boost Converter, Super Capacitor, Voltage Regulator-7805, Max232 IC, 6V Battery, PIC-16F877A Microcontroller, Single Phase Inverter using IGBT (FGA25N120ANTD), Single Phase Transformer and an LCD display.

2.4 Thermo Electric Generator((TEC 1-12706)

TEG uses one or more thermoelectric models as the primary components and either an active or passive cooling system, such as fan cooled, liquid cooled or open air heat sink, fabricated as a single functional unit. The block diagram is as shown in figure 2.

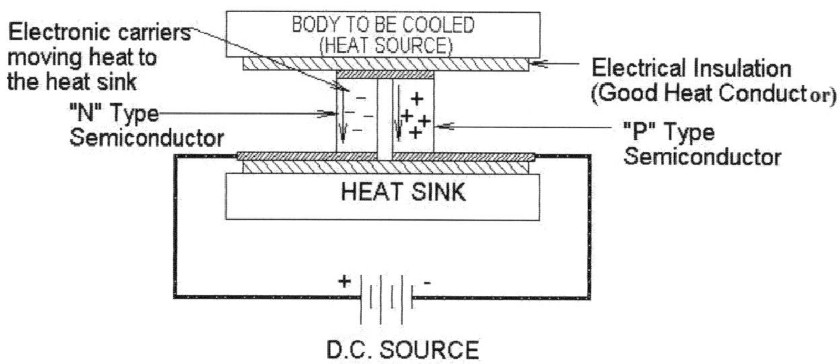

Figure.2 Thermoelectric Generator

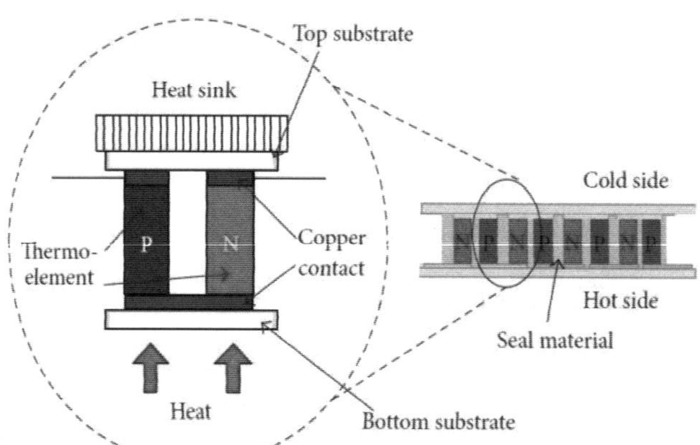

Figure.3 Thermoelectric Generator Internal View

Figure 3 shows the internal view of TEG module. When heat is given to the hot junction of a TEG, it produces electricity.TEC1- 12706 is a 40mm x 40mm sized, single stage module that is designed for heating and cooling applications. It can be widely used in industries, coffee maker, portable refrigerators, and medical instruments.

2.5 Super Lift Converter

Super lift converters are advanced DC/DC converters that provide high efficiency, high voltage transfer gain and minimized ripple voltage and current. Super lift method that is set up by split capacitors; amplifies the output voltage in geometric progression. It is a type of Switched-Mode Power Supply (SMPS) which consists of semiconductor switches and energy storage element.

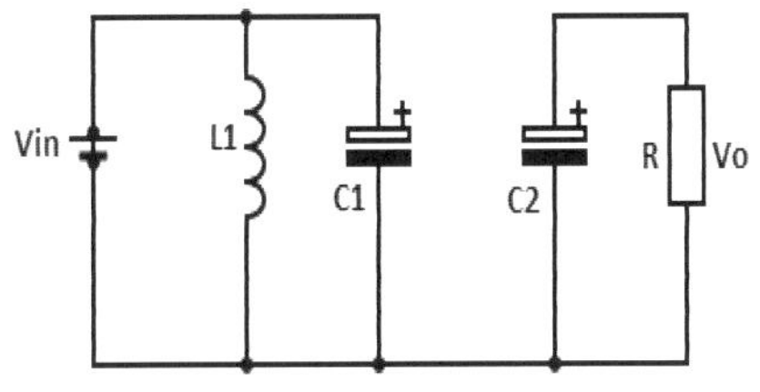

Figure.4 Super lift converter

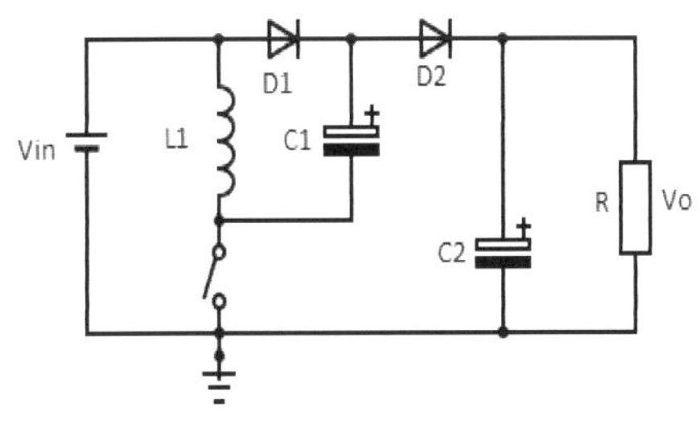

Figure.5 ON state working diagram

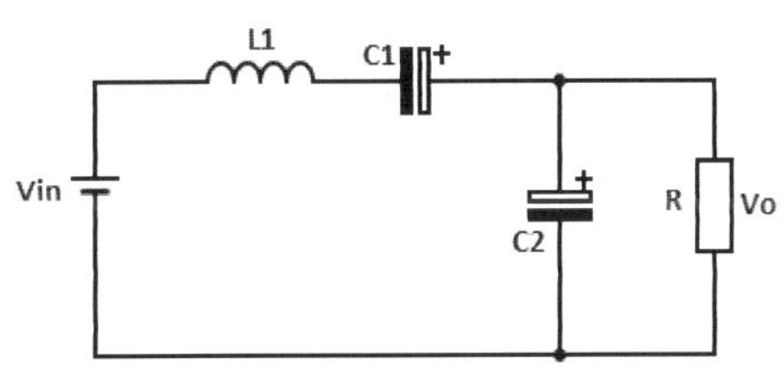

Figure.6 OFF state working diagram

Transfer gain of super lift converter is given by,

$$G = \frac{V_0}{V_{in}} = \frac{2 - K}{1 - K}$$

Where G – Transfer gain
V_0 – Output voltage
V_{in} – Input voltage

3. TEG Power And Efficiency Calculation

The conversion efficiency (η) is the ratio of the generated electrical power (oT) to the heat supplied to the module Qh
Qh – heat removed from the cooled side.

$$\frac{P_T}{Q_H} = \frac{P_T}{Q_C + P_T} (1)$$

There are four basic physical entities that are involved in the operation of Thermo Electric Generators (TEG) namely, the Thomson effect, the Joule effect, the Seebeck effect and the Peltier effect. Under steady state conditions, these entities together explain the phenomena of energy flow below. [8]

$$TJ\frac{da}{dx} + \tau J\frac{dT}{dx} - \rho J^2 - \frac{d}{dx}\left(k\frac{dT}{dx}\right) = 0 (2)$$

Where T - temperature
J – Electrical current density
α – Seebeck coefficient
τ – Thomson coefficient
ρ – Electrical resistivity
k – Thermal conductivity of the material

The equation of heat flow at hot side is:

$$Q_H = K_T.(T_H - T_C) + S_T.T_H.I - \frac{1}{2}I^2R_T (3)$$

Where $K_T = N(k_n + k_p)G$ – total thermal conductance of N couples.
$S_T = N(a_n + a_p)$ – Total Seebeck coefficient.
$R_T = N(\rho_n + \rho_p)/G$ – Total resistance.
G = area/length – Geometry factor.

Heat flow at cold side is:

$$Q_H = K_T.(T_H - T_C) + S_T.T_c.I - \frac{1}{2}I^2R_T (3)$$

Hence, the power generated by the module P_T is

$$P_T = Q_H - Q_C = S_T.(T_H - T_C).I - I^2R_T = [S_T(T_H - T_C) - IR_T].I = V_TI (4)$$

The output voltage of the module, V_T is

$$V_T = S_T(T_H - T_C) - IR_T \quad (5)$$

Therefore, the P_T, Q_H, Q_C, η can be found, for known material properties. In practice, it is impossible to measure the temperature of both the hot and cold junction.

$$W_H = (W_{T1} + W_{T2} + W_{T3} + W_{CR}) (6)$$

Where W_H - Total thermal resistance between T1 andT_H.
W_{T1},W_{T2},W_{T3} – Thermal conductivity resistances.
W_{CR} – Thermal contact resistance.

The hot junction temperature T_H is given by

$$T_H = T_1 + Q_HW_{TC} \quad (7)$$

Similarly, the hot junction temperature T_H is given by

$$T_C = T_2 - Q_CW_{TC} \quad (8)$$

Where W_{TC} – Total thermal resistance between T_C andT_2.

IOP Conf. Series: Materials Science and Engineering **623** (2019) 012009 doi:10.1088/1757-899X/623/1/012009

The following expressions show the relationship between the power output of Peltier module and the gain of the superlift converter.

$$P_O = P_T . G \quad (9)$$

$$P_O = V_T I . \frac{V_O}{V_{in}} \quad (10)$$

Where $\mathbf{P_O}$ – Stepped up output power of the superlift converter

4. Hardware Implementation

The hardware implementation of the proposed system using Peltier module and superlift converter is shown in figure 7.

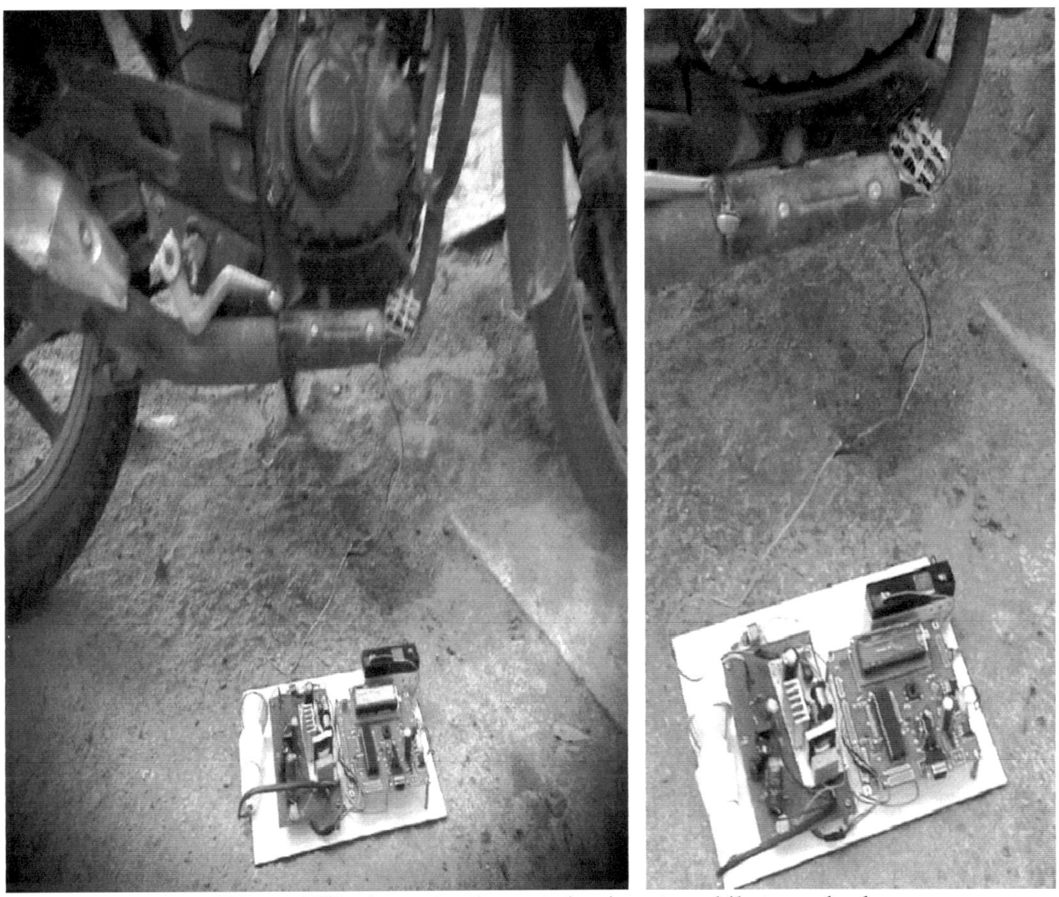

Figure.7 Hardware implementation in automobile two wheeler

Figure 9, circuit consists of Peltier device (TEC 1-12706), Boost converter with MOSFET switch IRFP460, PIC-16F877A Microcontroller interfaced with MAX232 and 16x2 LCD display, Voltage Regulator 7805 and X1 crystal Oscillator to maintain constant 20MHz frequency. In Figure 7TEC1-12706 is placed on the exhaust surface of automobile two wheeler muffler as shown. It is also used to convert waste heat generated in hot waste gas exhaust industrial furnace, air conditioner outdoor unit and laptops. This setup was implemented in an automobile to utilize waste heat and generated electrical energy is used to charge the battery of the vehicle. [9]

IOP Conf. Series: Materials Science and Engineering **623** (2019) 012009 doi:10.1088/1757-899X/623/1/012009

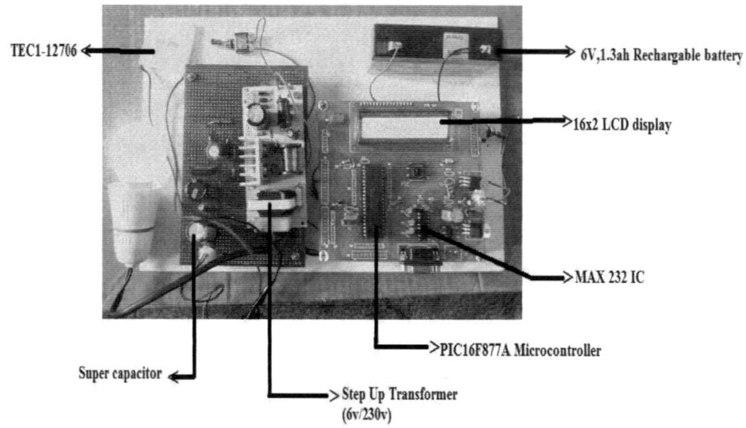

Figure.8 Hardware implementation

TABLE II Output Measurement

Temperature (°C)	Input Power (W)	Output Power (W)
42	1.684	2.873
56	1.928	4.973
87	2.45.95	
71	2.148	5.278
58	1.72	3.554

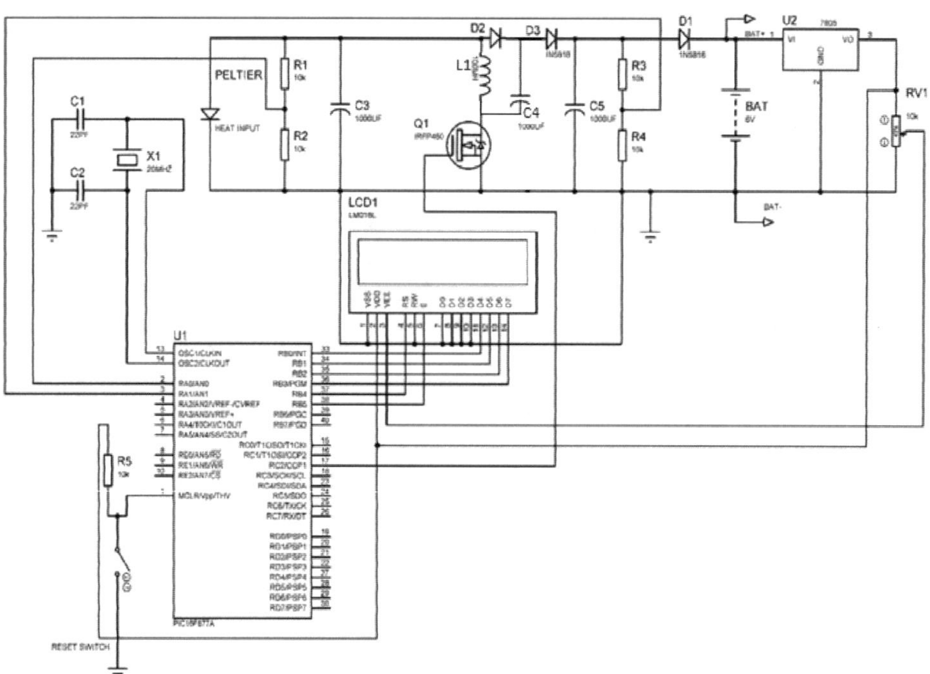

Figure.9 Schematic Diagram of Heat to Electrical Energy Converter

Figure.10 Temperature Measurement

5. Conclusion

The extraction and conversion of exhaust heat from automobile engine, into electrical energy, was successfully implemented on both hardware and software. From the results obtained, it can be concluded that Peltier device along with the concept of super lift converters can efficiently utilize the waste heat energy from automobile exhausts and convert it into electrical energy. This technique is more effective when the distance travelled by the automobile is longer. The Peltier device (TEC1-12706) can work more efficiently below 150 °C beyond which the power output starts to degrade. The scope of the proposed work can be extended to convert waste heat generated in air conditioner outdoor unit and laptops.

6. References

[1] Takashi Kyono, Ryosuke O. Suzuki, and Katsutoshi Ono, Conversion of Unused Heat Energy to Electricity by Means of Thermoelectric Generation in Condenser", IEEE transactions on energy conversion, 2003, vol. 18, no.

[2] G.Span, M.Wagner, S.Holzer2, T.Grasser, „Thermoelectric Power Conversion using Generation of Electron-Hole Pairs in Large Area p-n Junctions', IEEE transactions on thermo electronics, 2006, vol. 10, no. 5.

[3] R.J.M. Vullers, R.vanSchaijk, I.Doms, C.Van Hoof, R.Mertens, "Micropower energy harvesting", IEEE transactions on Solid-State Electronics, Kapeldreef, 2008, vol. 01, no. 6.

[4] Joao Paulo Carmo, Luis Miguel Goncalves, Jose HiginoCorreia , "Thermoelectric Microconverter for Energy Harvesting Systems", IEEE transactions on industrial electronics, 2010, vol. 57, no. 3.

[5] Jaydeep.V, Joshil, Thermoelectric system to generate electricity from waste heat of the flue gases", ISSN 0976-8610 Advances in Applied Science Research, 2012, vol. 3, no. 2.

[6] KoheiKawabuchi, Toshiaki Yachi, ‚Analysis of the Heat Transfer Characteristics in a Thermoelectric Conversion Device" , IEEE Transactions on Thermoelectric conversion ,2012, Vol. 2, no. 6.
[7] R.Saidur, M.Rezaei, W.K.Muzammil, M.H.Hassan, M.Hasanuzzaman, Technologies to recover exhaust heat from internal combustion engines", UMPEDAC on Renewable and Sustainable Energy Reviews 16, Level 4,2012
[8] J. S. Jadhao, D.G.Thombare, Review on Exhaust Gas Heat Recovery for I.C. Engine", International Journal of Engineering and Innovative Technology (IJEIT),2013, Vol. 2, Issue 12.
[9] S. Kodeeswaran, T. Ramkumar and R. J. Ganesh, "Precise temperature control using reverse seebeck effect," 2017 International Conference on Power and Embedded Drive Control (ICPEDC), Chennai, 2017, pp. 398-404. doi: 10.1109/ICPEDC.2017.8081122.

IOP Conf. Series: Materials Science and Engineering **623** (2019) 012010 doi:10.1088/1757-899X/623/1/012010

Solar powered air conditioner using BLDC motor

A Nazar Ali[1], D Sivamani[2], R Jaiganesh[3] and M Pradeep[4]

[1] Associate Professor, Department of EEE, Rajalakshmi Engineering College, Chennai, Tamilnadu, India.

[2] Assistant Professor, Department of EEE, Rajalakshmi Engineering College, Chennai, Tamilnadu, India.

[3] Assistant Professor, Department of EEE, K.Ramakrishnan College of Technology, Trichy, Tamilnadu, India.

[4] Assistant Professor, Department of EEE, Rajalakshmi Engineering College, Chennai, Tamilnadu, India.

Abstract. This paper present a load compressor fed Brushless DC motor for solar power air conditioner under variable air conditions. At current scenario, air conditioners are manufactured according to room size and space with available of objects present in the room.The output of the air conditioner cannot be alternatively changed. This problem is overcome by modify the fixed output of the air conditioner using BLDC motor. The BLDC motor performance is superior when compared to other motor. Solar power is used as the main source which is a renewable energy and readily available and also very useful for the future generation. This work deals with the output of the air conditioner can be alternatively changed according to the size of the room by varying the speed of Brushless DC motor.

1. Introduction

Air temperature varies according to climate and place, to control the temperature of air, air can be conditioned. It can be condition in two ways. They are

1. Forcing of air, e.g.: forced air network
2. Conditioning of air e.g.: air conditioner. Air conditioning introduced by maintaining the temperature of air at standard medium by circulating the air at closed space or room. In this project, air conditioning is done by varying gas compressor or open drive compressor, the gas used in this is proposed work is R134Aa i.e. tetrafluromethane gas which is a coolant gas mainly used in air conditioning. Since, it has high boiling point and low melting point and it turns into crystal at very low temperature. The compressor is driven by Brushless DC motor using belt drive. The compressed R134Aa gas has raised its temperature and pressure which is given as the input of condenser where its temperature pressure is reduced and the output of condenser is given to evaporator through drier and capillary tube and the cool air with 18 degree Celsius to 200. Celsius is given as output of air conditioner. The source for Brushless DC motor supplied from solar panel or photovoltaic cell which is a renewable energy [1, 2, and 3]. BLDC motor is suitable to produce flat torque characteristic than the other motor because of no brush fiction to reduce the torque and it relatively as a high speed range for supporting the torque. The BLDC motor also supports the design it varying character by producing low noise, since the design is based on room appliance.

2. Proposed System

IOP Conf. Series: Materials Science and Engineering **623** (2019) 012010 doi:10.1088/1757-899X/623/1/012010

The proposed work block diagram shown in fiqure1.It consists of Solar panel,Charge controller, Battery,Boost converter,BLDC motor and Air conditioner.

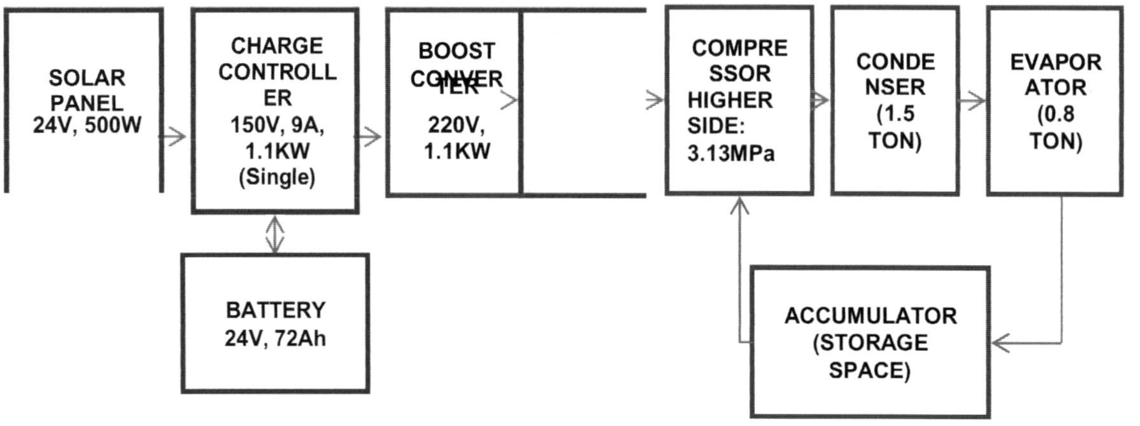

Figure.1 Block Diagram of Proposed system

2.1 Solar Panel

The photovoltaic cell Conversion of light energy in electrical energy is created on a portent called photovoltaic effect. When semiconductor materials are bare to light, the some of the photons of light ray are immersed by the semiconductor crystal which origins significant number of free electrons in the crystal. This is the basic reason of creating electricity due to photovoltaic effect. Photovoltaic cell is the basic unit of the system where photovoltaic effect is employed to harvest electricity from light energy. Silicon is the most extensivelysecond-hand semiconductor material for creating photovoltaic cell. In this design, 24V, 500W, 4 panels is used to 1500W.

2.2 Charge Controller

A solar charge controller is fundamentally a voltage or current controller to charge the battery and keep electric cells from overcharging. Here we use Pulse Width Modulation controller. This is the traditional type charge controller, for instance anthrax, Blue Sky and so on. These are essentially the industry standard now. The most essential charge controller basically controls the device voltage and opens the circuit, halting the charging, when the battery voltage ascents to a certain level. More charge controllers utilized a mechanical relay to open or shut the circuit, halting or beginning power heading off to the electric storage devices. In this design, we are going to use 24V, 50A PWM type. Since solar panel and battery voltage is 24V [4].

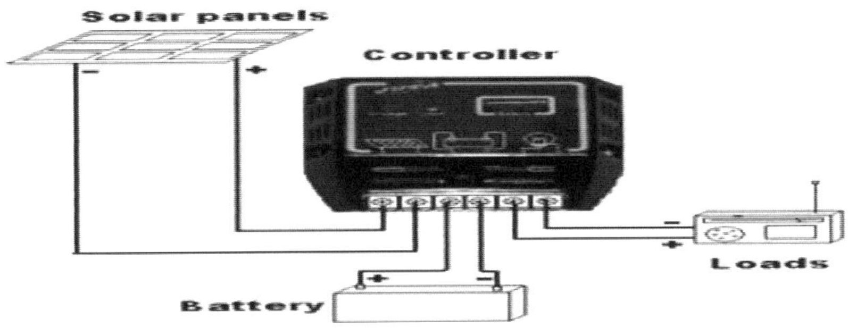

Figure.2 Charge Controller

2.3 Battery

In this paper, lead acid battery with the capacity of 24V, 72Ah can be used. Since, it is lower in cost it can be used in home appliances. It is capable to tolerate the overcharging. It can deliver high current.

2.4 Boost Converter

The low input DC voltage is converted into high output DC voltage using Boost converter. This project Boost converter given 48V from 24V .Its output is connected to the BLDC motor. [5, 6]

2.5 Brushless DC Motor

The Brushless DC motor is powered up by a DC electric source from a battery which is charged by the photovoltaic cell. Since, the rotor parts of BLDC motor consist of permanent magnet which rotates around a fixed armature develops a maximum torque when stationary, linearly decreasing when velocity increases. It has high efficiency and lower mechanical veer [7, 8]. Compared to other motors, BLDC motor is more efficient is to transfer electrical energy into mechanical energy. Hence, BLDC motor is applied in this concept with 2 hp, 48V, 33A, 8 pole, and Single phase with a driver

2.6 Compressor

The Varying compressor consists of a cylinder with a reciprocating piston of 1ton with 31.3 bar pressure. The forward and backward movement of the piston is conducted by the bearings according to the rotation of the snap ring which is rotated by the belt drive from the BLDC motor. When the piston takes its backward movement the R134Aa i.e., Tetrachlorofluro Methane gas enters the compressor and during the forward movement the entered R134Aa gas is compressed. The compressed gas with high temperature and pressure is passed through the outlet of the compressor.

2.7 Condenser

The outlet of the compressor is connected to the inlet valve of the condenser where the gas with high temperature and pressure is condensed by cooling the condenser with reduction of heat by a suction fan. During condensation the hot gas with high temperature and pressure is converted into liquid and partial vapour. After condensation both the temperature and pressure of the gas drops to a certain temperature from the room temperature and the outlet of the condenser is passed through the drier.

2.8 Drier

The outlet of the condenser is passed through the drier which is in the form of liquid and partial vapour. The drier filters the vapour and just allows the liquid to pass through. The vapour filtered liquid is then passed through the capillary tube.

2.9 Capillary Tube
The liquid outlet from the drier is passed through the capillary tube. The capillary tube works on the principle of Avogadro which states that when a liquid of certain temperature is passed through a pipe or valve which has a constant volume throughout the length will lose its temperature below the atmospheric temperature. Hence, the outlet of the drier which is passed through the capillary tube will lose its temperature below the atmospheric temperature at maximum level.

2.10 Evaporator
The liquid with low temperature from the outlet of the capillary tube is passed through the evaporator. When the liquid travels inside the evaporator it is blown with the atmospheric air by the blower. Hence, heat reduction takes place. According to the thermodynamics principle, when a hot and cold body gets in contact heat exchange takes place. Hence, by the thermodynamics principle the atmospheric air obtains the temperature of the liquid inside the evaporator and the liquid obtains the temperature of the atmospheric air and partially converted into vapour. The liquid which has been increased in temperature is passed through the accumulator.

2.11 Accumulator
Accumulator is storage or collecting tank that stores the outlet of the evaporator which is in the form of hot vapour and liquid. The vapour is again passed through the compressor by suction method. Hence, it is a cyclic process the compressor again compresses the outlet of the accumulator.

3.Circuit Diagram

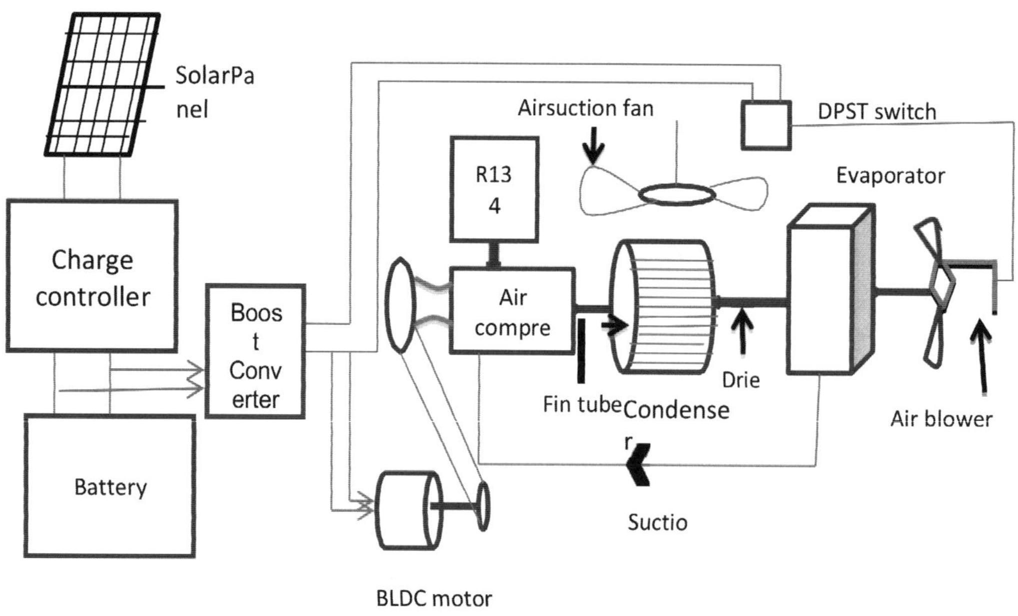

Figure.3 Circuit Diagram

4.Working of Air Conditioner Using BLDC Motor

The working of air conditioner is based on the principle that when air with definite temperature is circulated inside a closed room will lose its temperature to a certain level. The air inside the room can be cooled by using R134Aa which as high boiling point and low melting point, which is ecofriendly and it, is present in liquid state before it reaches the compressor. The working stages of the air conditioner consist of

1. Power source.

2. Driving unit.

3. Compression.

4. Condensation

5. Evaporation.

6. Cooling.

4.1 Power Source

Here, the power source obtained from the solar panel which is a variable source that depends on the climatic condition. The power source from the solar panel is given to the charge controller to give a regulated voltage, this output is given to the battery since charge controller has 6 terminals two for input two for battery where the regulated voltage is stored for many hours and other two terminals for boost converter because battery and charge controller are 24V but the load need 48 V. so, the output of charge controller is given as the input to boost converter where 24V is converted into 48V and 33 A . The output of boost converter is fed to BLDC motor.

4.2 Driving Unit

The driving unit in this setup is the BLDC motor. The BLDC motor drives the compressor with a belt transmission which leads to the working of overall setup. When the input source is given to the BLDC motor the electronic controller controls the charge controller output amplitude and waveform which is transmitted to the permanent magnet by means of boost converter[10,11,12]. The permanent magnet which rotates around a fixed armature connects the current to the moving armature. The electronic controller continuously switches the face to the windings to run the motor, which produces a uniform torque. The resultant torque and speed of the driving unit is enough to run the compressor with belt drive.

4.3 Compression

The compression process takes place inside the compressor when motor drives at high speed with required torque. Inside the compressor setup consist of cylinder with piston when the motor drives the compressor the piston moves in a reciprocating motion. The R134Aa gas is compressed by the forward and backward motion of the piston during backward movement the suction of gas takes place and during the forward motion compression of gas takes place which increases the temperature and pressure of the gas. Here, the load on the compressor depends upon the temperature of the atmospheric air. If the temperature of the atmospheric air increases, the temperature of the R134Aa gas increases. Hence, the load of the compressor increases and the power unit of the motor also increase with increase in torque.

4.4 Condensation

The condensation process is carried out in the condenser. According to the air conditioning principle, a condenser must be 0.5 ton higher than the capacity of the compressor. Since, our project is based on varying compressor the highest range of the compressor 0.9 ton i.e., approximately 1 ton. So, the condenser is of 1.5 ton capacity is used the condenser is of two row single pass type since it has a single passage for input and output transmission. Condenser consist of 32 fins arranged in a row by folded mode with vast face between one fold to another when compared to evaporator, the condenser fins in this manner for the fast reduction of heat by the suction fan. Since, the compressed gas entering into the with high temperature and pressure by heat reduction and changes into liquid state and passed through the condenser.

4.5 Evaporation
The evaporation takes place in the evaporator with the help of the blower. According to the air conditioning principle, the capacity of the evaporator depends upon the capacity of compressor where it varies between 0.2 to less than the compressor. The evaporator is also of two row single pass type with 35 fins arrangement. Here fins are folded in a row and closely packed than the condenser for forcing the air that passes through the evaporator with the help of the blower. Hence, evaporator of the air conditioner is takes place by the atmospheric air.

4.6 Cooling
Cooling is conditioning the air inside the room temperature. The cooled air or the air with reduced temperature compared with the room temperature is the output of the air conditioner. When the evaporator consisting of the cold liquid gets in contact with the atmospheric air heat transfer takes place. Therefore, the liquid gets heated up by the temperature of the atmospheric air and the atmospheric air gets the temperature of the liquid and spreads over the room as cooled wind.

5.Design Calculation For The Air Conditioner

5.1 Calculation for Charge Controller

Ampere =DC Connected Watts/DC System Voltage
= 1100/ 24 (24 V Battery)
= 45.83A=50A (approx.)

5.2 Calculation for Battery

Power = 1100Wh
Ah= 1100/24 (24 V Battery)
= 45.83Ah
Depth of discharge = 0.7
Ah= 45.83/0.7 = 65.47Ah = 72Ah (approx.)

5.3Calculation for Solar Panel

Battery Capacity = 72Ah
Sunrise-sunset= 5hours
Power = (Battery Capacity*solar panel voltage)/hour = (72*24)/5 = 432Watts= 500W (approx.)

5.4Calculation of Compressor

*To convert kg/cm^2 to MPa
For higher side:1 kg/cm^2 = 0.98MPa
The higher side ranges from 28 – 32 kg/cm^2
Assuming higher side value, 32 kg/cm^2 = 3.13 MPa

*To convert MPa to bar 1MPa = 10 bar
= 3.13*10 = 31.3 bar
*To convert bar to ton 1 bar= 10.19 tonnes per meter
= 31.3*10.19= 318 tonnes per meter
*Room size = 318/1000 = 0.31 ton

To convert ton to Kilowatts1ton = 3.516KW= 0.31 3.516=1.05 KW

IOP Conf. Series: Materials Science and Engineering **623** (2019) 012010 doi:10.1088/1757-899X/623/1/012010

*To convert kilowatts to horsepower 1KW=1.32hp =1.32*1.05 = 1.38hp

So, the BLDC motors need 1.38hp for a closed room, since the ac in open space in practical we need minimum 2hp,1.5kw motor because R134a gas temperature varies according to the atmosphere, so temperature increases. Therefore, weight increases it need high torque with high horse power.

5.5 Calculation of BLDC Motor

Since, we need high torque, so current must be high in 2 hp motor we can get about 33amps motor
So, I=33A, V=P/I = 1500/33= 48V
*So, the BLDC motor with 48V, 33A, 8 poles, with driver need for this design

5.6 Calculation for Compressor

The thumb rule is condenser must be 0.5 ton greater than the compressor.So, the compressor varies till 0.9 ton i.e., 1 ton. Therefore condenser is 1.5 ton.

5.7 Calculation for Evaporator

The thumb rule is evaporator must be 0.2 ton lesser than the compressor. So, evaporator is 0.8 ton.
Experimental Setup

Figure.4 Experimental Setup

6.Working Of Brushless DC Motor

IOP Conf. Series: Materials Science and Engineering **623** (2019) 012010 doi:10.1088/1757-899X/623/1/012010

The BLDC motor replaces the brushes of the conventional motor by employing an electronic commutator. The system consist of three coils of the motor arranged in a y formation i.e., a microchip PIC microcontroller, an insulated gate bipolar transistor (IGBT) driver, and a three phaseinverter comprising of 6 IGBT's. MOSFET (Metal Oxide Semiconductor Filed Effect Transistors) can also be used for the high power switching. When the input current is given to the motor the electronic commutator transmits it to the microcontroller [13, 14]. The output from the microcontroller comprises pulse width modulated signals (PWM) that determines the average voltage and average current to the coils.

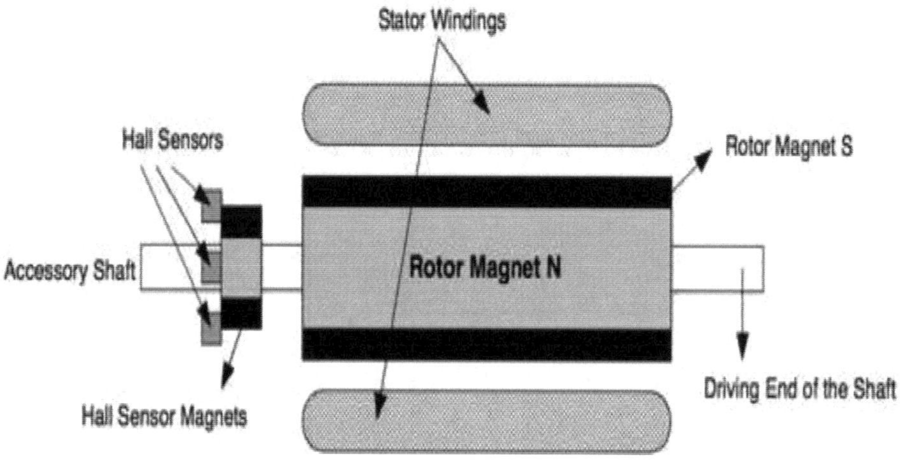

Figure.5 Constructional Diagram of Brushless DC Motor

The motor practices three Hall Effect sensors to specify the rotor locations.

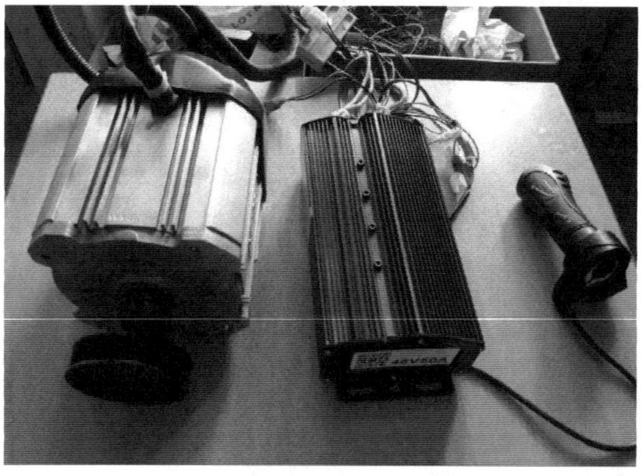

Figure.6 Brushless DC Motor

The rotor themselves uses two pairs of permanent magnets two make the magnetic flux. Conferring to Lenz's law, it gives growth to a current in the winding with a magnetic field that faces the unique change in magnetic flux. Henceforth the electric motor produces a voltage potential due to the instant of the windings done the related magnetic field. This potential is recognized as EMF (Electromotive Force). Meanwhile the motor consist of six step commutation arrangement for all electrical revolution with the assistance of two pairs of magnets, two electrical revolutions are compulsory two spin the motor once. Hence, the electrical response is converted into mechanical energy and this mechanical energy is transferred through belt drive to ride the compressor.

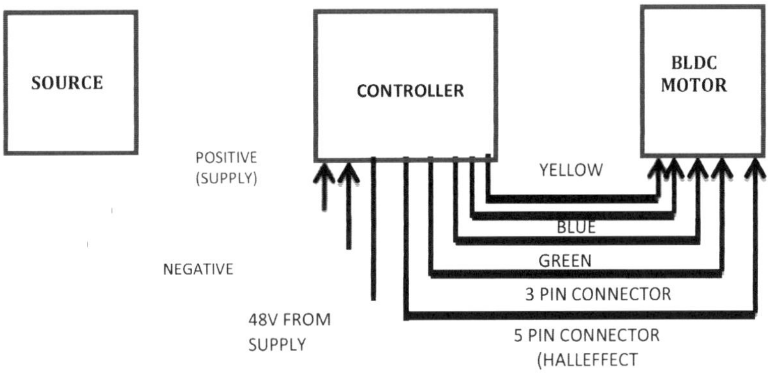

Figure.7 Wiring Diagram of BLDC Motor with Controller

7. Conclusion

The output of the air conditioner can be varied by varying the speed of the motor. The BLDC motor also supports the design it varying character by producing low noise, since the design is based on room appliance. The speed is constant even at high load and torque at low speed also high. The BLDC motor is very suitable to run the setup. This work can be implemented in hardware also. Based on the result, it has better thermal performance and withstand at high temperature. In BLDC motor the voltage drop on the electronic device which improves the efficiency of the motor compared to other As the BLDC motor is free from brushes and commutator their maintenance is less with efficient life time. Hence it is confined that the BLDC motor is more suitable for this design compared to other motors.

7. References

[1] Arora and Domkundwar, Refrigeration & Air Conditioning, Dhanpatrai& Company.
[2] Solar Electricity Handbook – 2014 Edition: A Simple Practical Guide to Solar Energy – Designing and Installing Photovoltaic Solar Electric System
[3] Solar Powered Variable Speed Power Factor Corrected BLDC Motor Drive, Muhammed Rajees.P.P
[4] C. L. Xia, Permanent Magnet Brushless DC Motor Drives and Controls, Hoboken, NJ, USA: Wiley, 2012.
[5] Comparison of Basic Converter Topologies For Power Factor Correction. Huai Wei, IEEE Member, and IssaBatarseh, IEEE Senior Member , University of Central Florida Orlando, FL 32816.
[6] H.Wei,I.Batarseh,"Comparison of basic converter topologies for power factor correction,"inProc.IEEE,pp 348-353,Apr.24-26 1998
[7] SanjeevSingh,BhimSingh,"Voltage controlled PFC zeta converter based PMBLDC drive for an air-conditioner,"in Proc.ICIIS.pp.550- 555,Jul./Aug. 2010

[8] Rajan Kumar and Bhim Singh, "Buck boost converter fed BLDC motor drive for PV array based water pumping," IEEE International Conference on Power Electronics, Drives and Energy Systems (PEDES), pp.1-6, 16-19 Dec. 2014.

[9] S. Veeraraghavan, M. Kumaravel, Krishna Vasudevan, and Ashok Jhunjhunwala, "Experimental studies and performance evaluation of Solar PV powered BLDC motor drive with an integrated MPPT in fan applications," 40th IEEE Photovoltaic Specialist Conference (PVSC), pp.3713-3718, 8-13 Jun. 2014

[10] Solar powered speed control of Brushless DC motor drive using PID FUZZY Controller, Volume 8, Issue 8, August 2017, pp. 1135–1147, Article ID: IJMET_08_08_113

[11] Simulation of Solar Powered PMBLDC Motor Drive, © 2014 IJEDR | Volume 3, Issue 1 | ISSN: 2321-9939

[12] Sanjeev Singh and Bhim Singh "PFC Bridge Converter for Voltage-controlled Adjustable-speed PMBLDCM Drive" by Journal of Electrical Engineering & Technology Vol. 6, No. 2, pp. 215~225, 2011

[13] PadmarajaYedamale "Brushless DC (BLDC) Motor Fundamentals" Microchip Technology Inc, 2003 Microchip technology

[14] Sanjeev Singh, Member, IEEE , and Bhim Singh "A Voltage-Controlled PFC Cuk Converter-Based PMBLDCM Drive for Air-Conditioners" IEEE TRANSACTIONS ON INDUSTRY APPLICATIONS, VOL. 48, NO. 2, MARC H/APRIL 2012

IOP Conf. Series: Materials Science and Engineering **623** (2019) 012011 doi:10.1088/1757-899X/623/1/012011

A Novel wearable therapeutic aid with intelligent information processing systems

P.Sabarish[1], A.T.Sankara Subramanian[1], A.Gayathri[2] and A.Anton amala praveen[1].

[1]Assistant Professor, Department of EEE, K.Ramakrishnan College of Technology, Trichy, Tamilnadu, India.

[2]Assistant Professor, Department of School of information Technology & Engineering, VIT, Vellore,Tamilnadu, India.

Abstract. This paper usually starts in primary childhood and its cruelty is associated with outcomes and cruelty of autism in teens and adulthood. Management is frequently behavioural. However, capable new concepts and indication are developing from neurobiology and development psychology that recognize neural adaptation, lack of environmental stimulation, arousal, and adaptive functions as key factors for the beginning and maintenance of Restricted and Repetitive Behaviours (RRBs). This paper consists of accelerometer sensor, tilt sensor, heart beat sensor, PIC microcontroller, motor with help of driver relays and IoT module. The accelerometer and tilt sensor are used to detect the child activity continuously. The sensor is fed to PIC (16F877A) microcontroller. If the sensor value crosses the threshold value, the vibration triggers the child and voice (music) will be played while in autism spectrum disorder. The IoT is used to monitor the child activity continuously.

1. Introduction

AUTISM spectrum disorders (ASD), characterized by debits in communication and social interaction organized with constrained, tedious, and labelled patterns of behaviour, represent a series of neurodevelopmental incapacities. The growing of technologies such as abundant computing and ambient intelligence are enlightening the eminence of health care and medicine treatments. Today the concept of patient in-the-loop effects the development of new system health oriented, for this reason, new difficulties and challenges are coming out. One comes out in those situations where the recognizing and the logging of patients' gestures are significantly important in the direction of improve the quality of healthcare provided.

2. Disadvantages of existing systems.

This allows two children to play a series of interactive games in a virtual reality environment by using simple and hard gestures in order to demonstrate the potential for fostering their communication & collaboration skills. [2,3] These only develop the skills of autism children and improve their interaction with others.Till now there is no first aid for epilepsy (ASD) in previous papers.[1]

IOP Conf. Series: Materials Science and Engineering **623** (2019) 012011 doi:10.1088/1757-899X/623/1/012011

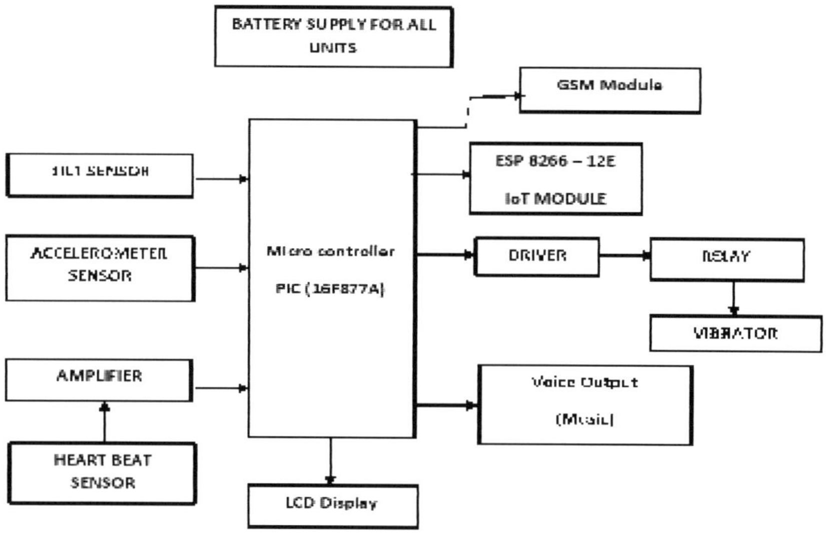

Figure 1.Block Diagram of proposed system

This proposed is mainly based on providing a first aid kit for rectifying epilepsy of an autism disorder people.This paper consists of accelerometer sensor,tilt sensor,heart beat sensor,PIC microcontroller,motor with the help of driver relays and Iot module.The accelerometer and tilt sensorsare used to detect the child activity continuously.[3]The sensors values are fed to the microcontroller.If sensor values crosses the threshold value,the vibration trigger the child and voice (music) will be played for autistic spectrum disorder.The Iot is used to monitor the child activity continuously.[4]

3.Advantages of Proposed system

According to literature survey they used the Iot module to improve the communication skillv of autism people. But in our proposed system we use to Iot module to rectify the epilepsy problem of autism disorder people.[5] The new biomedical technology named as digital pill is added as an enhancing feature to detect the intake of the medicine for this problem.[6]

4. Component description

4.1LEAD ACID BATTERY

A lead-corrosive battery-operated is an electrical stockpiling gadget that usages a reversible chemical reaction to store energy. It utilizes a blend of lead plates or frameworks and an electrolyte comprising of a weakened sulphuric corrosive to change over electrical vitality into potential compound vitality and back once more. [7,8] The electrolyte of lead-corrosive batteries is risky to your security and may produce consumes and other perpetual destruction in the event.

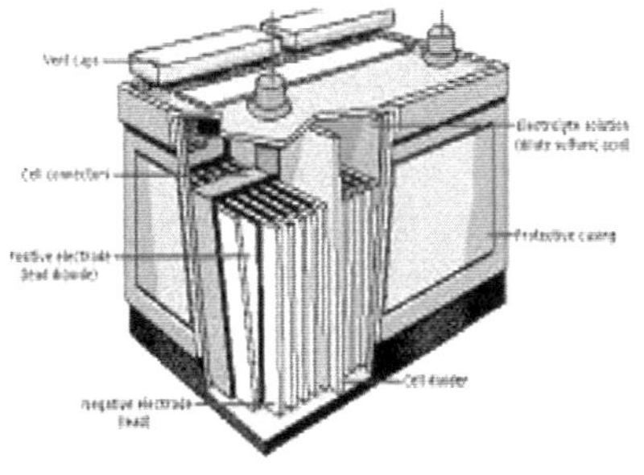

Figure 2.Lead Acid battery

4.2 TRIPLE AXIS ACCELEROMETER ADXL 335

Breakout board for the 3 hub ADXL335 from Analog Devices. This is the furthermost recent in a long, established line of simple sensors - The ADXL335 is a triple pivot MEMS accelerometer with very truncated clamour and power utilization - just 320uA.The included 0.1uF capacitors set the transfer speed of every pivot to 50Hz. also, locally available controller 3.3volts.[9]

An accelerometer is a device that estimate appropriate cumulative speed, legitimate quickening is not equivalent to organize increasing speed. For instance, an accelerometer very still on the outside of the Earth will gauge a speeding up because of Earth's gravity, straight upwards (by definition) of g $\approx$ 9.81 m/s2. [10]

4.3 HEART RATE SENSOR

This paper is based on the principle of photoplethysmography (PPG) which is a non-aggressive method of determining the distinction in blood volume in tissues using a light source and a sensor. Since the modification in blood volume is synchronous to the heart beat, this approach can be used to employed to figure the pulse. Transmittance and reflectance are two undeveloped types of photoplethysmography. As a result of the inhibited infiltration understanding of the light through organ tissue, the transmittance PPG is applicable to a limited body part, for example, the finger or the ear flap. As the light doesn't requirement to intrude the body, the reflectance PPG can be associated to any pieces of human body. In either case, the detected light replicated from or transmitted through the body part will vary as per the pulsatile blood stream brought about by the thumping of the heart.[11]

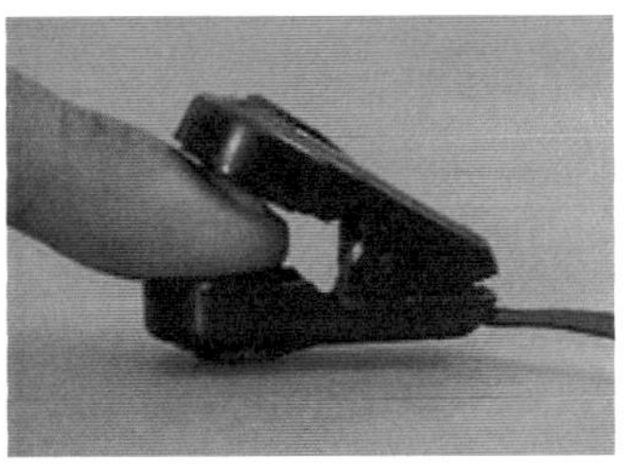

Figure 3 Heart Rate Sensors

IOP Conf. Series: Materials Science and Engineering **623** (2019) 012011 doi:10.1088/1757-899X/623/1/012011

4.4 MICROCONTROLLER

The microcontroller PIC16f877a is standout between the most famous microcontrollers in the industry. This regulator is exceptionally beneficial to operate, the coding or user interface design of this controller is as well simpler. [12] One of the primary points of the interest is that it very well may be combine eradicate whatever number occasions as could reasonably be expected on the grounds that it utilizes FLASH memory innovation. It has an absolute number of 40 pins and there are 33 pins for information also yield.

Figure 4 Microcontroller

PIC16F877A finds its applications in a massive number of devices. It is utilized in remote sensors, security and well-being gadgets, home mechanization and in numerous modern instruments. An EEPROM is additionally contained within in it which makes it imaginable to store a portion of the data for all time like transmitter codes and receiver frequencies and some other related information. The expenditure of this controller is short and its dealing with additionally simple. Its adjustable and can be operated in regions where microcontrollers have never been operated as in coprocessor applications and clock capacities so on.[13]

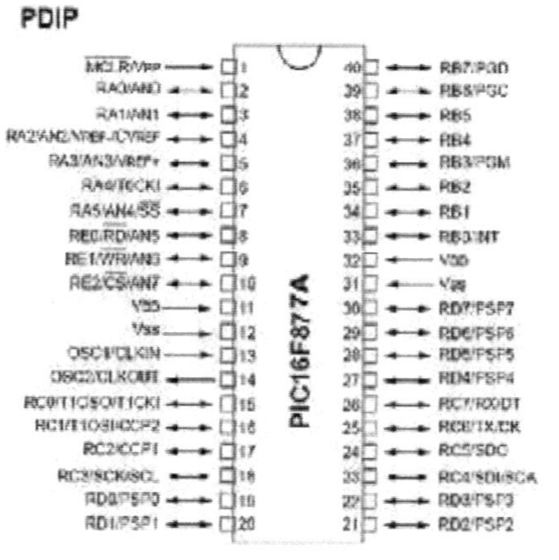

4.5 VIBRATOR

A vibrator is a mechanical gadget to create vibrations. The vibration is often generated by an electric motor with an unbalanced mass unit's driveshaft.

4.6 GSM

GSM, which represents global system for mobile exchanges, rules as the world's greatest broadly utilized phone innovation. Mobile phones apply a PDA administration transporter's GSM establish via scanning for PDA towers in the adjacent territory.

For functional and regular purposes, GSM offers clients more extensive global meandering abilities than different U.S. arrange innovations and can empower a wireless to be a "world phone". Future developed GSM consolidates the before TDMA standard.

5. Software Requirement

5.1 PROTEUS

The controller can understand a program inscribed in low level computing concept, it must be accumulated into a language of zeros and ones. The initial unique alludes in the direction of arrangement of standards utilized intended for constituting program for the microcontroller, while the later alludes in the direction of a program. accumulated program is similarly called Machine Code.[6]

In machine code, a similar order is spoken to through a 14-bit exhibit of ones justifiable by the microcontroller. All low-level computing construct directions are also arranged into the comparing cluster of ones. An information document utilized for putting away gathered program is called "official record, that is "HEX data file". The name comes from the hexadecimal exhibition of a data file and has a suffix of "hex" as well, for illustration "probe.hex". [8]

5.2 Embedded C

Glancing around, we wind up to be encompassed by different kinds of implanted frame work. Be it an advanced camera or a cell phone or a clothes washer, every one of them has sort of processor working inside it. Related with every processor is the installed programming. On the off chance that equipment shapes the body of an implanted framework, installed processor goes about as the mind, and inserted programming shapes its spirit. It is the installed programming which fundamentally oversees the working of implanted frameworks.[14]

During earliest stage long periods of microchip-based frameworks, programs were created utilizing constructing agents and melted into the EPROMs. There used to be no component to discover what the program was doing. LEDs, switches, and so forth were utilized to check right execution of the program. Some 'blessed' engineers had In-circuit Simulators (ICEs), yet they were excessively exorbitant and were not exactly solid also.[10]

5.3 MPLAB

Integrated tool set for the development of embedded application. It provides single integrated to develop code for embedded microcontrollers.

5.4 ISIS(Proteus)

It has wide range of components in its library. It is embedded with the foot prints of different category of components like ICs, transistor, connectors and other discrete components. It offers auto routing and manual options to the PCB designer.

IOP Conf. Series: Materials Science and Engineering **623** (2019) 012011 doi:10.1088/1757-899X/623/1/012011

6. Circuit Diagram

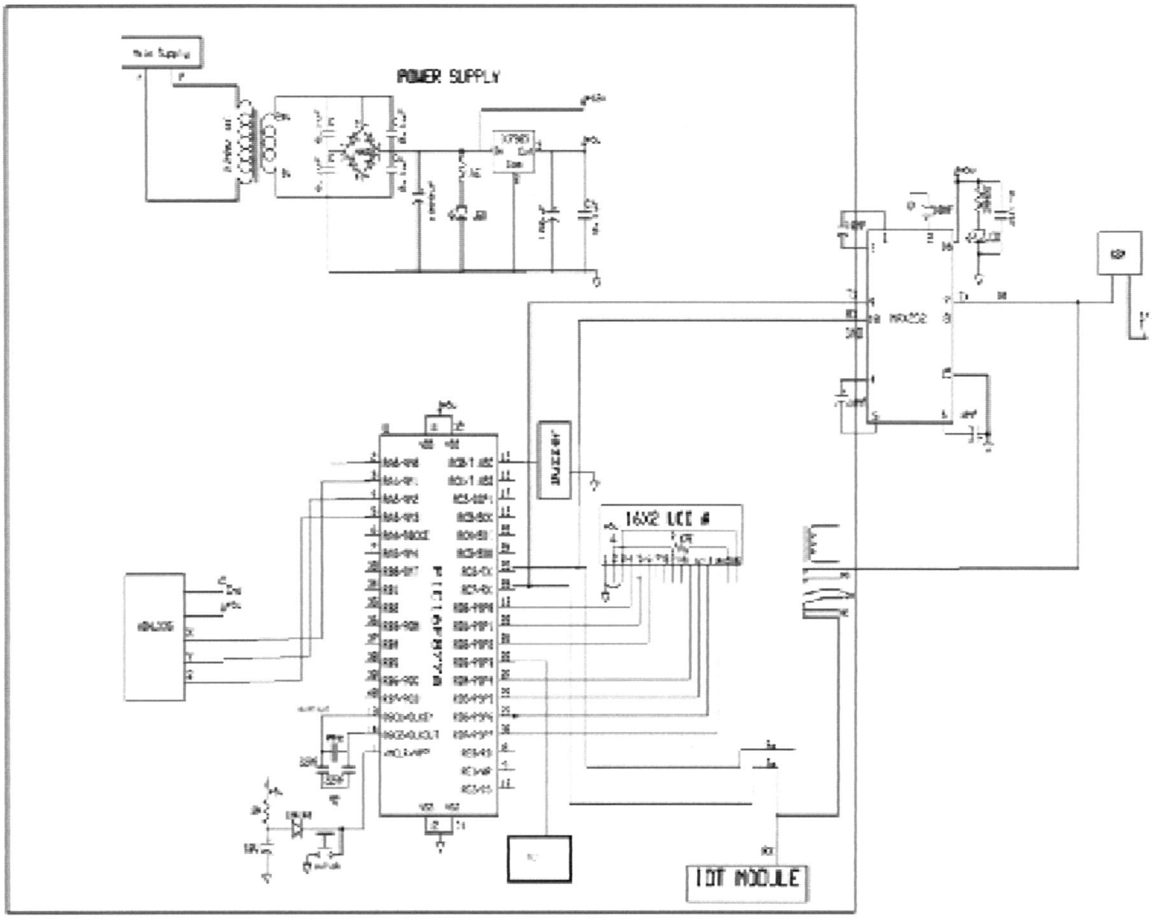

7. Future Work

Autism spectrum disorders embarrassment system will be further enhanced to support more naturalistic combined game play platform. We are now working on scheming the ASD with the haptic edges that is able to produce physical feedback to the user. We expect the haptic ASD system could growth the sense of cooperation between partners. Additionally, extra participants are desired in the future for the user study to evaluate the practical value of the system for children with ASD. In order to determine the influence of the system on the communication capability of the participants, we plan to continue with the analysis of the participants' conversations in terms of the game-oriented content and the social content and perform a statistical investigation of the modification in the content of the conversation.

IOP Conf. Series: Materials Science and Engineering **623** (2019) 012011 doi:10.1088/1757-899X/623/1/012011

8. Simulation Results

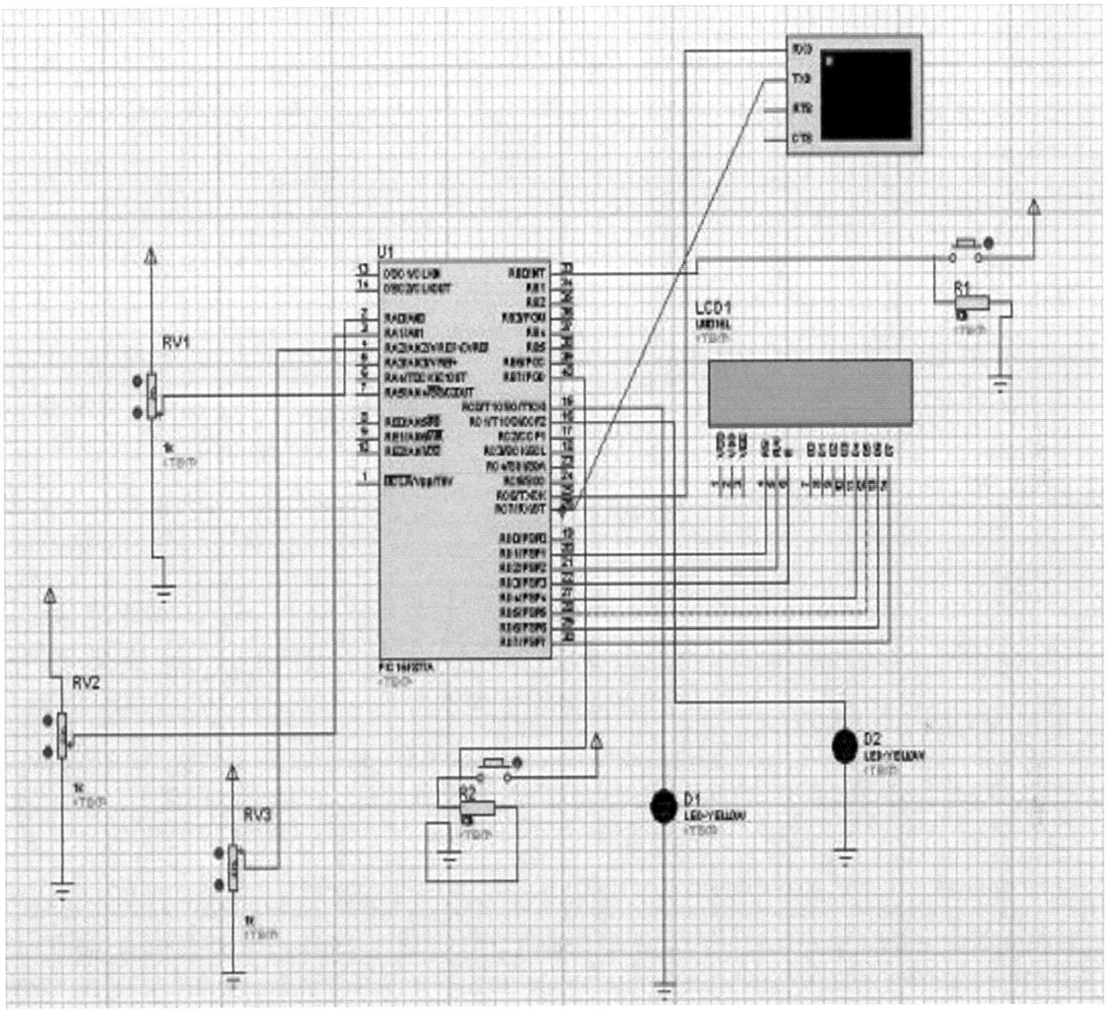

9. Conclusion

Technology-assisted systems can provide a quantifiable, customized therapy platform. Directly available frameworks are planned fundamentally to chain learning by means of parts of one's exposition alone confining individualization. System signals that were developed with an acceptable level of precision and thereby confirm the possibility of an anxiety-sensitive system to be used as a social communication skill learning platform for children with autism. This paper presents the expansion and estimation of the Autism spectrum disorders prevention system, which can provide a naturalistic social interface platform for children with ASD and their peers, growth the chances for communication and cooperation within the combined games and collect quantifiable data regarding combined and communicative performance of the participants. The possibility study tested the appropriateness of the system among children with ASD and attained an opening assessment of the system.

IOP Conf. Series: Materials Science and Engineering **623** (2019) 012011 doi:10.1088/1757-899X/623/1/012011

10. References

[1] J. Baio, "Prevalence of autism spectrum disorders: Autism and developmental disabilities monitoring network, 14 Sites, United States, 2008," MMWR Surveillance Summary, vol. 61, no. 3, pp. 1–19, 2012.
[2] Y. S. Kim et al., "Prevalence of autism spectrum disorders in a total population sample," Amer. J. Psychiatry, vol. 168, pp. 904–912, 2011.
[3] C. P. Johnson and S. M. Myers, "Identification and evaluation of children with autism spectrum disorders," Pediatrics, vol. 120, pp. 1183–1215, 2007.
[4] C. Home, "Prevalence of autism spectrum disorder among children aged 8 years—autism and developmental disabilities monitoring network, 11 sites, United States, 2010," MMWR Surveillance Summary, vol. 63, pp. 1–21, 2014.
[5] M. A. Maglione, D. Gans, L. Das, J. Timbie, and C. Kasari, "Nonmedical interventions for children with ASD: Recommended guidelines and further research needs," Pediatrics, vol. 130, pp. S169–S178, 2012.
[6] S. J. Blumberg, M. D. Bramlett, M. D. Kogan, L. A. Schieve, J. R. Jones, and M. C. Lu, "Changes in prevalence of parent-reported autism spectrum disorder in school-aged US children: 2007 to 2011–2012," Nat. Health Statist. Rep., vol. 65, pp. 1–7, 2013.
[7] J. P. Stichter, K. V. O'Connor, M. J. Herzog, K. Lierheimer, and S. D. McGhee, "Social competence intervention for elementary students with Aspergers syndrome and high functioning autism," J. Autism Developmental Disorders, vol. 42, pp. 354–366, 2012.
[8] E. W. Carter et al., "Promoting social competence and peer relationships for adolescents with autism spectrum disorders," Remedial Special Educ., vol. 35, pp. 91–101, 2013.
[9] J. Locke, E. H. Ishijima, C. Kasari, and N. London, "Loneliness, friendship quality and the social networks of adolescents with high-functioning autism in an inclusive school setting," J. Res. Special Educ. Needs, vol. 10, pp. 74–81, 2010.
[10] N. Bauminger and C. Kasari, "Loneliness and friendship in highfunctioning children with autism," Child Dev., vol. 71, pp. 447–456, 2000.
[11] S. M. Myers and C. P. Johnson, "Management of children with autism spectrum disorders," Pediatrics, vol. 120, pp. 1162–1182, 2007.
[12] K. Francis, "Autism interventions: A critical update," Developmental Med. Child Neurol., vol. 47, pp. 493–499, 2005.
[13] A.T. Sankara Subramanian, P. Sabarish and R Jai Ganesh."An Improved Voltage follower Canonical Switching Cell Converter with PFC for VSI Fed BLDC Motor."Journal of science and technology. ISSN:2456-5660 http://jst.org.in/Vol-2Issue-10October2017.html
[14] A. T. S. Subramanian, P. Sabarish, A. N. Ali, "A power factor correction based canonical switching cell converter for VSI fed BLDC motor by using voltage follower technique", 2017 IEEE International Conference on Electrical Instrumentation and Communication Engineering (ICEICE), pp. 1-8, 2017.

IOP Conf. Series: Materials Science and Engineering **623** (2019) 012012 doi:10.1088/1757-899X/623/1/012012

Intelligent system to monitor and diagnose performance deviation in Industrial Equipment

A T Sankara Subramanian[1], M D Udayakumar[2], V. Indragandhi[3], R.Ramkumar[4]

[1,2 & 4]Assistant Professor, Department of EEE, K.Ramakrishnan College of Technology, Trichy, Tamilnadu, India.
[3]Associate Professor, Department of EEE, VIT, Vellore, Tamilnadu, India.

Abstract. As per ISO norms, industries used to calibrate the meters which they are using, but performance deviations of the working equipment which are not affecting the economy or production of industries are not given significant importance. So, this work aims at concentrating on fine performance deviation in induction machine which if avoided without taking corrective action may even stop production. This is done by taking a set of reference parameters and the newly taken values are compared with the reference ones and error margin is determined. Various sensors are interfaced for measuring the physical parameters of the machines. Thus, with all these the performance of industrial induction motor with different loading conditions are carried out and the possible damage that could occur if the fault in the machine is not cleared, and also will show the preventive measures to be taken to avoid the failure of the drives are framed. In industries the machines used will be of higher ratings. And if there occurs any error or fault then its production will be greatly affected, which in turn will create huge production and economic losses, so to avoid that the proposed system to forecast the performance deviations of the drives will be of great commercial significance. With this system, the production and economic losses can be reduced to greater extent and also can prevent the machines from damage.

1. Introduction

Condition monitoring of electric machines has received considerable impetus from the industry to ensure consistent and reliable operation of modern industrial systems. The principal aim is to develop schemes for reliable detection of faults at an early stage, which will allow controlled and scheduled maintenance instead of sudden failure, thus reducing production losses, outage time, and damage to the equipment [2].

Induction motors are the Literal spine for any industry. Anyway like some other machine, they will in the end fall flat due to heavy and continuous duty periods, poor workplace, establishment and assembling factors, and so on. With heightening requests for dependability and productivity, the field of fault determination in induction motors is picking up significance. Be that as it may, numerous strategies have been proposed for issue identification and conclusion, yet the majority of the techniques require a decent arrangement of mastery to apply them effectively. Less difficult methodologies are expected to empower even unprofessional administrators with ostensible information of the framework to examine the issue condition and settle on solid choices[1].

2. Existing system

A 3 phase induction motor which is suitable for pumping application is fed with a supply of 50Hz and a separate loading arrangement which is required to test the machine at different loading conditions is attached and for different loading conditions test data of voltage and current are measured and recorded.

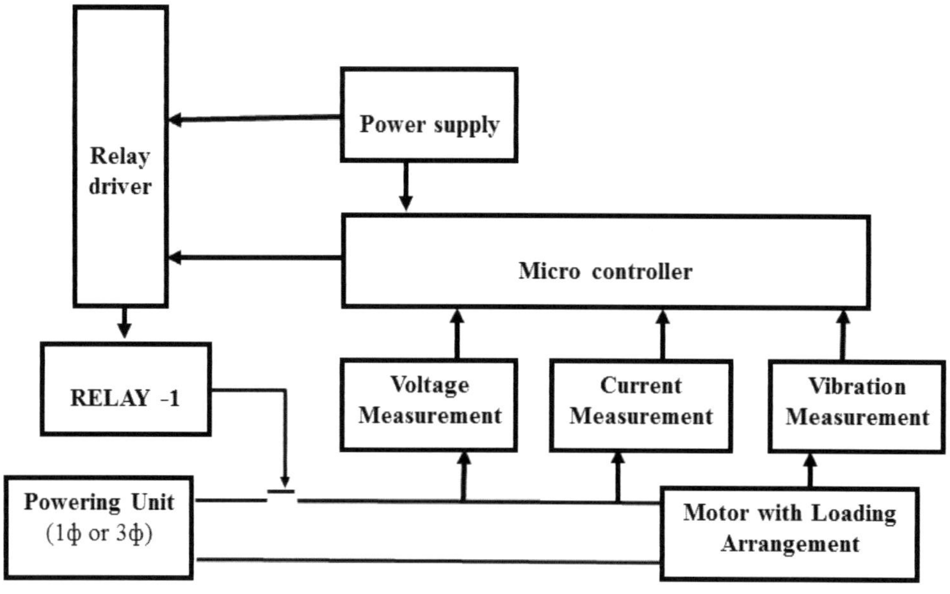

Figure. 1. Block diagram of existing system

If there occurs any vibration in the machine due to the faults then it will be measured by the vibration measuring unit. All the fault signals from the various measuring units are passed to the micro controller. This micro-controller will produce a control signal which is passed to the relay driver circuit [3]. The relay driver circuit will operate the relay 1 which in turn will open the switch that will disconnect motor from the supply and thus eliminating the fault and preventing the machine from damage.

The machine faults either electrical or mechanical are fuzzy in nature or typically it is very difficult to analyze as the parameters associated with such analysis is closed linked to one another. There are two efficient condition monitoring techniques to effectively deploy the strategies for the predictive maintenance. Basically many harmonic analysis techniques are employed and elimination is done at specific characteristic frequencies generated by specific faults as the unique rotating flux [4]. Fuzzy logic system has also been utilized, which assesses severity of the fault and operating condition of machine.

The fault detection can be conducted using vibration signals and/or stator current signals. The machine oscillation on the bed will cause mechanical vibration under high speed conditions which can be measured using vibration sensors which has more signal to Noise ratio. But installation of such sensor is really a tough job and need experienced engineers to do it. Measuring current values of induction motor is not a tough job it can be done simply using simple current sensors [3].The projected current monitoring procedure there is no supplementary sensors required. This is because for machines with larger power ratings associated with different industrial applications are wired with current transformer and potential transformer for voltage and current measurement [5]. Lab automation system is a means that allow users to control electric appliances. Many existing, well established lab automation systems are based on wired communication. Despite that the evolution of wireless communication system in the last decade and its massive growth and concurrence in the field of industrial automation makes it more ideal and cheap [6]. The faults that occur in the machines are corrected by analyzing various parameters in various stages of fault correction methods [8].

IOP Conf. Series: Materials Science and Engineering **623** (2019) 012012 doi:10.1088/1757-899X/623/1/012012

In this work, the major focus is on forecasting all the major faults that may occur in any machine and also to provide the necessary details required for correcting or eliminating the faults. This work also helps to determine or understand the fault occurrence much before the fault occurs so that can one can avoid the post fault problems.

The Powering unit consists of source and converters, the source can be either a single phase or three phase supply which is connected to the machine and the converters are used for the requisite conversions.

The voltage measuring unit is connected across the lines in parallel to measure the supplied voltage, and the current measuring unit is connected to a line in series to measure the current consumed by the machine.

The speed measuring unit is joined to the shaft to estimate the speed at which it rotates or to mark the change in speed, along with this a load measuring unit is also attached to estimate the load applied. A temperature measuring unit is also attached to the machine.The output of this measuring unit is connected to the Arduino controller. The Arduino controller processes the input based on the coding uploaded

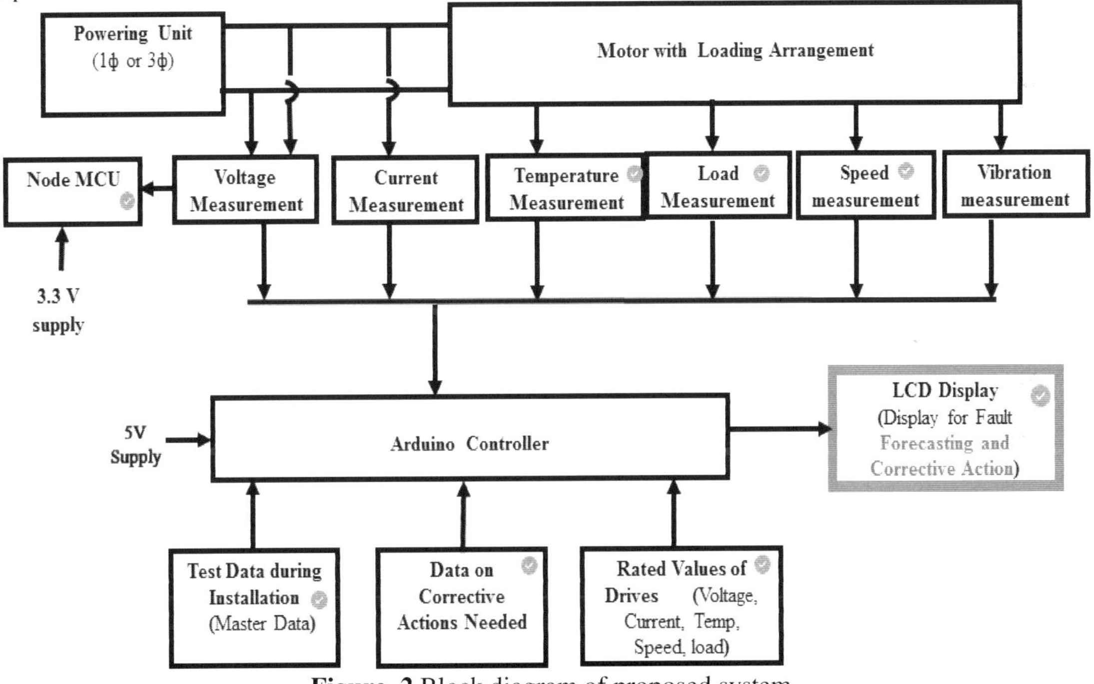

Figure. 2 Block diagram of proposed system

The input to the Arduino controller is the test data during installation which consist of all the master data that are calculated manually, the rated values are referred for the comparison of voltage, current, speed, temp and load for performance deviations and then data on Corrective Actions if needed is also given to it. Its output is fed to the 20x4 LCD display which displays the fault forecasting and corrective measures.

The hardware is made with a small rating Induction of 1 KW and the specifications of the testing machine is noted down clearly to fix the reference values as per standard.

3. Software calculation

The Arduino controller interfaced with the hardware can be fed with the micro C coding based on the following calculation,

Voltage = Sensed value * (3.0 / 1023.0) * 76.0

3.0 – Sensor output voltage.
1023.0 – Controller bit value.
76.0 – Multiple factor (to convert 1ϕ voltage).
Current = ((A_{dc} voltage – offset voltage) / sensitivity)
A_{dc} voltage = (A_{dc} value / 1023.0)*5000.0.
Sensitivity = 30 Amp – 66 MV/Amp.
A_{dc} value = Input sensed current.
Speed = No. of times sensor input becomes high for every 5 seconds.

4. Hardware Description

This system is mainly focused on the performance deviation of induction motor. So the main hardware will be the induction motor. Here various sensors such as current sensor, voltage measuring unit, temperature sensor, speed sensor, load cell along with load amplifier module and vibration sensor are used to measure the physical properties of Induction motor. All the output sensed signals are passed to the Arduino board. Using the Arduino IDE software the output is displayed using the LCD display.

The entire setup is fixed on a hard surface, and also the motor is placed over a bed as shown in the fig.3. The load arrangement is also provided so that the load can be applied externally and also an arrangement is given to measure the load applied.

The setup is tested for its mechanical and electrical stability and the same has been recorded as test data and compared with the simulated results of the PROTEUS software. The loading arrangement and load measuring setup.With the help of a rope tightened upward will help to measure how much load is applied. And with help of a rope lying downward load can be applied. The voltage measuring unit consists of a transformer, a pot, bridge rectifier IC, capacitor and Zener diode. The voltage measurement is done so that if it exceeded the prescribed values then the system is not under healthy condition.

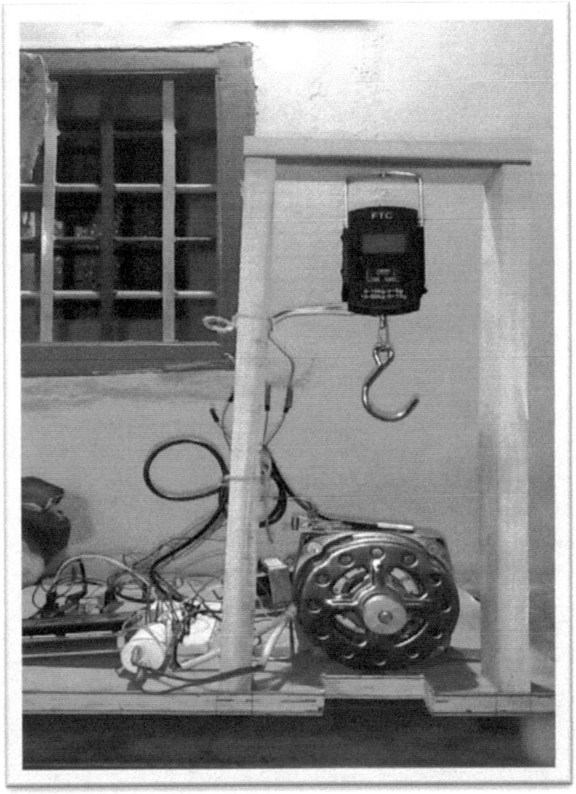

Figure. 3 Hardware setup of proposed system

The setup for measuring the current consists of the current sensor ACS712. Arrangement for speed measurement consists of the speed sensor. A metal piece is attached to the rotor shaft and with this setup the speed is measured using the sensor.
The setup of measuring temperature and vibration consists of the temperature sensor DS18B20 and vibration sensor.The Vibration sensor shows the changes in the Mechanical movements in terms of frequency changes and the same if exceeds the reference frequency it will raise and indication as a warning.

5. Simulations of the Proposed System

5.1 Proteus Simulation for Voltage Measurement

Voltage measuring unit consist of bridge rectifier ic, Zener diode, capacitor, potentiometer and transformer. 230V ac is stepped down to 12V ac using the step down transformer, then it passed through the rectifier and to get 12V dc. This is further passed through the filter circuit and to obtain 3V dc supply. This is given to the Arduino analog pin.
The coding is written in micro C. one can also use Java language to write the coding. The controller processes the coding and provides output. The main advantage of using Arduino is that the programs can be changed even in the run time without flashing out the controllers. It gives more easiness and flexibility in writing the coding. The calculation made earlier can be directly converter into coding. There are specific libraries need to be used in Arduino if not available in the basic package, then that can be easily downloaded free from many websites. Figure 4 shows the simulation circuit for voltage measurement. It is quite easier to draw and simulate in Proteus.

IOP Conf. Series: Materials Science and Engineering **623** (2019) 012012 doi:10.1088/1757-899X/623/1/012012

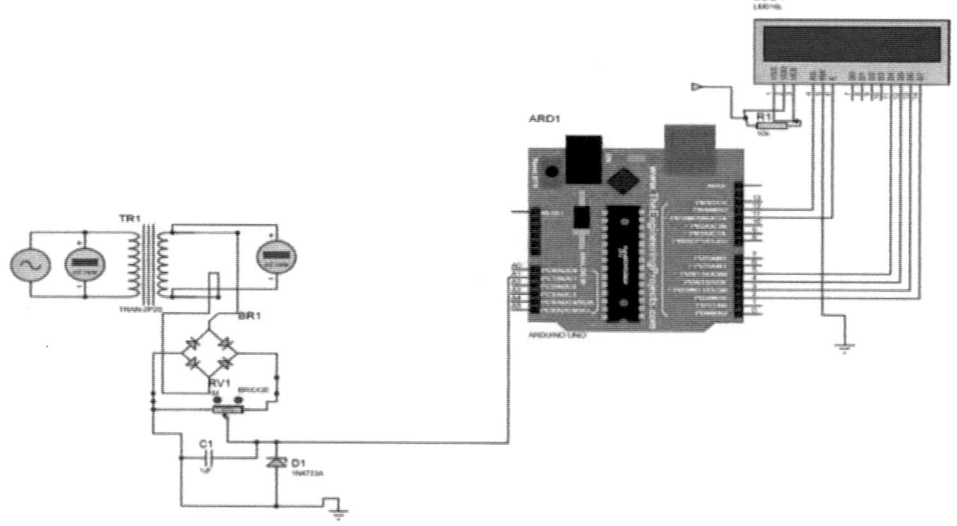

Figure. 4 Simulation circuit of voltage measurement.

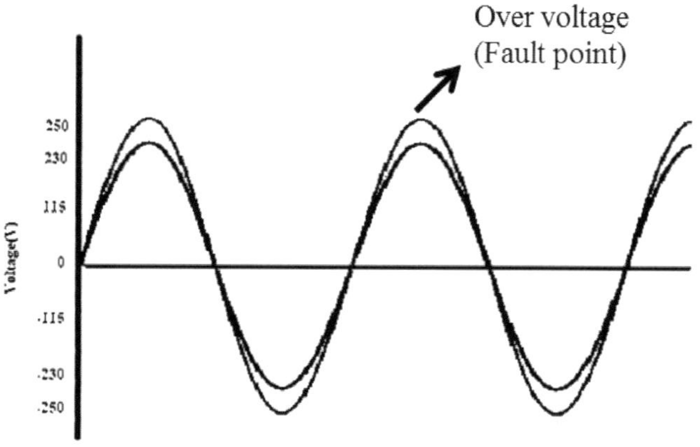

Figure.5 Output voltage waveform with fault point.

Figure.5 consists of two sinusoidal waves of voltage. One is the normal waveform which is obtained before the fault occurred. The other one is obtained when there occurs any fault and the current value rises beyond the rated value, this is depicted by the upper wave showing the over voltage at the faulty point.

5.2 Proteus Simulation for Current Measurement

ACS712 current sensor is used here. The sensed value is given to the digital pin of the Arduino. Figure 6 shows the simulation of current measurement.

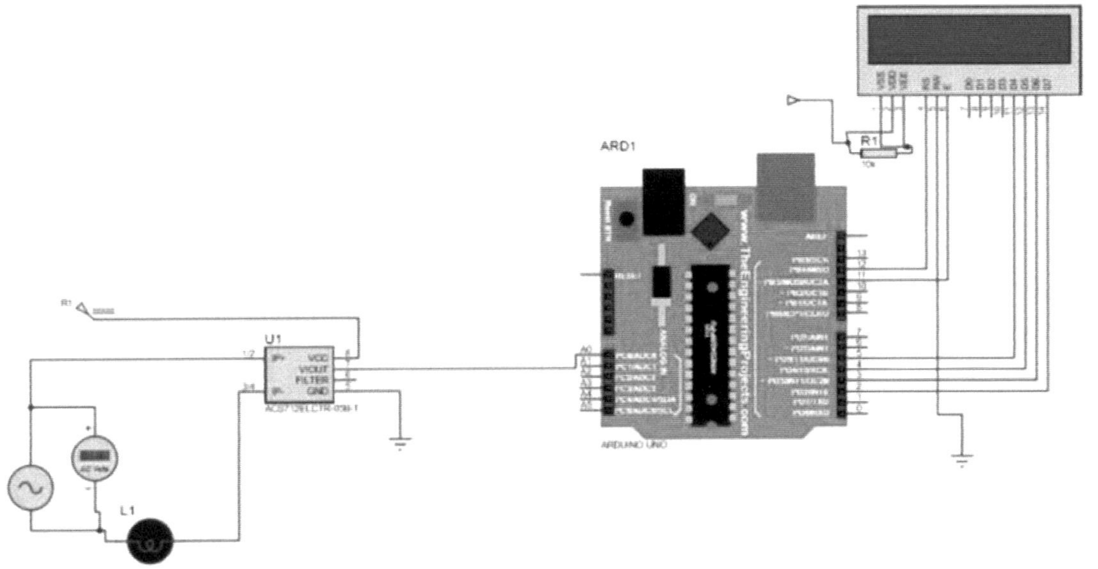

Figure. 6 Simulation circuit of current measurement.

In the programming part the current value is obtained after certain calculations. The values sensed by the current sensor are used for the calculations purpose.

At the fault point the current flowing will be more than the rated current. This is clearly shown in the figure7. The dotted lines in the graph depict the faulty region. The sudden increase in current may occur due to transient switching or increase in the load applied. This type of situation leads to severe damage of the machine and hence it needs to be identified and corrected as early as possible.

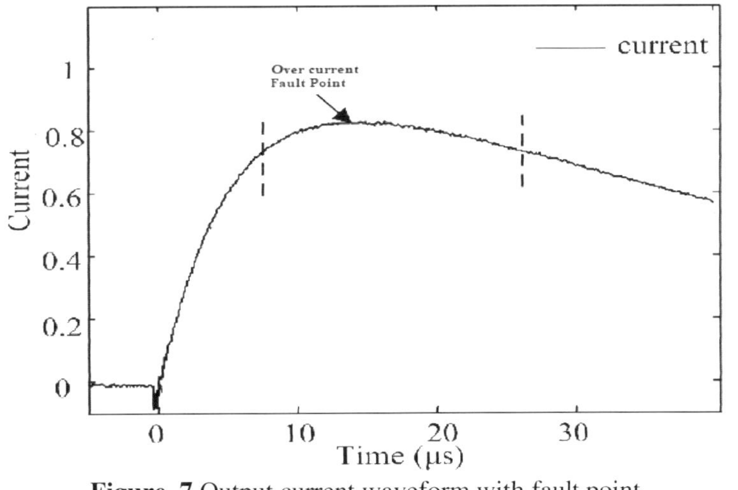

Figure. 7 Output current waveform with fault point.

5.3 Proteus Simulation for Temperature Measurement

DS18B20 digital temperature is used to measure the temperature variation. This is shown in figure 8.

IOP Conf. Series: Materials Science and Engineering **623** (2019) 012012 doi:10.1088/1757-899X/623/1/012012

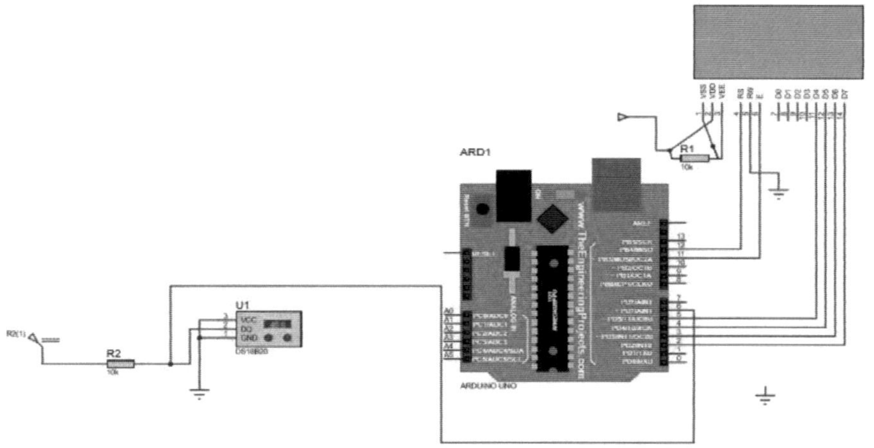

Figure.8 Stimulation circuit of temperature measurement.

The temperature sensor is places close to the motor so that the temperature rise in the motor can be easily determined. This temperature value is used to check whether there is temperature rise in the motor winding or any other part of the system. This is so accurate that it will show minute increase in the temperature.

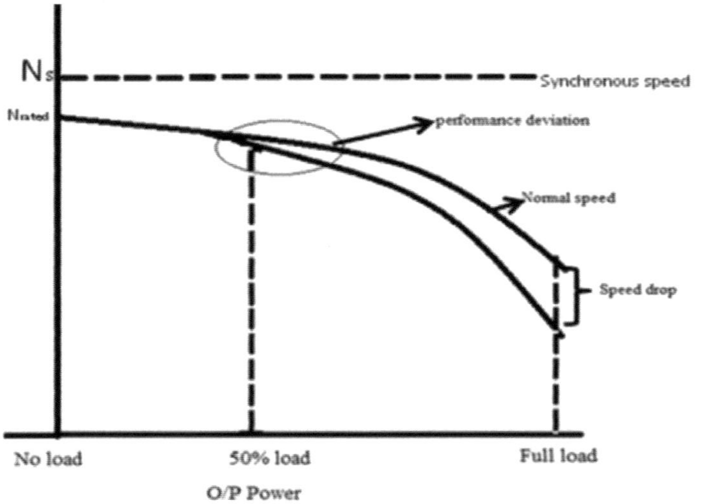

Figure. 9 Identification of Speed deviation in drives

Figure 9 show the speed deviation in the induction motor. Ns represent the synchronous speed while N_{rated} represents the induction motor speed. It is clear from the graph that induction motor runs at lesser speed than the synchronous motor. One shows the normal speed reduction as the load increases but the other shows rapid reduction of speed as the same load is increased this shows that there is some problem or fault occurred in the system.

IOP Conf. Series: Materials Science and Engineering **623** (2019) 012012 doi:10.1088/1757-899X/623/1/012012

6. Software Results

The software result in figure 10 shows the output values that are obtained through the coding used in the Arduino IDE software.

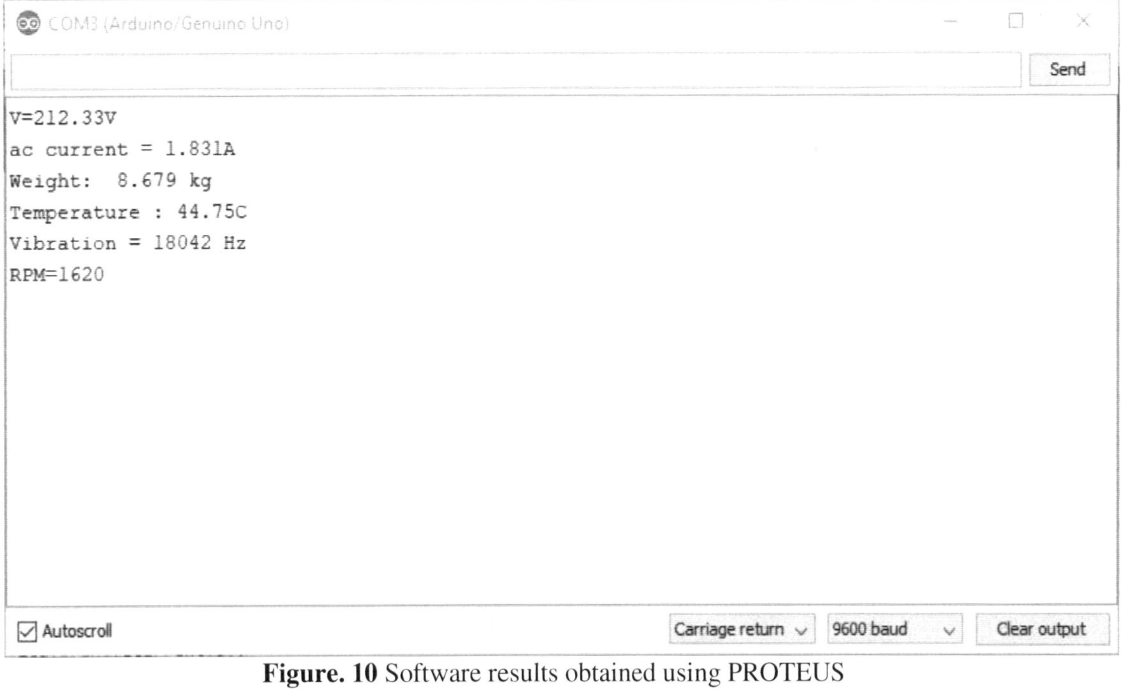

Figure. 10 Software results obtained using PROTEUS

7. Conclusion

This workpays way for a novel way of forecasting the fault much before the fault occurs and the permanent damage takes place. Any fault or damage does not occur instantly, it happens due to the lack of proper observation and lack of awareness. It focus on forecasting or predicting the future damage that may occur if the small deviation occurring is neglected. This method is not available in any places at present, which is much necessary to avoid huge collateral damage. This provides a great opportunity for the engineers to expand their knowledge in the area of fault forecasting. The high cost machines can be prevented from getting damaged by preventing the fault to precede further, this can be achieved only with our proposed system. And also this can also be incorporated with IoT based system for enhanced controlling and forecasting purpose.

8. References

[1] Partha Sarathee Bhowmik, Sourav Pradhan, Mangal Prakash, May 2013, "Fault Diagnostic And Monitoring Methods Of Induction Motor". Volume 1.
[2] Ayaz Ahmed Soomro, Imtiaz Hussain Kalwar, Kamran Kazi, Sarang Karim Khoso and Shahzeb Ansari, Year: 2016, "A Hybrid Monitoring Technique For Diagnosis Of Mechanical Faults In Induction Motor", Volume: 9
[3] De Z. Li, Wilson Wang, Year: 2015, "An Intelligent Monitor for Electric Motor Fault Diagnosis", Volume: 5.
[4] Pero Ostojic, Arijit Banerjee, Dhaval C. Patel, Wrichik Basu, Shahid Ali, Year: 2014, "Advanced Motor Monitoring and Diagnostics", Volume: 5.

IOP Conf. Series: Materials Science and Engineering **623** (2019) 012012 doi:10.1088/1757-899X/623/1/012012

[5] Prof. V. P. Kaduskar, Nitish Gupta, Yash Bhardwaj, Shivam Kumar, "IOT BASED LAB AUTOMATION SYSTEM".

[6] R.Megala, Dr. V. Eswaramoorthy, "Fault Detection and Prediction Of Failure Using Vibration Analysis".

[7] Jaiganesh, R., et al.,(2017), "Smart Grid System for Water Pumping and Domestic Application Using Arduino Controller." International Journal for Modern Trends in Science and Technology 3.05: 385-390.

[8] Sagayaraj, R., and A. Nazar Ali, (2018), "Performance Analysis of Quazi Z-source inverter Fed Induction Motor under Semiconductor Failure Condition."

[9] D Sivamani, Dr. R Sagayaraj, R Jai Ganesh and Dr. A. Nazar Ali,September 2018, "Smart Incubator using Internet of Things", International Journal for Modern Trends in Science andTechnology, Vol. 04, Issue 09, pp.-23-27.

[10] A.T. Sankara Subramanian, P. Sabarish and R Jai Ganesh."An Improved Voltage follower Canonical Switching Cell Converter with PFC for VSI Fed BLDC Motor."Journal of science and technology. ISSN:2456-5660 http://jst.org.in/Vol-2Issue-10October2017.html

[11] A. T. S. Subramanian, P. Sabarish, A. N. Ali, "A power factor correction based canonical switching cell converter for VSI fed BLDC motor by using voltage follower technique", 2017 IEEE International Conference on Electrical Instrumentation and Communication Engineering (ICEICE), pp. 1-8, 2017.

[12] Subramanian Sankar, "A PV SYSTEM BASED HIGH STEP UP CONVERTER WITH A VOLTAGE MULTIPLIER MODULE USING FUZZY LOGIC CONTROLLER", International Journal of Applied Engineering Research 10.9, pp. 7441-7446, 2015

[13] Arunraj, S., et al. "A Novel Zeta Converter with Pi Controller for Power Factor Correction in Induction Motor'." International Journal of Scientific Research in Science and Technology (IJSRST), Print ISSN (2017): 2395-6011.

[14] Nagarajan, L. "Star Delta Starter using Soft Switch for Low Power Three Phase Induction Motors." Australian Journal of Basic and Applied Sciences 9.21: 175-178.

[15] Vikram, A. Arun, et al. "Solar PV Array Fed BLDC Motor Using Zeta Converter For Water Pumping Applications." Journal of Science and Technology (JST) 2.11 (2017): 09-20.

[16] A.Rajkumar and B.Karthikeyan., Fuzzy Logic Controller for Fast Transient Response DC - DC Converter. Adv. in Nat. Appl. Sci., 8(21): 65-71, 2014

[17] Rajkumar, B. Karthikeyan and S. Senthil Kumar , Fuzzy Logic Control of Fast Transient Response DC-DC Converter. Middle-East J. Sci. Res., 23 (9): 2256-2263, 2015.

[18] A Rajkumar, B Karthikeyan, S Senthilkumar, Reliability Analysis of Fuzzy Logic Controlled Fast Transient Response DC- DC Converter. International Journal of Applied Engineering Research 10 (85), 728-730.

[19] A Rajkumar, B Karthikeyan, S Senthilkumar, Fault Tree Based Reliability Analysis of Fast Response DC-DC Converter. International Journal of Applied Engineering Research 10 (85), 731-737.

[20] A Rajkumar, B Karthikeyan, S Senthil Kumar, Performance Evaluation Fast Transient Response DC-DC Converter using Soft Computing Analysis with PI Controller, Australian Journal of Basic and Applied Sciences 9 (33), 373-379.

IOP Conf. Series: Materials Science and Engineering **623** (2019) 012013 doi:10.1088/1757-899X/623/1/012013

An IBSMF Optimization Technique used Distribution system with Renewable Energy Sources

J Suresh[1], V Ganesh[2]

[1]Ph.D Scholar, Department of EEE, JNTUA, Anantapuramu, Andhra Pradesh, India
Suresh55j@yahoo.co.in
[2]Professor, Department of EEE, JNTUA College of Engineering, Pulivendula, Andhra Pradesh, India
ganivg@gmail.com

Abstract. In this paper, the interconnection of renewable energy sources is proposed for power quality improvement of renewable energy sources (RESs) with an energy storage system. The proposed system is the combined execution of Improved Bat Search Algorithm (IBSA) through Moth Flame Optimization Algorithm (MFOA) named as IBSMFO. The searching behaviour of the bats is modified with efficient neighborhood search functions like crossover & mutation. In the proposed method, the MFOA has handled the searching behavior of the IBSA in perspective of the minimum error objective function. The objective of the proposed IBSMFO approach was to enhance the power quality as for the real and reactive power variations. To accomplish the objective, MFOA is optimized for minimize power variations as well as the operational cost of the RESs in light of weekly and daily forecast of data for grid electricity charge, electrical load and environmental parameter. The proposed IBSMFO procedure deals with the performance of the energy storage apparatus as far as the enhancement of power in the entire system. By at that point, the proposed system is executed in MATLAB/SIMULINK working model and the execution is surveyed with the current procedure.

Index Terms-Power Quality Enhancement (PHE), control scheme, IBSA, MFOA.

1. Introduction

Energy deficiency problem concerted by contemporary high petroleum cost has resulted in a severe impact to several technological aspects [1]. Efficiency development of high power apparatus, research and improvement of alternative energy, the study of integrated various renewable energy resources, etc. have been eagerly progressing [2]. All through the past several decades, a huge quantity of natural resources of the earth has been unlimitedly consumed and our living environment has been severely destroyed as well as contaminated [3]. Global environmental protection concept and concern have been widely excited with several innovative forms of renewable energy sources such as solar photovoltaic systems (PV) and wind power generation systems (WPGS) to supplement fossil fuels have been examined, integrated and developed in the entire world [4-6].

A Renewable energy storage system (RES) system can combine all different kinds of existing renewable energy linked with available energy storage systems [7]. The necessary power for the connected loads can be efficiently delivered and supplied through the hybrid power energy storage system with suitable control with efficient coordination among different subsystems [8]. Due to the global technological development and promoted experience on the Wind Power Generation System, the charge of generating electricity from the WPGS has been reduced and the price of electricity-generating of the WPGS might close to one of the conventional energies [9-11]. With the enhancement of semiconductor manufacture technology, Solar PV has enhanced efficiency and the cost of Solar PV becomes much lower with the installed capacity becomes much higher in recent years [12]. However, the Solar PV contains lower energy conversion efficiency, lower power density and higher cost compared with WTGs. Huge Solar PV may generate sufficient electricity for supplying isolated loads or delivering energy to a utility grid through DC–AC converters [13].

RES system for power quality (PQ) enhancement is done by using several controllers [14]. Several researchers are worked with the most modern controllers such as a predictive controller, sliding mode controller, H-infinity controller used for better steady-state and transient response of systems. This control technique depends on complex mathematical analysis. In order to avoid the problems in controller designing, intelligent controllers are used [15, 16]. For enhanced result, intelligent controller is now applied to different hybrid energy system problems. An application of fuzzy logic-based controller (FLC) for inverter voltage and frequency control works thoroughly even after variation in system constraint and operating circumstances [17, 18]. Here an Fuzzy Logice Controller used for battery charging or discharging is implemented for system Power Quality improvement to suppress the power oscillation and to supply a quality power to load [19]. In spite of all the system parameter are recognized, there may be parameter variations through the operation of the system. So it is difficult to design controller parameters and more time is required [20].

This paper proposed an IBSMFO control scheme for the power quality improvement in the micro grid using the RES. The main objective of the present work is to improve the power quality as for the real and reactive power variations. The error function is minimized by using the searching technique named as MFO. The remaining section of the paper is depicted in the section underneath. Section 2 delineates the recent research works about the PQ enhancement. Section 3 depicts a summary of the proposed system. Section 4 and section 5 includes the experimental results and the conclusion.

2. Literature survey

Abundant research works have previously existed in the literature which was based on the optimal control strategy for power quality improvement of the renewable energy sources using different techniques and various aspects. Some of the works are reviewed here.

G. Talapur et al. [21] have presented a reliable micro grid used for a residential population with modified control techniques to attain improved operation during grid connected, islanded and resynchronization mode. A modified power control technique was developed such that, local load reactive power demand, harmonic currents and load unbalance compensated by particular residential local DG. A supplementary modified control technique was also having developed to achieve a seamless transition of micro grid between grid connected mode and islanded mode. M. Mosaad and H. Ramadan [22] have presented the PQ enhancement of the Fuel Cell (FC) integrated to the power network through a chopper and an inverter with the conventional PI controller. Two PI controllers tuned by three recent dissimilar evolutionary computing techniques namely Harmony Search (HS), Modified Flower Pollination Algorithm (MFPA) and Electromagnetic Field Optimization (EFO) methods are measured by the author. The two PI controllers are used for driving the inverter connected the on-grid Fuel Cell in order to govern the PCC voltage between the Fuel Cell and the power network. The control of an interface inverter for the application of a hybrid solar photovoltaic and battery energy storage system (BESS) was planned by S. Mousazadeh Mousavi et al. [23]. A control approach for battery state of charge and power management was working for improving the performance of BESS and load leveling purpose.

For controlling the shunt active power filter (SAPF), S. Prince et al. [24] have proposed a Kalman filter (KF) based proportional integral (PI) current control strategy. To achieve accuracy in the Kalman Filter performance, Particle Swarm Optimization (PSO) algorithm was adopted for tuning parameter. Thus, the tuning and parallel resonance problems are eliminated associated through the existing proportional integral (PI) technique. V. Lopez-Martinet al. [25] has employed a modification for PFC controllers by adapting the operation mode depending on the measured THDV. As a result, the PFCs operate either in a low current Total Harmonic Distortion (THD) mode or in the conventional resistor emulator mode and contribute to the regulation of the THDV and the PF at the distribution feeders. To verify the concept, the modification was applied to a current sensor less Non Linear Controller (NLC) applied to a single phase Boost rectifier.

2.1. Background of the Research Work

The review of the recent research work demonstrates that the power quality (PQ) improvement of renewable energy sources (RESs) utilizing an energy storage system is an essential contributing variable. Be that as it may, the RESs unsettling influences, for example, frequency drop, environmental dependency, noisy operation in wind energy systems, harmonics, intermittency and variation of solar irradiation through the sun intensity are the absolute most ruling PQ problems in RES. These disturbances result in malfunctions decreased duration and failure of electrical apparatus. Be that as it may, there are several procedures have been actualized for the power flow control, for example adiabatic compressed air energy system (ACAES), flywheel energy storage system (FESS), proton exchange membrane (PEM), fuel cell (FC), model predictive control (MPC) technique etc., ACAES has been connected used for enlarging the operational ranges yet the primary downsides are poor maintenance of transmission and distribution system prompts loss of energy, inefficient control and leakages prompt loss of compressed air. The flywheel energy storage system has the capacity to accelerating a rotor (flywheel) to high speed with maintaining the energy in the system as rotational energy though it displays constraints like short discharge time. MPC technique can improve the control execution and inherent compensation for dead times. However, impediment like computation complexity and the solutions are not ensured. Despite the fact that the exceeding techniques are utilized for attractive the power quality in RES, the complexity of the algorithm is high because of the increased number of samples required. To overcome these challenges, optimal control using advanced technology is necessary. In interconnected works few control techniques are exhibited to tackle the RES issue. The earlier mentioned impediments have inspired to do this research work.

3. Outline of the System with Proposed system

Figure 1 illustrate the proposed IBSMFO control scheme of RES to improve the Power Quality (PQ) in the MG system. Here, the control strategy based proposed method is the parallel implementation of the IBSA and MFDOA technique. In the proposed approach the RES are the compilation of the Solar PV, wind turbine, fuel cell and battery. By using these techniques the energy is transmitted to the DC bus. Furthermore to improve the dynamic optimal PQ and also to study the electrical energy, the load and the grid side parameters are determined optimally in this system. The architecture also specify the Voltage Source Inverter (VSI) based RES unit with the control modules for the Power Quality Enhancement (PQE). Here the VSI is used to implement the suitable power control mode due to the active and reactive power of the HRES unit must be regulated under grid connected mode based on their reference values. At the time of load demand changes the power supplied by the main grid and RES must be suitably changed. The power delivered commencing the main grid, Solar PV array as well as Fuel Cell must be coordinated to meet the load demand. Here the constant active and reactive power is essential to control Power Quality issues between the HRES and the utility grid. The active and reactive power control strategy of the grid connected operation mode is clearly depicted in the part underneath.

IOP Conf. Series: Materials Science and Engineering **623** (2019) 012013 doi:10.1088/1757-899X/623/1/012013

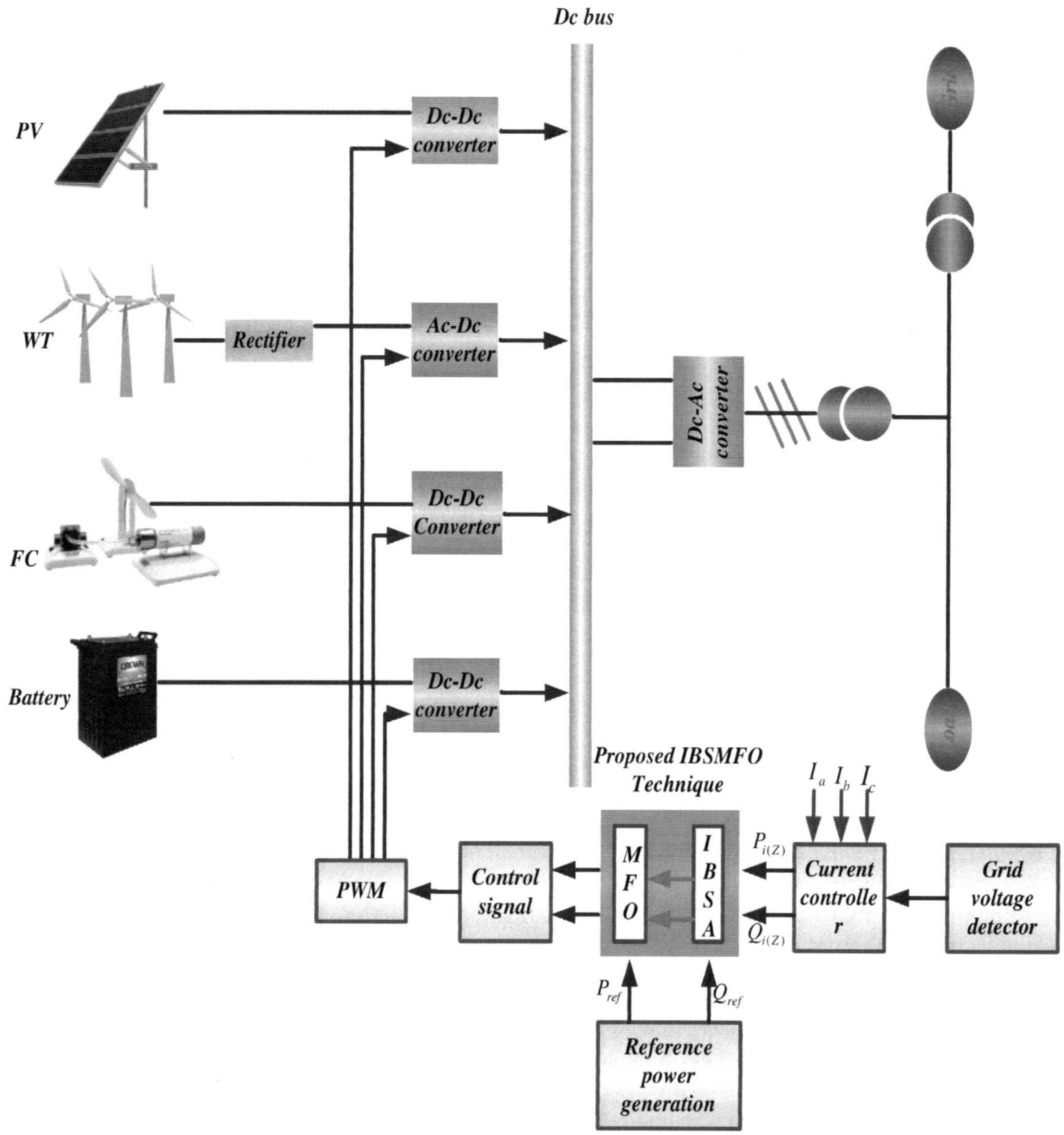

Fig. 1 Control Structure of HRES with the proposed controller

3.1. Active and Reactive Control strategy for Grid Connected Power Quality Enhancement

This part analyzed the power quality improvement with the gird and load of the HRES. The power generated from the main grid and HRES should be changed when the load demand changes. Furthermore the power generated from the main grid, Solar PV array and Fuel Cell must be synchronized to meet the load demand. In order to contribute with constant active and reactive power it is required to control the power flow between the HRES as well as the utility. The power balance from sources to AC bus and to storage devices are controlled to satisfy the active and reactive power demand in the load. Here the power balance should be satisfy both at the DC link and at the Point of Common Coupling (PCC) which is expressed in equation (1).

$$P_{HRES}(t) = \sum_{i} P_{WT}(t) + P_{PV}(t) + P_{FC}(t) + P_{Bat}(t) \qquad (1)$$

From the available renewable energy sources at the time , the generation of total power is denoted as $P_{HRES}(t)$. The need of the load and the output of the HRES are used to derive the power balancing of the system and also for the active and reactive power control process, $P_{Grid}(t)$ and $P_{Load}(t)$ in the grid as well as load connected mode is determined from the following equation,

$$P_{Grid}(t) = P_{Load}(t) - P_{HRES}(t) \tag{2}$$

$$P_{Load}(t) = P_{HRES}(t) + P_{Grid}(t) \tag{3}$$

$$Q_{Grid}(t) = Q_{Load}(t) - Q_{HRES}(t) \tag{4}$$

$$Q_{Load}(t) = Q_{HRES}(t) + Q_{Grid}(t) \tag{5}$$

In the entire time due to the uncertainty of the RES with the variation of non linear load demand. The above mentioned power balanced conditions are not satisfied. The result of the power control mode is approximately based on the high performance operation of the HRES. Moreover the control mode should be dedicated to improving the quality of power supply as PQ issue like voltage, frequency deviation and harmonic distortion rises. Her the active and reactive power should be computed as follow,

$$P_I(t) = \frac{3}{2}\left[V_d^I * i_d^I + V_q^I * i_q^I\right] \tag{6}$$

$$Q_I(t) = \frac{3}{2}\left[V_q^I * i_d^I - V_d^I * i_q^I\right] \tag{7}$$

In this case in order to inject the preset active and reactive power values the amplitude and phase angle of the inverter current are controlled which can be defined locally by the grid. In the grid connected mode the active and reactive power control strategy is achieved based on the voltage regulation and frequency regulation. To examine the HRES units the real and reactive power process is considered and with the help of irradiance the Solar PV is varied and in the WT parts the Battery is utilized. For the grid connected operation mode the following subsection presents the most relevant power and current control strategy.

3.1.1. Current and Power Control Strategy

The purpose of using this strategy is to control the supply of active and reactive power to the load and also control the PQ issues between the grid and the utility. Here, Figure 2 illustrate the proposed controller scheme for enhancing the Power Quality in the Micro Grid system based on the two PI regulators. This control scheme employed the outer control loop to generate the reference current vectors ID and IQ representing that the control objective has been achieved a high quality of the inverter output power. In addition which is ensured by relatively slow changes of the reference current trajectory consequently. And also the VSI based DG unit and PQ control strategy based on the IBSMFO technique is proposed. This IBSMFO is an efficient process to release qualified reference current vectors which also provides optimum control parameters. The active and reactive power have to be achieved when the DG unit connects to the grid or at the load change condition. The active and reactive power outputs of the inverter are regulated by the controller based on their reference values ($P_{ref}(t)$ and $Q_{ref}(t)$)while the decoupling between the active and reactive power which has been achieved by using equation (8) and equation (9). Here DQ reference vectors current expressed as

$$I_D^*(t) = \left(P_{ref}(t) - P_I(t)\right) * \left(\delta_p^P + \frac{\delta_I^P}{t}\right) \tag{8}$$

$$I_Q^*(t) = \left(Q_{ref}(t) - Q_I(t)\right) * \left(\delta_p^Q + \frac{\delta_I^Q}{t}\right) \quad (9)$$

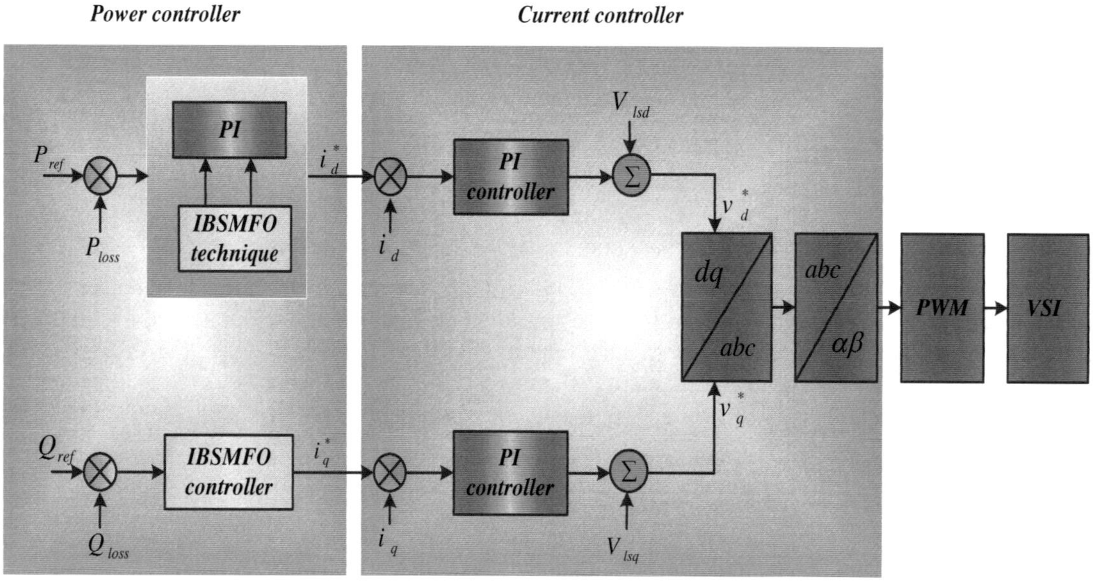

Fig. 2 Proposed Controller for PQ enhancement

The objective function of proposed controller is taken as to track accurately and to reduce inverter output current transients. To develop the steady state and dynamic performance both the inverter current loop and the grid voltage feed forward loop are employed by both converter as well as the current error is eliminated by using two PI regulators. Thus the output signal of the controller is represented by the reference voltage signals in the DQ frame. Moreover, the controller technique provides the desired output voltage vectors with less harmonic distortion by the use of the Pulse Width Modulation technique. Here the reference voltage signal is expressed as follows,

$$\begin{bmatrix} V_D^* \\ V_Q^* \end{bmatrix} = \begin{bmatrix} -K_p & -\omega L_t \\ \omega L_t & -K_p \end{bmatrix} \cdot \begin{bmatrix} I_D \\ I_Q \end{bmatrix} + \begin{bmatrix} K_p & 0 \\ 0 & K_p \end{bmatrix} \cdot \begin{bmatrix} I_D^* \\ I_Q^* \end{bmatrix} + \begin{bmatrix} K_I & 0 \\ 0 & K_I \end{bmatrix} \cdot \begin{bmatrix} X_D \\ X_Q \end{bmatrix} + \begin{bmatrix} V_{tD} \\ V_{tQ} \end{bmatrix} \quad (10)$$

$$\frac{dX_D}{dt} = I_D^* - I_D \text{ and } \frac{dX_Q}{dt} = I_Q^* - I_Q$$

As shown in the following equation (10) and (11) can be transformed into α, β a stationary frame Using Clarke's transformation.

$$\begin{bmatrix} V_\alpha \\ V_\beta \\ V_0 \end{bmatrix} = \frac{2}{3} \begin{bmatrix} V_a \\ V_b \\ V_c \end{bmatrix} \begin{bmatrix} 1 & -0.5 & -0.5 \\ 0 & 0.866 & -0.866 \\ 0.5 & 0.5 & 0.5 \end{bmatrix} \quad (11)$$

The inductor current can achieved bya low pass filter Equation (12) and it is given in a first order transfer function.

IOP Conf. Series: Materials Science and Engineering **623** (2019) 012013 doi:10.1088/1757-899X/623/1/012013

$$f \cdot \frac{1}{1+tT_i} = f_i \quad (12)$$

where the filter input value is given as f the filtered value as f_i and the time constant as T_i.

4. IBSMFO Based Power Quality Enhancement of HRES

This section presented the new system for the PQ enhancement of HRES in the MG system. For this improvement of PQ the proposed work employed the combined execution of the Improved Bat Search Algorithm (IBSA) and Moth Flame Optimization (MFO) algorithm [26]. To improve the PQ the optimal gain parameters are obtained by the IBSA and the searching behavior of the IBSA is improved by the MFO algorithm. The Bat algorithm is a meta heuristic nature inspired algorithm which was invented by Yang [27]. The BA is formulated idealizing bats characteristics in hunting their prey. The bat algorithm is formulated by idealizing the echolocation behavior of bats at first which includes the behavior of micro bats and Acoustics of Echolocation. As per this work by the better BA is used to reduce the error function to improve the PQ of HRES system [28]. The error of the system was reduced and the optimal PQ enhancement is achieved based on the PI parameters tuning. To find the objective function the step by step process of bat algorithm is analyzed which are described below.

4.1. Gain Parameter Optimization of PQ Enhancement using IBSMFO

Step 1: Initialization

In the initialization process the real and reactive power values are the input of the algorithm. In the first step the gain parameters K_p and K_i values are randomly generated and which is expressed as follows,

$$X_i = \begin{bmatrix} K_p^{11} K_i^{11} & K_p^{12} K_i^{12} & \cdots & K_p^{1n} K_i^{1n} \\ K_p^{21} K_i^{21} & K_p^{22} K_i^{22} & \cdots & K_p^{2n} K_i^{2n} \\ \vdots & \vdots & \vdots & \vdots \\ K_p^{m1} K_i^{m1} & K_p^{m2} K_i^{m2} & \cdots & K_p^{mn} K_i^{mn} \end{bmatrix} \quad (13)$$

where, K_p and K_i is the gain parameters.

Step 2: Fitness Evaluation

The error of the system is minimized by the objective function which can be expressed through the fitness evaluation and which can be expressed as,

$$Obj\, F_j = Min\ E(x) \quad (14)$$

$$E(x) = P_{ref}(t) - P_l(t), \text{ for the real PQ enhancement}$$

$$E(x) = Q_{ref}(t) - Q_l(t), \text{ for the reactive PQ enhancement}$$

where, E(x) is the error function of the system. Here once the minimum objective function is achieved the process gets optimized and the corresponding Kp and Ki parameters are tuned [29].

Step 3: Solution Generation:

In this step the new solution is generated by adjusting the pulse frequency and keeping wavelength as constant [30]. During the optimization process, the position (x_i) and velocity (v_i) of each bat should be defined and updated. The new solutions x_i^t and velocity v_i^t at time step t are generated by using the following equations,

$$F_i = F_{\min} + (F_{\max} - F_{\min}) \times \eta \tag{15}$$

$$v_i^t = v_i^{t-1} + (x_i^{t-1} - x^*) \times F_i \tag{16}$$

$$x_i^t = x_i^{t-1} + v_i^t \tag{17}$$

where η is denoted as the range of [0, 1] randomly drawn the best location is defined as X*. The velocity increment is represented by the product of Fi and θi. According to a problem the velocity increment can be adjusted by changing one and keeping fixed another.

Step 4: Local search process by using MFOA
The MFOA is a novel inspired optimization paradigm and the main inspiration of this algorithm is the navigation method of moths in nature called transverse condition [31-33]]. The use of MFOA in this paper is to develop the searching behavior of the IBSA. In MFOA the moth is the search agent and the flame is the best position of the moth. Here the current solution is selected among the available solution and which is given in the following equation,

$$X_{new} = (-1) + X_{old} * \left(-\frac{1}{T}\right) \tag{18}$$

Step 5: Updating
The position of the moth is updated by using the following equation,

$$X_{i,j} = S(X_i, F_j)$$

where, the ith moth is represented as X_i and the jth flame is represented as F_j and the spiral function is represented as S. The spiral function of a moth is computed as follows,

$$S(X_i, F_j) = D_i.e^{bt}\cos(2\pi t) + F_j$$

where the distance of ith moth for the jth flame is represented as D_i, the shape of the spiral motion is defined as b and t is the random number in the range {-1, 1}and which is expressed as follows,

$$D_i = |F_j - X_i|$$

The best position is updated by using the number of flames and which can be expressed as follows,

$$U = round\left(N - l * \frac{N-1}{T}\right)$$

where the present number of iteration is l, T is the maximum number of iteration and N is the maximum number of flame.

Step 6: Termination

This procedure of computation is terminated if the maximum count of iterations is reached which satisfy the stopping criterion or else repeat the process going to steps 2 and 4 [34, 35]. The error minimization output for PQ enhancement is given as follows,

$$\begin{bmatrix} E^{11} & E^{12} & \cdots & E^{1n} \\ E^{21} & E^{22} & \cdots & E^{2n} \\ \vdots & \vdots & \vdots & \vdots \\ E^{m1} & E^{m2} & \cdots & E^{mn} \end{bmatrix} = \begin{bmatrix} G_p^{11} G_i^{11} & G_p^{12} G_i^{12} & \cdots & G_p^{1n} G_i^{1n} \\ G_p^{21} G_i^{21} & G_p^{22} G_i^{22} & \cdots & G_p^{2n} G_i^{2n} \\ \vdots & \vdots & \vdots & \vdots \\ G_p^{m1} G_i^{m1} & G_p^{m2} G_i^{m2} & \cdots & G_p^{mn} G_i^{mn} \end{bmatrix}$$

After completion of above steps shown in algorithm the system achieves improvement in PQ of proposed system by making the optimal gain parameters with possible limit error. The flowchart justification of the proposed IBSMFO is illustrate as follows.

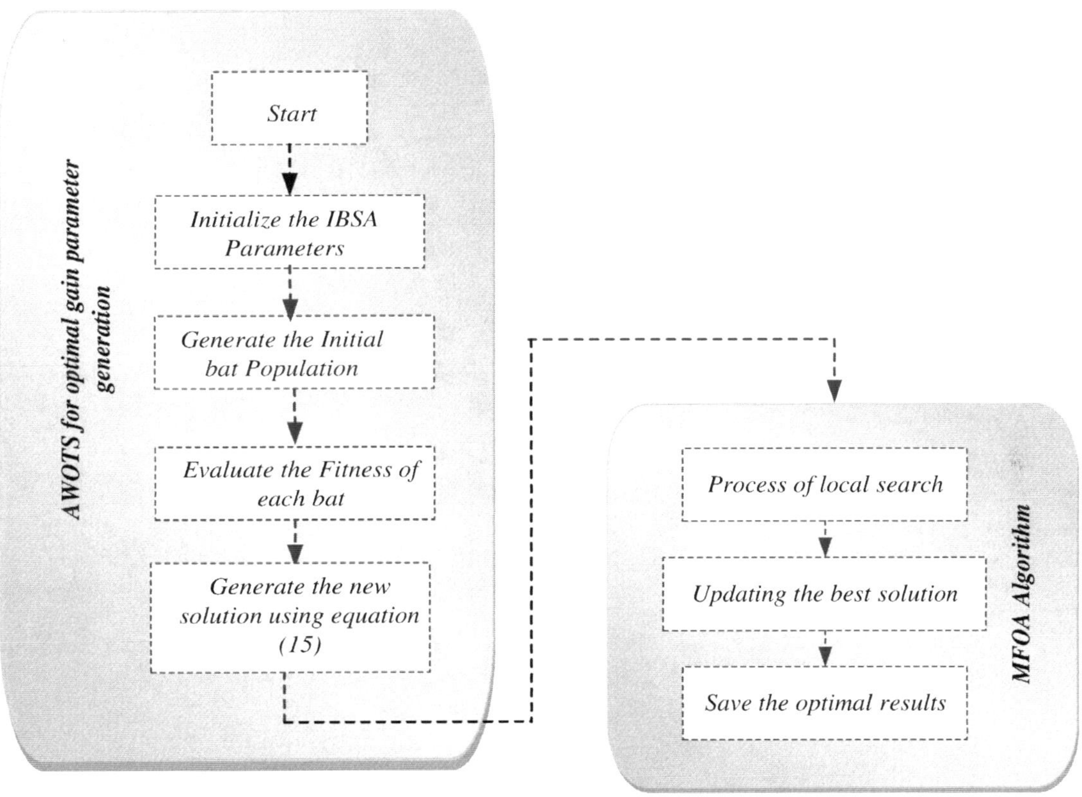

Fig. 3 Flowchart of the Proposed IBSMFO technique

5. Experimental Results and Discussions

In this section the IBSMFO method based PI controller is proposed and implemented in MATLAB/SIMULINK platform. The proposed system is used to improve the power quality of HRES unit. At this point the projected process is rooted in HRES unit which is picked to organize for constructing orientation signals of the DC/DC converter. Through this projected technique the real and reactive power quality optimization is done through the IBSMFO method. Subsequently the prohibited signals are generated from the projected technique that can construct the scheming pulses for developing the presentation of DC/DC converter. So, the projected process is utilized to progress the presentation of the converter also to recompense the PQ troubles. The projected process is experienced and its presentation is demonstrated. In order to estimate the performance of the proposed tuning controller, the simulation results are compared with those of Artificial Bee Colony (ABC) algorithm [36-39], Gravitational Search Algorithm (GSA) [40-42], the Firefly Algorithm (FA) [43-45] and IBSMFO controllers.

5.1. Performance Analysis

In the subsection the performance of the proposed method is analyzed by using the combined IBSA and MFO for the enhancement of PQ of the HRES unit in the MG system. Here the design of the proposed controller is developed in light of the PQ control mode for the HRES. For the viability of the proposed work, this experimental section of this paper is analyzed by using two cases named as,

Case 1: Balanced supply with the unbalanced load condition
Case 2: Unbalanced supply with the balanced load condition

In the above cases, the case study 2 has two modes;
They are given as follows:
Mode A: Step response in the Solar system
Mode B: Zero response in the Solar system

The unbalanced supply arises when there is a sudden deviation in voltage or current signal from the normal sinusoidal waveform. Here, the unbalanced PQ issue of Solar PV arises during the fault condition and it is evaluated by the proposed method. In this section the performance of the proposed IBSMFO technique is compared with the existing techniques for PQE such as ABC algorithm, GSA and the FA. The two case studies analyzing the performance of the proposed technique are beneath section.

Case 1: Balanced power supply with the unbalanced load condition

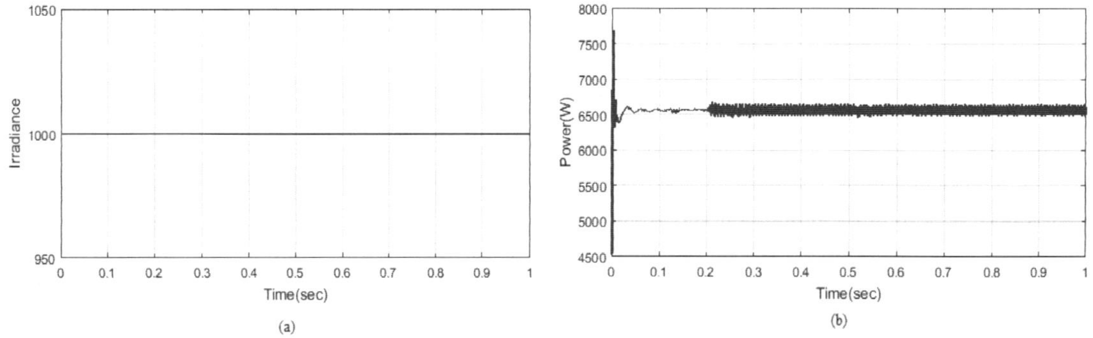

Fig. 4 Analysis of (a) Solar PV irradiance (b) Solar PV power

In the first case the PQ enhancement of HRES is analyzed under balance supply with unbalance load condition. The main objective of this case is to evaluate the response of the power controller when the necessity of the load is unbalanced. The performance of Solar PV irradiance and power is shown in figures 4 (a) and (b). At the range of 1000 within t= 0 to1 sec, the irradiance under Solar PV runs at a constant level. The Solar PV power performance, where the power starts at zero range is shown in figure 4 (b). The output of the Solar PV power is performed at constant level after the time period t=0.25 sec. The power analysis of the HRES such as Solar PV, WT, FC, battery, grid and the load is analyzed by figures 5 (a) to (d). Here, the performance of the HRES is satisfied with the balance power condition. In the analysis of figures 5(a) to (d) the power (w) is represented in the x-axis and the time in a sec is represented in the y-axis. At time t=0.023 sec the power of Solar PV is kept in peak variation at 3500 kW and suddenly falls at 0.08 sec. The power of WT is kept in peak variation at 4200 kW at time t=0.01 sec and suddenly falls at 0.023 sec. At time t=0.01 sec the power of FC is kept in peak variation at 1500 kW and suddenly falls at 0.03 sec. The power of the battery is kept in peak variation at 14000 kW at time t=0.01sec suddenly falls at 0.02 sec. In the particular time period the system starts to operate in a grid connected mode at that time the power generated from the grid is illustrated in figure 5(e). Based on the designed the control operation of the HRES system is expected that the load would be shared. The performance of the load and grid under the unbalanced condition is analyzed from figures 5(e) and (f). At time t= 0.25 sec the power of the grid is 2kW after that it goes constantly till the end of the operation. At time t= 0.25 sec the power of the load is 3500kW after that it goes constantly till the end of the operation.

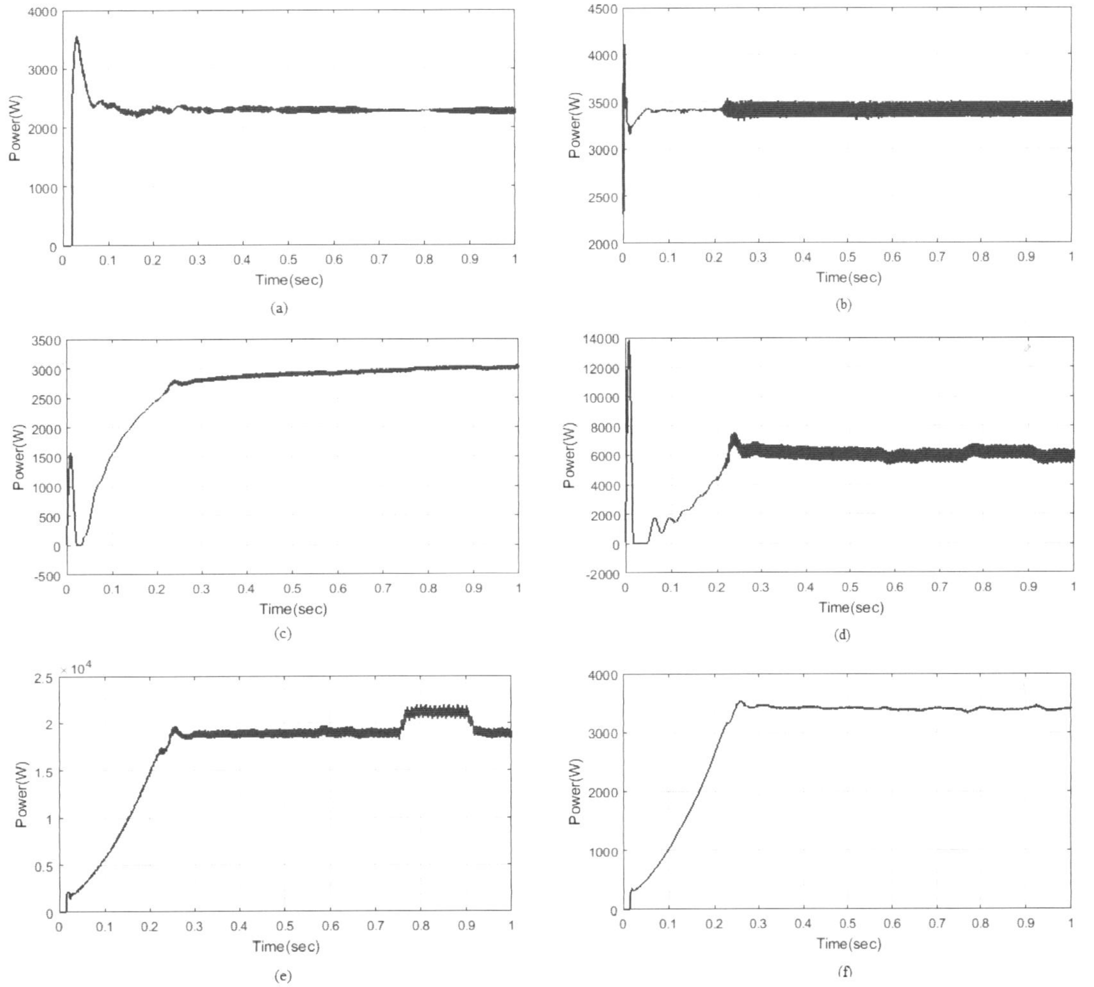

Fig. 5 Power analysis of (a) Solar PV (b) WT(c) FC (d) Battery (e) Grid and (f) Load

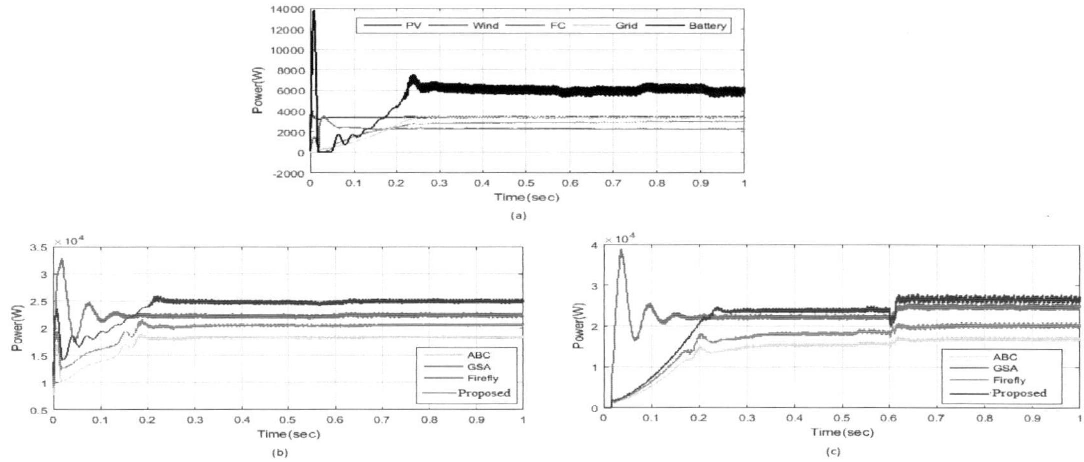

Fig. 6 Analysis of (a) Total power in HRES using proposed method (b) Total power comparison (c) Load power comparison

Figures 6(a) to (c) illustrates the total power of the HRES using a proposed controller, total power comparison and the Load power comparison. In figure 6(a) the power variation of Solar PV, wind, FC, Grid and Battery are 4000 MW, 3800 MW, 1800 MW, 3900 MW and 7800 MW respectively. The effectiveness of the proposed method is compared with the existing techniques such as ABC, GSA and FA respectively. The comparison of total power and load power under load changes is shown in figures 6(b) and (c). In this section the peak overshoot time is zero by using the proposed method. The settling time of the proposed ABC, GSA and the FA method is t=0.22sec, 0.28 sec, 0.32 sec and 0.4 sec respectively and likewise, the rising time of the proposed, ABC, GSA and the FA method t= 0.04 sec, 0.03sec, 0.028 sec and 0.03 sec respectively. The above results of the comparison analysis shows the settling time, rising time and peak overshoot time of the proposed method is highly less when compared with the existing method.

Case 2: Unbalanced supply with the balanced load condition

In this section the optimal power quality of the proposed method is analyzed. To achieve the optimal PQ the active and reactive power balanced is controlled by using the proposed system. The high quality output power of the inverter and the control objective has been achieved by a relatively slow change of the reference current/voltage trajectory consequently. An Active and reactive power control strategy is analyzed with the HRES unit connecting to the grid or at the load change condition. In this research both the inverter current loop and two PI regulators are used to eliminate the current error. To ensure the PQ enhancement the unbalanced supply and balanced load condition is considered. Based on the Solar PV variations two sets of analysis is presented here. (i) Mode A: Step response in the Solar system (ii) Mode B: Zero response in the Solar system.

Mode A: Step response in the Solar system

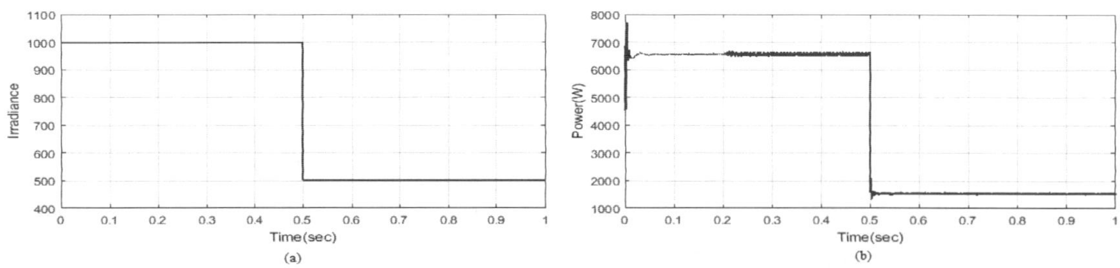

Fig. 7 Performance of (a) Solar irradiance (b) Solar power

IOP Conf. Series: Materials Science and Engineering **623** (2019) 012013 doi:10.1088/1757-899X/623/1/012013

In the subsection the power analysis of Solar PV, WT, FC and battery is analyzed. Figures 7(a) and (b) specified the irradiance and power of Solar PV under unbalanced condition. In the time instant t=0 to 0.5 sec the irradiance of Solar PV is 1000w/m2 power is produced by the Solar PV under the step irradiation conditions which is then reduced to 500w/m2 at the period of 0.5 second to 1 second and which is shown in figure 7 (a). Figure 7(b) shows the Solar PV power generation under unbalanced load condition using the proposed methodology. Figures 8 (a)-(f) shows the maximum power of Solar PV, WT, FC, battery, grid and load. The maximum power of 3500kW is produced by the Solar PV at the time of t=0.02 sec, after that, the maximum power is reduced to 2500kW in the period of 0.06 sec. Figure 8 (b) shows the power of WT, here the maximum power WT is 4100kW at the starting time and it can be reduced to 3250kW at the time period of 0.2 sec. Figure 7(c) shows the extraction of FC power1500kW power is extracted from the FC at the initial stage by the proposed method without any oscillations. At normal conditions 2700kW power has been delivered and changes slightly at the time of unbalanced load conditions. The power of the battery is kept in peak variation at 14000 kW at time t=0.01 sec and suddenly falls at 0.02 sec and which is shown in figure 8(d). The performance of the load and grid under the unbalanced condition is analyzed from figures 8(e) and (f). At time t= 0.25 sec the power of the grid is 2kW after that it goes constant till the end of the operation. At time t= 0.25 sec the power of the load is 3500kW after that it goes constant till the end of the operation.

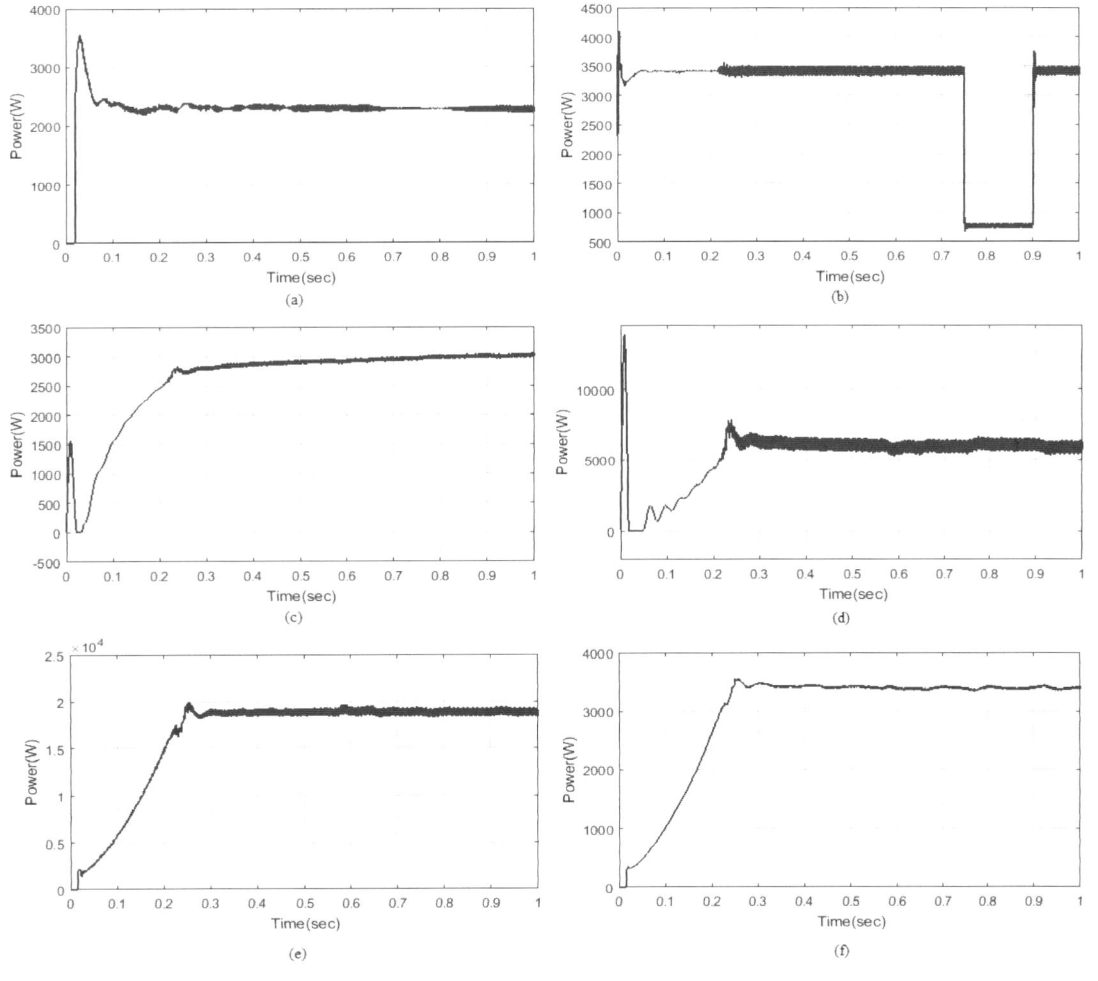

Fig. 8 Power representation of (a) Solar (b) WT(c) Fuel cell (d) Battery (e) Grid & (f) Load

Figure 9(a) shows the total power in the HRES unit and figure 9(b) shows the comparison of total power generation by using various PQE scheme. The extraction of the required power by the proposed control scheme is obtained from the renewable energy sources, energy storage and grid. Due to the unawareness of the environment 1.5*104 kW power generated by the ABC is not a required power. The GSA based controller develops the power generation from the sources by slightly observing the need of the system. Constant power generation is maintained until the end of the operating condition by the FA methodology but it does not meet the required criteria. Figure 9(c) shows the load power comparison using different techniques. Due to low power generation the ABC does not meet the load requirements and slight deviations are found in the GSA based controller. From the high power generation of the FA technique it is observed that the load variation is not identified by this method. The figure describes how effectively the proposed method identifies the situation and solves the power requirement of the load. To improve the PQ of the energy storage connected smart grid system the proposed method has much more efficient when compared to the other techniques.

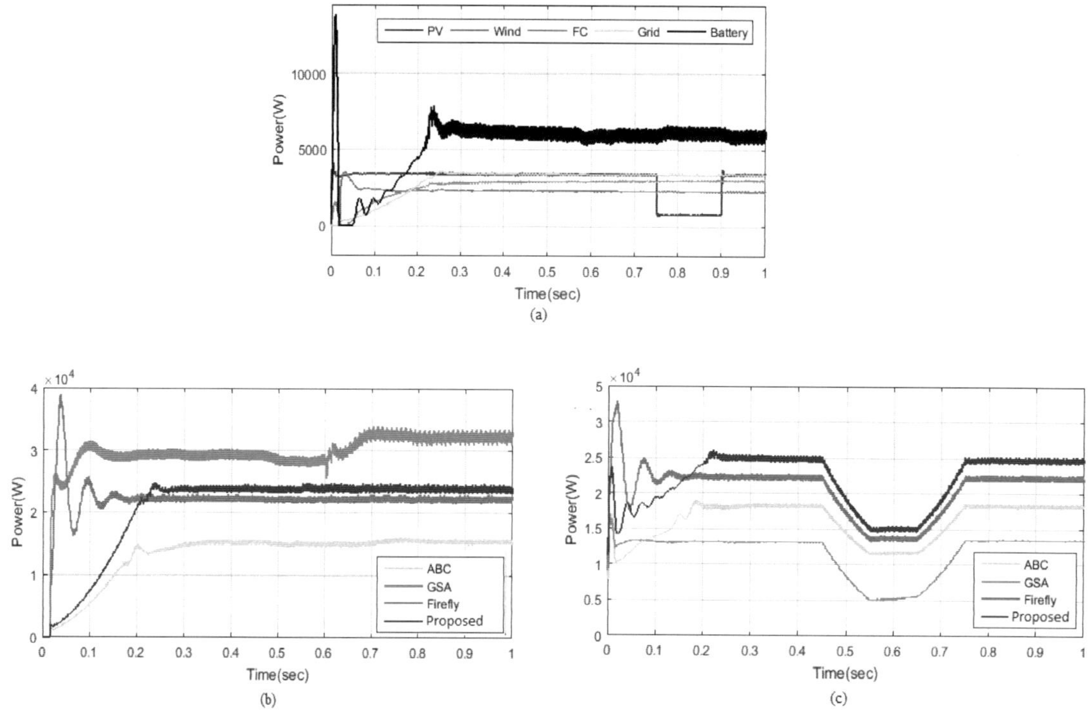

Fig. 9 Analysis of (a) Total power in HRES using proposed method (b) Total power comparison (c) Load power comparison

Mode B: Zero response in the Solar system

The zero response analysis of Solar PV is illustrated in figures 10 (a) to (c). It shows the resulting controlled total and Solar PV power flows in the HRES system. It obvious that the PQ issues between the grid and the HRES unit is controlled perfectly. While the HRES unit still injects a sustained output power the unbalanced power is supplied by the grid in this case. In this instance the HRES unit supplies the load automatically and its excess power is fed to the grid. With the proposed and existing methods the total power and battery power is analyzed from the graphical representation. The maximum power and optimal PQE is achieved in the proposed method.

IOP Conf. Series: Materials Science and Engineering **623** (2019) 012013 doi:10.1088/1757-899X/623/1/012013

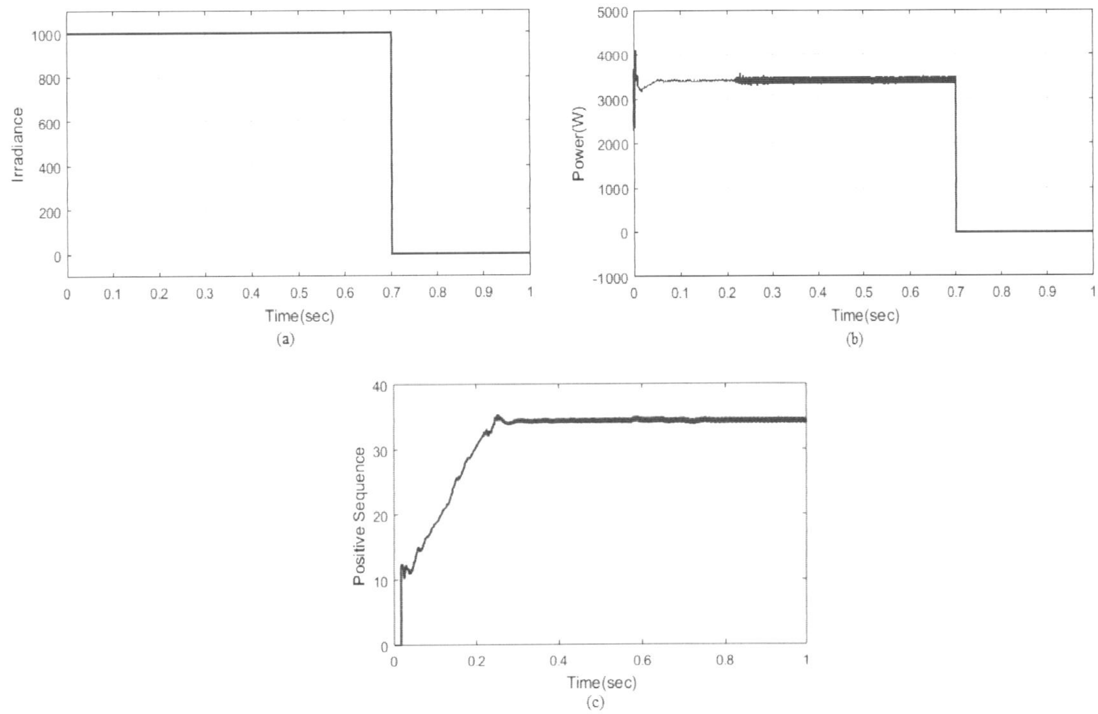

Fig. 10 Performance of Solar (a) Irradiance (b) power (c) Inverter voltage

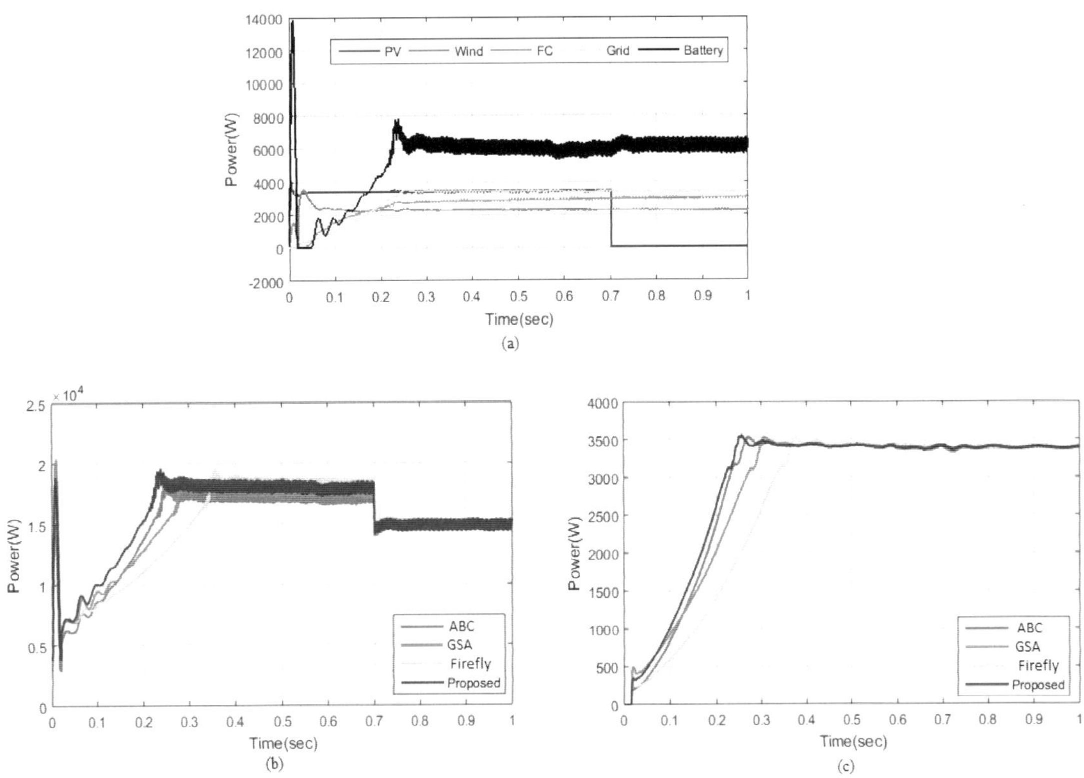

Fig. 11 Analysis of (a) Total power in HRES using proposed method (b) Total power comparison (c) Load power comparison

IOP Conf. Series: Materials Science and Engineering **623** (2019) 012013 doi:10.1088/1757-899X/623/1/012013

Figures 11(a) to (c) illustrates the total power of the HRES using a proposed controller, total power comparison and the load power comparison. In figure 11(a) the power variation of Solar PV, wind, FC, Grid and Battery are 4000 MW, 3800 MW, 1800 MW, 3900 MW and 7800 MW respectively. The effectiveness of the proposed method is compared with the existing techniques such as ABC, GSA and FA respectively. The comparison of total power and load power under load changes is shown in figures 11(b) and (c). In this section the peak overshoot time is zero by using the proposed method. The settling time of the proposed, ABC, GSA and the FA method is t=0.22sec, 0.28 sec, 0.32 sec and 0.4 sec respectively and likewise, the rising time of the proposed, ABC, GSA and the FA method t= 0.04 sec, 0.03sec, 0.028 sec and 0.03 sec respectively. The above results of the comparison analysis show, the settling time, rising time and peak overshoot time of the proposed method is highly less when compared with the existing method.

Table 1 controller parameters after optimization

Parameters		ABC	GSA	FA	IBSMFO method
Active power controller parameters	K_{pp}	8.2270	9.6325	9.2545	8.8554
	K_{pi}	0.0020	2.3020	1.6580	1.8569
Reactive power controller parameters	K_{qp}	5.0978	8.3699	7.6691	8.1657
	K_{qi}	0.0020	2.3652	1.6559	1.5698

Table 2 Statistical analysis for hundred iterations (P)

Solution techniques		Mean	Median	Std. deviation
GSA	Case 1	0.8493	0.8496	0.0257
	Case 2 (Mode- A)	0.8620	0.8623	0.0050
	Case 2 (Mode-B)	0.8993	0.8989	0.0058
FA	Case 1	0.7478	0.7443	0.0291
	Case 2 (Mode - A)	0.7787	0.7789	0.0248
	Case 2 (Mode -B)	0.8101	0.8098	0.0055
IBSMFO method	Case 1	0.7023	0.6976	0.0264
	Case 2 (Mode- A)	0.7235	0.7534	0.0323
	Case 2 (Mode -B)	0.7634	0.7568	0.0167

IOP Conf. Series: Materials Science and Engineering **623** (2019) 012013 doi:10.1088/1757-899X/623/1/012013

Table 3 Statistical analysis for Hundred iterations (Q)

Solution techniques		Mean	Median	Std. deviation
GSA	Case 1	0.9212	0.9211	0.0276
	Case 2 (Mode - A)	0.9314	0.9319	0.0051
	Case 2 (Mode -B)	0.9621	0.9625	0.0056
FA	Case 1	0.8731	0.8701	0.0270
	Case 2 (Mode - A)	0.8966	0.8966	0.0025
	Case 2 (Mode -B)	0.9071	0.9073	0.0029
IBSMFO method	Case 1	0.8143	0.8143	0.0242
	Case 2 (Mode - A)	0.8543	0.8526	0.0063
	Case 2 (Mode -B)	0.8735	0.8653	0.0034

Using various methods the real and reactive powers are tabulated in both cases from the above table 1. Using other methods, the maximum real power is achieved about (8.2270, 0.0020) while by using the proposed method it is (8.8554, 1.8569). The other methods are analyzed and the reactive power is also determined similarly. Table 2 and 3 tabulates the mean, median and standard deviation of the attained real and reactive power values from different algorithms. The voltage and current values as the input are elegantly utilized by the proposed method. In the event, to generate the optimal control pulses of the HRES DC-DC converter the proposed method becomes well-geared the procedures come to an end. Better results, when compared to the other methods, are given by the proposed method in the efficiency analysis. To their corresponding real and reactive PQE, the efficiency of the converter is analyzed by utilizing the proposed controller. The proposed method has PQE, reduced the disturbance of the system and maximizes the optimal performance based on the output response of the performance of the system. Similarly, to obtain the maximum power the existing controllers are analyzed with the converter. The proposed controller accomplished optimal PQE while contrasted with different control strategies. Drawn by the same burdens the nature of HRES power, load power, grid power and inverter voltage, PV voltage is watched. The two cases are balanced and an unbalanced supply performance is determined from the overall analysis. The proposed controller achieves better results while all the control schemes are efficient in optimal operation of DC-DC converter performances.

6. Conclusions

In the work, a novel optimization algorithm used and is known as IBSMFO by this control an optimal Power Quality enhancement in the interconnected renewable system with DC-DC converter. The proposed work, the active and reactive powers are regulated for distribution generator fed grid connected system. In this control scheme there are two loops i.e., inner current control loop and outer power control loop. In the proposed control algorithm is having two individual controls one is IBSA algorithm to locate gain parameters used for objective function with minimum error. The MFO technique is used to improve the searching performance of optimal solution. Finally, the control signal generated to enhance Power Quality of HRES. The performance analysis of proposed control technique is evaluated for 2 different load change conditions and are balanced and unbalanced supply. The simulation results are highlighted in terms of the settling time, rise in time and overshoot time of the control algorithm. The results are highlighted with statistical performance of HRES unit is analysed with the mean, median and standard deviations of the proposed controller with the existing techniques. The outcomes of the simulation results demonstrate the proposed algorithm works effectively for the renewable energy sources interconnected with grid with quick settling time, overshoot time and hence the stability of the proposed system is maintained successfully.

7. References

[1] H. Nian and B. Pang "Stability and Power Quality Enhancement Strategy for DFIG System Connected to Harmonic Grid with Parallel Compensation", IEEE Transactions on Energy Conversion, pp. 1-1, 2018.

[2] Q. Tabart, I. Vechiu, A. Etxeberria,and S. Bacha, "Hybrid Energy Storage System Microgrids Integration for Power Quality Improvement Using Four-Leg Three-Level NPC Inverter and Second-Order Sliding Mode Control", IEEE Transactions on Industrial Electronics, vol. 65, no. 1, pp. 424-435, 2018.

[3] Y. Zheng, B. Jenkins, K. Kornbluth and C. Træholt, "Optimization under uncertainty of a biomass-integrated renewable energy microgrid with energy storage", Renewable Energy, vol. 123, pp. 204-217, 2018.

[4] R.Agarwal, , I. Hussain and B. Singh, "Application of LMS-Based NN Structure for Power Quality Enhancement in a Distribution Network Under Abnormal Conditions", IEEE Transactions on Neural Networks and Learning Systems, vol. 29, no. 5, pp. 1598-1607, 2018.

[5] K. Fouad, H. Boulouiha, A. Allali, A. Taibi and M. Denai, "Multivariable control of a grid-connected wind energy conversion system with power quality enhancement", Energy Systems, vol. 9, no. 1, pp. 25-57, 2016.

[6] S.Agalar and Y.Kaplan, "Power quality improvement using STS and DVR in wind energy system", Renewable Energy, vol. 118, pp. 1031-1040, 2018.

[7] K. Rao and K. Srikanth, "Improvement of Power Quality using Fuzzy Logic Controller In Grid Connected Photovoltaic Cell Using UPQC", International Journal of Power Electronics and Drive Systems (IJPEDS), vol. 5, no. 1, 2014.

[8] B. Naga Pavan Kumar and D. Seshi Reddy, "Power Quality Enhancement of Integrated Grid Connected off-Shore Wind Farm and Marine Current Farm using Statcom", Indian Journal of Science and Technology, vol. 8, no. 29, 2015.

[9] R. Rouabhi, R. Abdessemed, A. Chouder and A. Djerioui, "Power Quality Enhancement of Grid Connected Doubly-Fed Induction Generator Using Sliding Mode Control", International Review of Electrical Engineering (IREE), vol. 10, no. 2, p. 266, 2015.

[10] V. Dash and P. Bajpai, "Power management control strategy for a stand-alone solar photovoltaic-fuel cell–battery hybrid system", Sustainable Energy Technologies and Assessments, vol. 9, pp. 68-80, 2015.

[11] Q. Tabart, I. Vechiu, A. Etxeberria and S. Bacha, "Hybrid Energy Storage System Micro grids Integration for Power Quality Improvement Using Four-Leg Three-Level NPC Inverter and Second-Order Sliding Mode Control", IEEE Transactions on Industrial Electronics, vol. 65, no. 1, pp. 424-435, 2018.

[12] M. Rahmani-Andebili, "Stochastic, adaptive, and dynamic control of energy storage systems integrated with renewable energy sources for power loss minimization", Renewable Energy, vol. 113, pp. 1462-1471, 2017.

[13] A. Mohamed, M. Elshaer and O. Mohammed, "Control enhancement of power conditioning units for high quality PV systems", Electric Power Systems Research, vol. 90, pp. 30-41, 2012.

[14] S. yafaruddin, E. Karatepe and T. Hiyama, "Performance enhancement of photovoltaic array through string and central based MPPT system under non-uniform irradiance conditions", Energy Conversion and Management, vol. 62, pp. 131-140, 2012.

[15] Z. Zeng, H. Yang, R. Zhao and C. Cheng, "Topologies and control strategies of multi-functional grid-connected inverters for power quality enhancement: A comprehensive review", Renewable and Sustainable Energy Reviews, vol. 24, pp. 223-270, 2013.

[16] J. Torreglosa, P. García, L. Fernández and F. Jurado, "Energy dispatching based on predictive controller of an off-grid wind turbine/photovoltaic/hydrogen/battery hybrid system", Renewable Energy, vol. 74, pp. 326-336, 2015.

[17] S. Upadhyay and M. Sharma, "A review on configurations, control and sizing methodologies of hybrid energy systems", Renewable and Sustainable Energy Reviews, vol. 38, pp. 47-63, 2014.

[18] V. Dash and P. Bajpai, "Power management control strategy for a stand-alone solar photovoltaic-fuel cell–battery hybrid system", Sustainable Energy Technologies and Assessments, vol. 9, pp. 68-80, 2015.

[19] C. Chang, Y. Lin, Y. Chen and Y. Chang, "Simplified Reactive Power Control for Single-Phase Grid-Connected Photovoltaic Inverters", IEEE Transactions on Industrial Electronics, vol. 61, no. 5, pp. 2286-2296, 2014.

[20] C. Cecati, F. Ciancetta and P. Siano, "A Multilevel Inverter for Photovoltaic Systems With Fuzzy Logic Control", IEEE Transactions on Industrial Electronics, vol. 57, no. 12, pp. 4115-4125, 2010.

[21] G. Talapur, H. Suryawanshi, L. Xu and A. Shitole, "A Reliable Micro-grid with Seamless Transition between Grid Connected and Islanded Mode for Residential Community with Enhanced Power Quality", IEEE Transactions on Industry Applications, pp. 1-1, 2018.

[22] M. Mosaad and H. Ramadan, "Power quality enhancement of grid-connected fuel cell using evolutionary computing techniques", International Journal of Hydrogen Energy, vol. 43, no. 25, pp. 11568-11582, 2018.

[23] S. Mousazadeh Mousavi, A. Jalilian, M. Savaghebi and J. Guerrero, "Power quality enhancement and power management of a multifunctional interfacing inverter for PV and battery energy storage system", International Transactions on Electrical Energy Systems, p. e2643, 2018.

[24] S. Prince, K. Panda, V. Kumar and G. Panda, "Power quality enhancement in a distribution network using PSO assisted Kalman filter — Based shunt active power filter", 2018 IEEMA Engineer Infinite Conference (eTechNxT), 2018.

[25] V. Lopez-Martin, F. Azcondo and A. Pigazo, "Power Quality Enhancement in Residential Smart Grids Through Power Factor Correction Stages", IEEE Transactions on Industrial Electronics, vol. 65, no. 11, pp. 8553-8564, 2018.

[26] B. Venkateswara Rao and G. Nagesh Kumar, "Optimal power flow by BAT search algorithm for generation reallocation with unified power flow controller", International Journal of Electrical Power & Energy Systems, vol. 68, pp. 81-88, 2015.

[27] E. Ali, "Optimization of Power System Stabilizers using BAT search algorithm", International Journal of Electrical Power & Energy Systems, vol. 61, pp. 683-690, 2014.

[28] A. Oshaba, E. Ali and S. Abd Elazim, "MPPT control design of PV system supplied SRM using BAT search algorithm", Sustainable Energy, Grids and Networks, vol. 2, pp. 51-60, 2015.

[29] R. Svečko and D. Kusić, "Feedforward neural network position control of a piezoelectric actuator based on a BAT search algorithm", Expert Systems with Applications, vol. 42, no. 13, pp. 5416-5423, 2015.

[30] A. Oshaba, E. Ali and S. Abd Elazim, "PI controller design for MPPT of photovoltaic system supplying SRM via BAT search algorithm", Neural Computing and Applications, vol. 28, no. 4, pp. 651-667, 2015.

[31] S. Mirjalili, "Moth-flame optimization algorithm: A novel nature-inspired heuristic paradigm", Knowledge-Based Systems, vol. 89, pp. 228-249, 2015.

[32] D. Allam, D. Yousri and M. Eteiba, "Parameters extraction of the three diode model for the multi-crystalline solar cell/module using Moth-Flame Optimization Algorithm", Energy Conversion and Management, vol. 123, pp. 535-548, 2016.
[33] S. Khalilpourazari and S. Pasandideh, "Multi-item EOQ model with nonlinear unit holding cost and partial backordering: moth-flame optimization algorithm", Journal of Industrial and Production Engineering, vol. 34, no. 1, pp. 42-51, 2016.
[34] M. Aziz, A. Ewees and A. Hassanien, "Whale Optimization Algorithm and Moth-Flame Optimization for multilevel thresholding image segmentation", Expert Systems with Applications, vol. 83, pp. 242-256, 2017.
[35] W. Yamany, M. Fawzy, A. Tharwat and A. Hassanien, "Moth-flame optimization for training Multi-Layer Perceptrons", 2015 11th International Computer Engineering Conference (ICENCO), 2015.
[36] D. Karaboga and B. Akay, "A comparative study of Artificial Bee Colony algorithm", Applied Mathematics and Computation, vol. 214, no. 1, pp. 108-132, 2009.
[37] R. Roy and H. Jadhav, "Optimal power flow solution of power system incorporating stochastic wind power using Gbest guided artificial bee colony algorithm", International Journal of Electrical Power & Energy Systems, vol. 64, pp. 562-578, 2015.
[38] H. Jadhav and P. Bamane, "Temperature dependent optimal power flow using g-best guided artificial bee colony algorithm", International Journal of Electrical Power & Energy Systems, vol. 77, pp. 77-90, 2016.
[39] K. Roy, K. Mandal and A. Mandal, "Modeling and managing of micro grid connected system using Improved Artificial Bee Colony algorithm", International Journal of Electrical Power & Energy Systems, vol. 75, pp. 50-58, 2016.
[40] B. Shaw, V. Mukherjee and S. Ghoshal, "Solution of reactive power dispatch of power systems by an opposition-based gravitational search algorithm", International Journal of Electrical Power & Energy Systems, vol. 55, pp. 29-40, 2014.
[41] A. Bhowmik and A. Chakraborty, "Solution of optimal power flow using non dominated sorting multi objective opposition based gravitational search algorithm", International Journal of Electrical Power & Energy Systems, vol. 64, pp. 1237-1250, 2015.
[42] M. Narimani, R. Azizipanah-Abarghooee, M. Javidsharifi and A. Azizi Vahed, "Enhanced gravitational search algorithm for multi-objective distribution feeder reconfiguration considering reliability, loss and operational cost", IET Generation, Transmission & Distribution, vol. 8, no. 1, pp. 55-69, 2014.
[43] H. Shareef, A. Ibrahim, N. Salman, A. Mohamed and W. Ling Ai, "Power quality and reliability enhancement in distribution systems via optimum network reconfiguration by using quantum firefly algorithm", International Journal of Electrical Power & Energy Systems, vol. 58, pp. 160-169, 2014.
[44] U. Vadivu and B. Keshavan, "Power quality enhancement of UPQC connected WECS using FFA with RNN", 2017 IEEE International Conference on Environment and Electrical Engineering and 2017 IEEE Industrial and Commercial Power Systems Europe (EEEIC / I&CPS Europe), 2017.
[45] N. Gowtham, S. Shankar and K. Rao, "Enhanced firefly algorithm for PQ improvement of wind energy conversion system with UPQC", TENCON 2017 - 2017 IEEE Region 10 Conference, 2017.

An IoT Based Home Automation System

Sridip Paul[1], V Indragandhi[2*], N K Kumar[3], R Raja Singh[4], V Subramaniyaswamy[5]

[*]Corresponding Author: indragandhi.v@vit.ac.in
[1] UG Scholar, Vellore institute of technology, Vellore
[2, 4] Associate Professor, Vellore Institute of Technology, Vellore
[3]Assistant Professor, Vellore Institute of Technology, Vellore
[5]Associate Professor, Sastra University, Thanjavur

Abstract: Industry experts, corporate heads, and chiefs of businesses all agree that we are on the verge of what is to be the Fourth Industrial Revolution – which will be more disrupting and impactful than any we've seen in the history of mankind. Automation is without a doubt, one of the pillars of this Industrial Revolution. Consumers worldwide prefer automated systems over manual ones. Today the Internet has become a part and parcel of daily life. We cannot imagine a life without the internet. In this paper we present to you an Automation System which is a real-world application of a fast- growing emerging Internet Technology, The Internet of Things (IoT). A Home Automation System, based on the IOT, makes the consumers life easier, more comfortable, faster and more secure. This cutting-edge technology allows the consumer to have complete control over his or her home appliances and devices from anywhere in the world, via a web browser with an internet connection. The simulation designed by us, in this paper, is a small approximation of a much larger real-world system, the concepts, approach and methodology however, remaining the same.

Index Terms: Fourth Industrial Revolution, Automation System, Internet of Things, Home Automation

1. Introduction

Problem Statement: To successfully design, implement and simulate an IoT based Home Automation System(HMS), using NodeMCU, which is an open source IoT platform, thereby enabling full control of all the loads in the system from anywhere in the world, via a public web application, with an internet connection.

Home Automation gives us access to control devices in our homes from a mobile device anywhere on Planet Earth. The term may be used for individual programmable gadgets, like thermostats and sprinklers, but "home automation" more accurately is associated with homes in which almost everything – from lights, fans to electrical outlets to heating, cooling and ventilation systems – are connected to a remotely controllable network. From a home safety outlook, this may also include any alarm systems, all the doors, windows, locks, smoke sensors, surveillance cameras and any other sensor, that may be attached to it.

2. Literature Survey

2.1 Bluetooth based HMS using mobile phones

In this sort of HMS, the home machines are associated with the Arduino BT board at information and yield ports utilizing relay. The program of Arduino BT board is based on the C language. The association is made by means of Bluetooth.

2.2 Zigbee based home automation framework, utilizing PDAs

To screen and control the home machines the framework is structured and actualized utilizing Zigbee. The gadget execution is record and store by system facilitators. For this a Wi-Fi organize is used, which utilizes the four-switch port standard remote ADSL current switch.

2.3 Home Automation utilizing RF modules

The essential objective of a HMS is to manufacture a home mechanization framework utilizing a RF controlled remote. In order to achieve this, a RF remote is consolidated to the smaller scale controller on transmitter side that sends ON/OFF signs to the recipients where gadgets are associated.

3. Proposed Methodology

3.1 Primary Objectives – Hardware

Successful design of a circuit with 5 loads, with a power adapter and relay switching circuits, and being connected to the IoT platform – NodeMCU[1-2]. Successful and complete implementation of the circuit with the necessary requisite material viz., NodeMCU, Power Adapter Circuit, Relay Modules, etc. on a plywood base. Design of the relay modules using sugar cube relays, on Printed Circuit Boards (PCBs). Writing the code to control the ESP8266 Wi-Fi module using Arduino IDE, and uploading the same onto the board, via a USB cable. This is followed by testing the code and all the circuit connections for loose/faulty connections, and also clamping/soldering certain parts of the circuit as and when required[3].

3.2 Primary Objectives – Software

Designing a web application that meets the necessary requirements of controlling the loads in the HMS. Choosing hosting server, selecting web hosting plan, buying server space, and registering the domain name. Designing the website using HTML, CSS and JavaScript, and using PHP, a server-side scripting language to process inputs/outputs and dynamic content on the webpage[3-4]. Creating and developing the necessary databases using MySQL, a relational database management tool, and linking the same with the PHP code. Publishing the web pages and testing the same before final deployment.

3.3 Secondary Objectives – Additional Features

1. *Login Feature:* The web application will be secured with a login feature. Only bonafide users who possess valid login credentials, username, password, will be able to login and access the web application.
2. *Password Change Feature:* Once successfully logged in, users will be able to proceed to change their passwords for the existing username. Once the password is successfully changed, a message stating the same will be presented to them on the screen.
3. *Run-time monitoring Feature:* Users will be able to monitor the time for which each device is On, after each On-Off period, simply by refreshing the webpage. The cumulative data of run-times of each of the 5 devices (loads) will get stored in the "Current Info" Tab, for each day, on a daily basis. This will get reset to zero at the start of each day. The previous usage, of each day right from the time of the development of the app can be viewed under the "Historical Info" Tab.

New user Registration feature**:** Once successfully logged in, users will be able to go on to register new users – that is, create new valid user name, password combinations, to allow other individuals/families to have access to the HMS. It is important to note that this feature can be accessed only after a successful login by clicking on the 'Settings' Tab. This ensures that the feature

IOP Conf. Series: Materials Science and Engineering **623** (2019) 012014 doi:10.1088/1757-899X/623/1/012014

is not misused or the safety and security of the home is not compromised, as new users can thus only be created after authorization from a bonafide current user.

4. DESIGN APPROACH & DETAILS - HARDWARE

4.1 Components Required - NodeMCU, AC Power Adapter, IC 7805 Regulator, Relay Modules, Switching Circuit for fan, The Devices – 1 Bulb, 1 Tube light, 1 DC Fan, 2 AC Sockets(Figure.1.).

4.2 Technical Specifications & Details

NodeMCU is a firmware, based on the open source LUA platform which is developed for a Wi-Fi module ESP8266. Since it is an open source platform, its hardware can be easily built, can be easily modified and can be easily edited. The ESP8266 is a low-price Wireless Fidelity chip developed by Espressif Systems. The Power Adapter Module is a general-purpose adapter, especially designed for Arduino and NodeMCU. It is used for power supply in small Robotics papers and other such Do It Yourself (DIY) Kits. It converts AC input voltage in the range of 100 V to 230 V, of frequency between 50 to 60 Hz to 12 V DC voltage. The Sugar Cube relays are superior quality 12 V Single Pole Double Throw (SPDT) relays(Figure.2.). They are rated up to 7 A at 240 V AC. They are completely enclosed, or sealed relays.

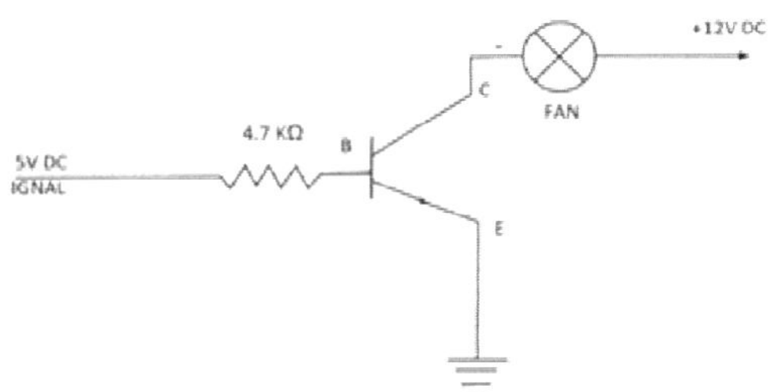

Fig. 1. Switching Circuit for FAN

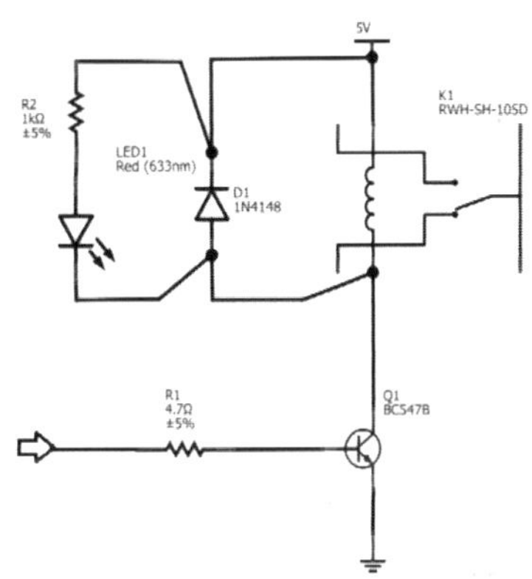

Fig.2.Relay Circuit

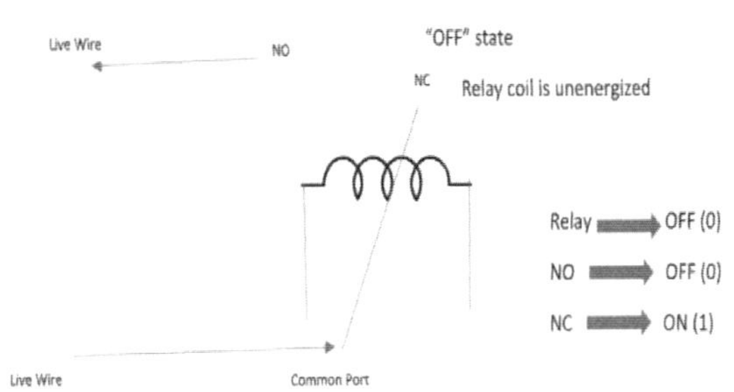

Fig.3.Unerengized Relay Circuit

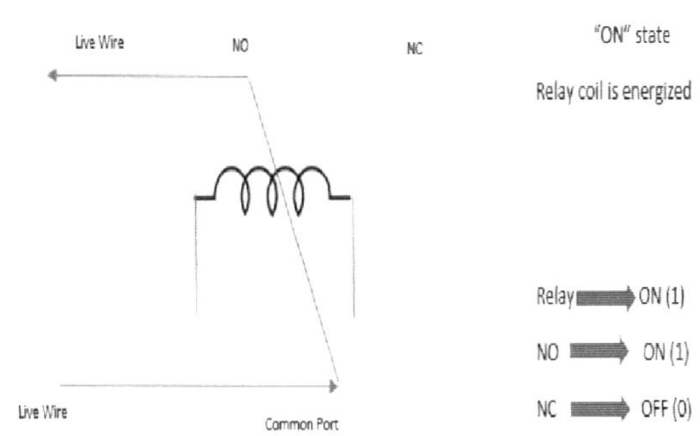

Fig.4.Energized Relay Circuit

The 5 V DC supply acts as a signal voltage for the relay. The NodeMCU cannot directly drive the relay, as it consumes a lot of current. Therefore, the NodeMCU drives the transistors which in turn drive the relay. The Live wire is connected between the "Normally Open (NO)" port and the "Common (COM)" port. In the 'OFF' state the relay switch is connected between the "Normally Closed (NC)" and the "Common (COM)" ports (Figure.3).

On getting energized it will deflect and close the NO and COM ports, thus completing the live wire circuit, acting like an electronic switch. A diode may be placed across the relay to counteract the effects due to Lens's Laws. All the components have been placed and soldered on a custom-made etched PCB (Figure.4).

Load #1 of the HMS is a Light Emitting Diode (LED) Tube light having a rating of 6 W. Load #2 is a 9 W LED Bulb, rated at 220-240 V at 50 Hz. Load #3 is a 6 W DC Fan, rated at 12 V and 0.15 A. It requires its own switching circuit. The switching circuit is as shown here:
This switching circuit is absolutely necessary as the DC Fan works on 12 V DC supply while all the other four loads in the HMS, work on normal household supply of 230 V AC.
Loads #4 & #5 are sockets / plug points, that supply the regular Indian household (domestic) supply voltage of 230 V with 50 Hz frequency. All regular household devices / gadgets can therefore be used or charged with these plug points, from smart phones and tablets to laptops, personal care devices, irons etc. Each of these sockets therefore symbolize a normal household power outlet.

4.3 Complete Circuit Design

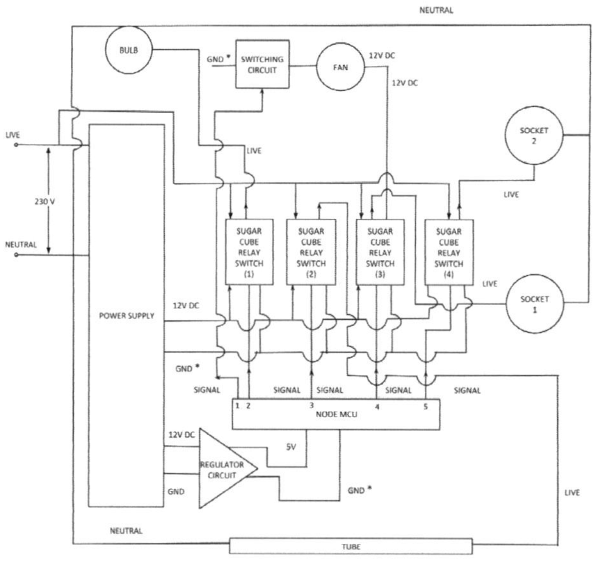

Fig.5. Proposed Circuit Design

The Power Supply Block shown in Figure.5. is actually the Power Adapter Module, which converts 230 V AC to 12 V DC. All the devices except DC Fan have a common ground (neutral). The household supply of 230 V AC will be converted to 12 V DC by the power adapter module. A regulator circuit, containing the transistor IC 7805, will further convert this 12 V DC to 5 V DC, and this in turn will be fed to the Node MCU, to be used as a signal voltage, toggled between 'High' and 'Low' , or 5 V and 0 V , as per the command given in the software. Each of the 4 relay modules, will be connected to the Node MCU, so that the 5 V signal voltage can be provided as and when required. Further each of the 4 relay modules will be given the 12 V DC supply to keep the relay coil energized. The Live wire connection will be made to pass through the ports of the relay, the relay acting as an electronic switch. When the signal voltage is 'High (+5V)' the relay will be energized, moving to the 'ON' state, closing the Live Wire path, to the corresponding device. When the signal is low (0V) the relay will be un energized, in the 'OFF' state, making the Live wire path open. The DC Fan will work on the 12 V DC supply, it will have its own switching circuit. The transistor will be connected through the port D8 of the Node MCU.

4.4 Configuring the Node MCU – The Arduino Code

The program includes the header files 'ESP8266WiFi' and 'ESP8266HTTPClient' with the help of the '#include' statement with the name of the package being enclosed within"<" and ">". Including the 'ESP8266WiFi' header file gives us access to specific Wi-Fi routines which we are calling to connect to the network. The real connection to Wi-Fi is initialized by calling: "WiFi.begin ("network name", "pass-to-network"). Here the network name is the SSID of the Wi-Fi network (which will be running in the home, or the Home Network) and the pass-to-network its password. Connection process can take time and we check for this to happen in this loop:

```
While(WiFi.status()!=WL_CONNECTED) {
delay(1000);
Serial.print("Connecting…" );
}
```

The While loop continue to loop till "WiFi.status" not equal to "WL_CONNECTED". The loop exits only when the status changes to "WL_CONNECTED".

Code: HTTP Client http ;

```
String url = http://www.sridip.online/files/bulb; // Dynamic API
http.begin(url);
int httpCode = http.GET();;
String payload = ""; // Response
If( httpCode > 0) {
payload = http.getString();

Serial.println(payload);
If(payload[0]=='1')
digitalWrite(relay1,1);
else
digitalWrite(relay1,0);
….                              … and so on.
```

The HTTP Client instance, in the subsequent lines, acts as a session to send HTTP requests and receive appropriate responses. The HTTP Client instance is a set of settings which are applicable to all requests executed by that instance only. Also, every HTTP Client instance uses its very own connection pool, thereby separating its requests from requests of other instances.

The String 'url' holds the URL of the web application of this HMS, with the extension 'files/bulb'. The 'Green/Red' button on the web application interface has an associated data attribute (Green = 1, Red = 0). This is configured in the Web Application using HTML and JavaScript. These data attributes corresponding to each 'ON/OFF' command stored as a string variable, which gets continuously concatenated. This is stored in the 'files/bulb' file extension. The NodeMCU is programmed to fetch this.

In Arduino, C/C++ coding, a string is treated simply as an array of characters. The 'get String' function returns this string. The string gets continually concatenated, with the first five characters (index positions 0-4) having the latest ON/OFF values. The 'http.getString()' function returns this string, which gets stored in the 'payload' string variable(which is actually treated as an array of characters).

The first five index positions correspond to the state (ON/OFF or 1/0) of each of the 5 devices in the HMS. Therefore, 5 If-Else statements are required, one for each device, if the corresponding index position has the value '1', the NodeMCU is programmed to send the signal voltage (+5 V) from the corresponding output pin (D0/D2/D4/D6/D8, which have been named as relay1/relay2/… relay5 respectively and have been previously configured in the program as output pins), with the function 'DigitalWrite'. If the signal voltage is zero, the NodeMCU is correspondingly programmed, to not send the signal voltage.

The Code is written in the Arduino IDE and uploaded onto the NodeMCU via a USB cable.

5. DESIGN APPROACH & DETAILS - SOFTWARE

5.1 Stage One – Brainstorming & Design

(The Goal) The web application designed should be able to facilitate smooth control of all the loads in the HMS. The user should be able to control (i.e., switch ON and/or switch OFF) all the 5 devices (loads) in the HMS, at will and without any delay/interruption. In addition, the webpage should also support the additional features viz., password protected interface, password change/modification, multi-user registration and the dynamic monitoring of usage of each device, and the ability to store a log of the same.

(The Solution) Web Hosting – it is essentially a service that is provided by many hosting companies, that hosts a website's files so that they can be accessed on the internet.
Reference may be taken from this analogy: We have got a large number of files on our computer, and if we want to access them, we just open up the file. But, if we wanted to share this file with someone else, they would have to be in the same room as us, or we would have to transfer the files to them by some means.
In this analogy, the web host is the sender. All of our site's files are kept in storage on our host company's servers. Now, as and when a visitor types in our domain name, the server transfers these files onto the web browser. Our website is nothing but a repository of various files and media elements, such as photographs. So we rent a space on a web server and store all of our files there.
Advantages include improved site performance, outstanding technical support, and improved web security.
Stages of web hosting are – Deciding on the nature of the website, deciding on the type of hosting server, selection of a web hosting plan, changing the Domain Name Server (DNS) address, and uploading the website.
For web applications, usually the hosting company rents out server space to the paying customer. These servers are housed inside datacenters. The hosting service company used by us is HostGator India. The type of hosting used is Shared Linux based Hosting. In Shared Hosting, one's website is placed on a common server along with many others, ranging from a few hundred to a few thousand. All domains share some common resources such as the CPU and RAM. In Linux Hosting, Linux is used as the operating system on the server. Linux is an OS like Unix, which is free and open source.
Registering a certain name for a website, on the internet, is called domain name registration. Domain names are used to differentiate between one or more Internet Protocol (IP) addresses. They are a substitute for the IP Address. For our website, the name choses was 'Sridip's Paper', URL 'https://www.sridip.online/index.php', and domain name 'sridip.online'.

5.2 Stage Two – Coding

Language Used – PHP (Hypertext Preprocessor– originally called "Personal Home Pages", hence the name)
Web pages designed using – HTML (Hyper Text Markup Language), CSS (Cascading Style Sheets), and JS (JavaScript).
Databases managed using – MySQL (A relational database management tool. Role of PHP – PHP is a server-side programming medium, which allows the developer to perform complex processes and display the requisite code. PHP lends functionality to the website, such as saving data in the databases, responding to inputs given by the user, etc.
Role of HTML – HTML allows us to put content (in the form of images, text, videos, etc.) in the web page, and define how the content would be placed/presented. Anything that the user can read/view on the webpage is being presented, by means of a browse, with the help of HTML.
Role of CSS – CSS is used to improve and enhance the layout, style and presentation of the webpage(s) involved. It allows the developer to define the overall look and feel of the website. While HTML predominantly deals with content and structure, CSS deals with presentation related elements such as fonts, layouts and colors. Before CSS came into use, all of this had to be defined in the HTML file itself, which made the code very complex and cumbersome.
Role of JS – JavaScript (JS) is a client-side scripting medium, which makes web pages more dynamic as well as interactive. It makes the web-pages smoother, user-friendly and helps bring about easy navigation for the user. If the developer so wishes, they can also implement visual effects to their web pages using JavaScript. JavaScript, along with HTML and CSS make a very powerful combination for web application development. Unlike PHP, JS is a client-side programming language while PHP is a server-side scripting medium.

Role of Relational Databases – A relational database is a group of multiple data sets which are streamlined by tables, records and columns. RDBs create a well-defined relationship between database tables. Tables interact with each other and share information, which allows smooth data search ability, management and presentation. There are several Relational Database Management Systems (RDBMSs) in use such as MS SQL Server, My SQL, Oracle, PostgreSQL, Oracle Database to name a few. The database management tool used for our web application is – My SQL, which is provided by the hosting service company, as an in-built feature.

5.3 Stage Three – Testing & Debugging

The final stage of any web application development is testing and debugging. There are several aspects of the web application which had to be tested, both with appropriate as well as extraneous inputs. Below is the complete list of items which had to be tested (at the time of development, and also at later stages, to ensure no issues developed later) and the results of testing.

6. EXECUTION & COMPILATION OF RESULTS

Fig.6. Hardware Setup

The final hardware circuit, containing the NodeMCU, power adapter circuit, sugar cube relays, the switching circuit for DC Fan, as well as the 5 loads(Figure.6.). The Tube light, the lamp (bulb), and a socket (charging a smart phone) has been switched on, using the mobile web application. The verified results are given in Table1 and Table2.

Table1 Primary Objectives

DEVICE	ON	OFF
LOAD #1 – LAMP	✓	✓
LOAD #2 – TUBE	✓	✓
LOAD #3 - DC FAN	✓	✓
LOAD #4 - SOCKET1	✓	✓
LOAD #5 – SOCKET2	✓	✓

Table 2 Secondary Objectives

FEATURE	RESULT
Login feature	Success
Password	Success

validation	
Change of password	Success
On time of each device	Success
Date wise log of on times	Success
New user registration	Success

7. CONCLUSION

This paper is special as effective yet, simplistic simulation of a complete HMS has been realized in this paper, using components and raw materials easily available commercially, at reasonable prices, and using design and implementation concepts of Electrical and Electronics Engineering. The system automates 5 different devices/loads, via a webpage which can be accessed from anywhere in the world via an internet connection. The webpage thus provided is secured with a login feature. Innovative usage monitoring feature has been provided, which measures the duration of usage of each device after each ON-OFF period. This will allow the user to monitor usage, and strategically reduce usage, energy consumption and thus utility bills.

8. REFERENCES

[1] Adrian McEwen and Hakim Cassimally, *Designing the Internet of Things,* Wiley.
[2] Peter Waher, *Learning Internet of Things,* Packt Publishing.
[3] Arshdeep Bahga, *Internet of Things : A Hands-On Approach,* ISBN:978-0996025515
[4] Jennifer Robbins, *Learning Web Design,* O'Reilly.
[5] Jon Duckett, *HTML and CSS – Design and Build Websites,* John Wiley and Sons.

IOP Conf. Series: Materials Science and Engineering **623** (2019) 012015 doi:10.1088/1757-899X/623/1/012015

A Review on Dynamic Voltage Restorer in power systems concerned to the issues of Power Quality

Bandla Pavan Babu[1], Indragandhi V[2]

[1]ResearchScholar, School of Electrical Engineering, VIT Vellore
[2]Associate Professor, School of Electrical Engineering, VIT Vellore

Abstract. Power Quality issues are the important concepts for all types of electricity usage consumers due to increase in power demand at different levels. For achieving power demanded, the power grid is incorporated with single or integrated renewable energy sources which resembles in rise of additional power quality issues than in regular isolated system. In this, a detailed review on different nature of power quality issues, monitoring devices of power quality & power quality mitigation especially by the dynamic voltage restorer-DVR has been explained .A review on DVR for power quality issues like voltage swell, sag, flickers, imbalance, reactive power imbalance and harmonics with distinct power circuit structure and the detailed control strategies for mitigating the mentioned issues is presented in this paper.

1. Introduction

By considering limitations of standards, it is very crucial in maintaining the electrical power quality. If the quality of power is not maintained, then it leads additional power losses, increase in harmonic content, interference with the telecommunication lines, reduction in system efficiency and also overall stability. With the increase in development of power electronics, quality of power can be maintained from different power quality issues particularly the in industrial customers voltage disturbances are the common. A power electronic converter is nothing but a DVR, connected in series for mitigation of disturbances/issues related to power quality. This DVR used in the scheme of flexible AC Transmission systems (FACTS) at transmission level as well as at level of distribution for custom power devices. Both voltage problems in steady state condition and dynamic state condition can be compensated by DVR and is also efficient in generating or consuming the reactive power by the energy storing elements used in DVR. DVR can also enhance the transient stability by compensating the reactive power along with this, disturbances like voltage sag/swell/flickers/imbalance can also be mitigated which are may be caused due to dynamic load changes. In the DVR, series injection of voltage can be achieved by voltage source converter VSC (inverter)-VSI's which consists of IGBT's (Insulated Gate Bipolar Transistor) with anti parallel connected diodes , which is supported by a capacitor stored by DC. This entire schema is linked between point of common contact (PCC) and at the receiving point. The study of DVR has been reviewed emphatically for past few years and a tremendous composition, which is available on the distinct topology and control methods. In this paper a review on DVR on different power circuit structures and control strategies has been clearly explained [1], [11].

2. Different Types of Power Quality Problems & its Mitigating Devices

2.1. Voltage Sag or Voltage dips

As per the IEC standard, the voltage Sag or voltage dips are defined as the sudden voltage reduction in electrical systems at a point pursued by a recovery voltage later a short recovery of time from a 0.5 cycle to a few seconds as shown in Figure. 1. In IEEE standards it is represented as the 20% variations of normal voltage.

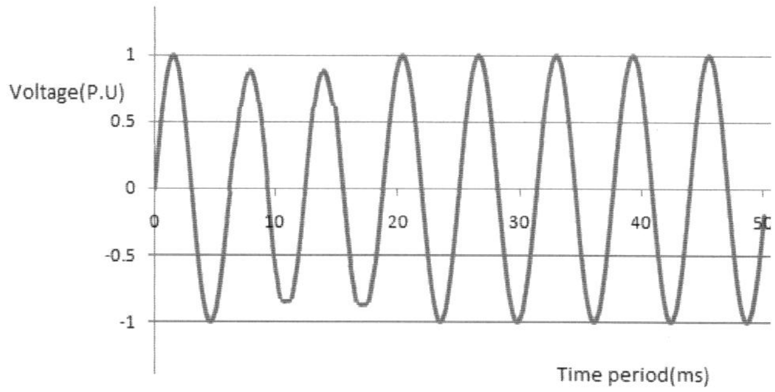

Figure. 1. Voltage sag or Voltage dips

2.2. Voltage Swell or Rise

Voltage swell is the inverse of voltage sag. With sudden removal of load under normal operating conditions, at terminal voltage may be higher than the input voltage which is nothing but Ferranti effect as displayed in Figure. 2.

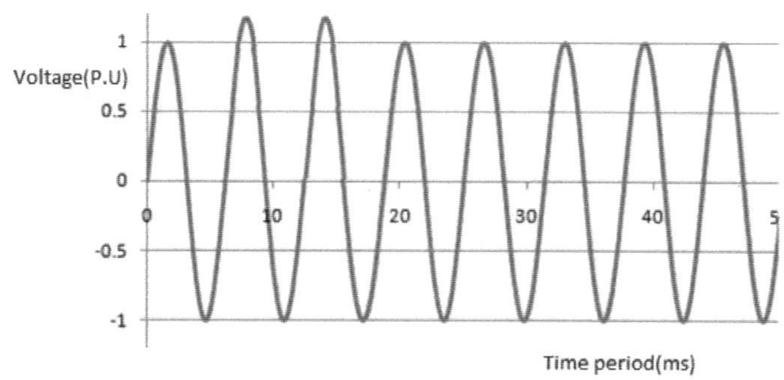

Figure. 2. Voltage Swell or Rise

2.3. Voltage Spikes

Generally voltage spikes are very short pulses of energy in power lines which contains very high voltages (it may also thousandths or millionths of a second high voltages) and it cause damage to the power electronic device easily which are incorporated in the system as represented in Figure. 3.

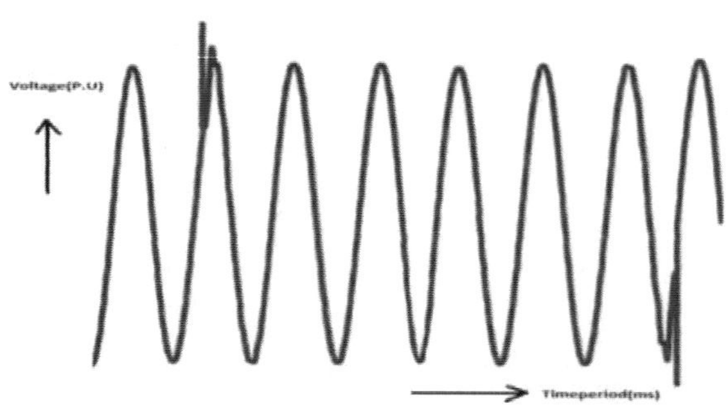

Figure. 3. Voltage spikes

2.4. Voltage Fluctuations and Voltage Flickers

As represented in Figure. 4, voltage fluctuations and voltage Flickers may be due to the surges or lightning's at the power grid but whereas coming to the grid connected renewable energy sources, this voltage fluctuations or power fluctuations are appeared due to solar irradiation which are caused by passing clouds. These voltage fluctuations and voltage flickers are comparatively very small i.e., shorter than $\pm 5\%$ variations in the RMS line voltage.

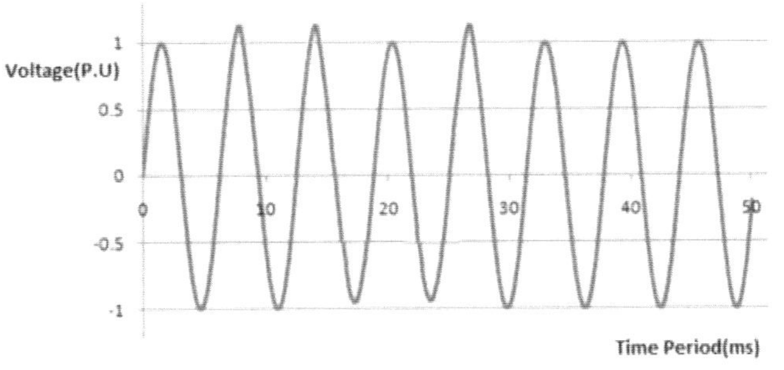

Figure. 4. Voltage Fluctuations and Voltage Flickers

2.5. Voltage Unbalance

Voltage unbalance or voltage imbalance is the change in magnitudes of single/three phase voltages in approximation among one another as displayed in Figure. 5.

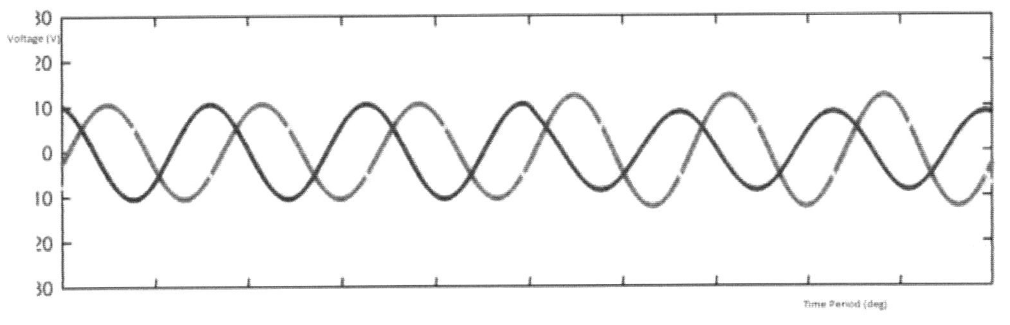

Figure. 5. Voltage Unbalance

2.6. Frequency Variations

One of the ultimate significant power quality problem is frequency variations which are appeared due to different reasons like harmonic instability, dynamic load changes etc.,. The permissible range of frequency variations is $\pm 5\%$ variations of fundamental frequency as displayed in Fig. 6.

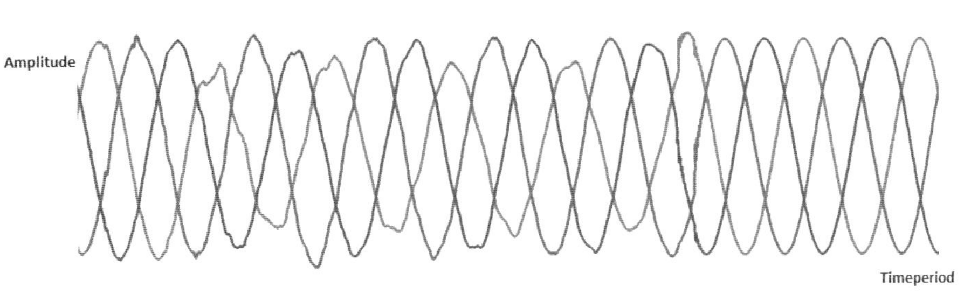

Figure. 6. Frequency Variations

2.7. Harmonics & Harmonic Distortions

Harmonics are defined as the Integral multiple of fundamental frequency. Harmonic Distortions or Total Harmonic Distortions (THD) are to be considered because it affects mainly the power quality specifically in grid connected renewable energy sources. Due to the choice of inverters in renewable energy sources for DC to AC conversions, which results in injection of voltage as well as current harmonics as represented in Figure. 7.

THD represented as

$$THD = \frac{\sqrt{\sum_{m=2}^{\infty} V_m^2}}{V_1}$$

Vm – Voltage Magnitude of Harmonic Component (For m = 2 to ∞)

V_1 – Fundamental Voltage

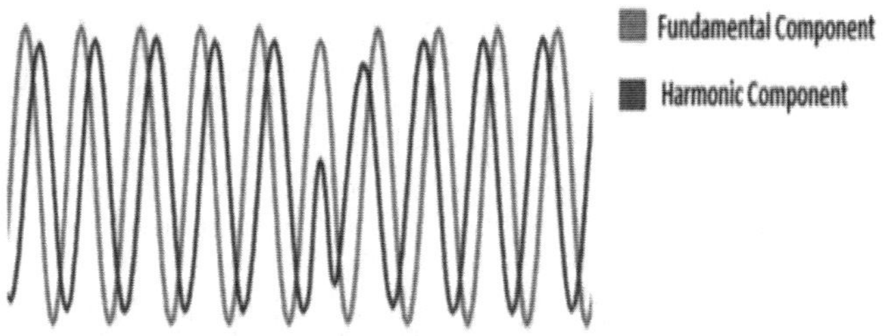

Figure. 7. Harmonic Distortions

Table 1. Mitigating equipments for power quality issues

Nature of Disturbances / Device	Power Quality Isssues							
	Transient	Swell	Sag	Frequency Deviations	Flicker	Harmonic Distortion	Noise	Interruption
SPS	×	×	×	×	×	×	×	✓
Voltage Regulator	×	✓	✓	×	×	×	×	×
Isolation Transformer	✓	×	×	×	×	×	✓	×
Filter	×	×	×	×	×	✓	✓	×
UPS	✓	✓	✓	✓	✓	✓	✓	✓

IOP Conf. Series: Materials Science and Engineering **623** (2019) 012015 doi:10.1088/1757-899X/623/1/012015

Passive Filter-PF	×	×	×	×	×	✓	×	×
Fixed Capacitor-FC	×	×	✓	×	×	×	×	×
Active power filter-APF	×	×	×	×	×	✓	×	×
Surge Suppressor	✓	×	×	×	×	×	×	×
Hybrid Active Power Filter-HAPF	×	×	×	×	×	✓	×	×
Power Conditioner	✓	✓	✓	×	×	×	✓	×
Static Var Compensator-SVC	✓	✓	✓	✓	×	×	×	×
UPQC	✓	✓	✓	✓	✓	✓	✓	✓
Static Synchronous Compensator-STATCOM	✓	✓	✓	✓	×	×	×	×

3. DVR System Configuration & Its Operation

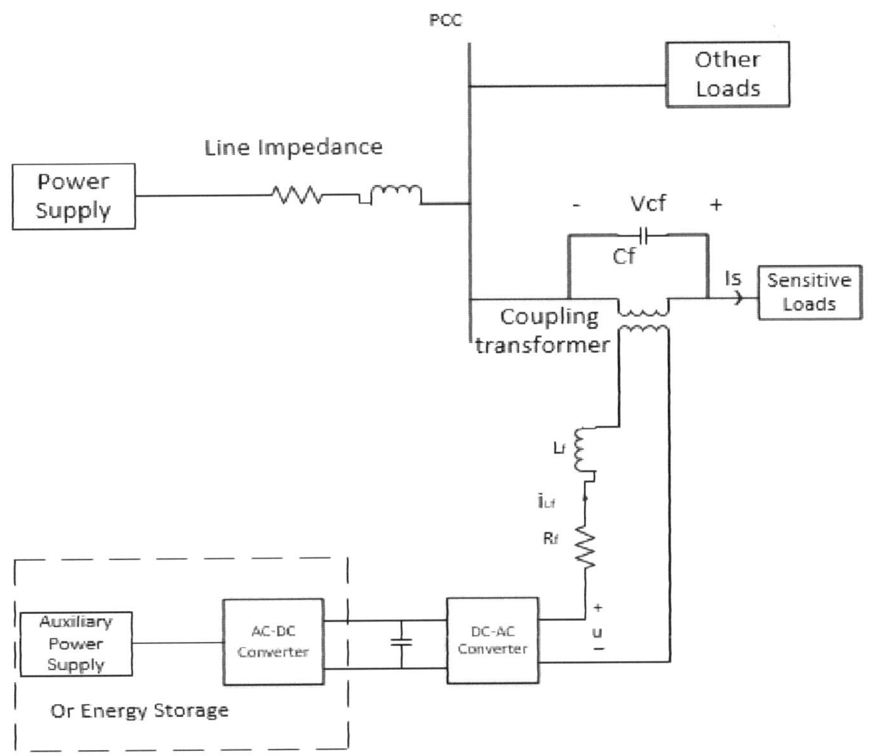

Figure. 8. DVR with a power system configuration

IOP Conf. Series: Materials Science and Engineering **623** (2019) 012015 doi:10.1088/1757-899X/623/1/012015

The compensator which inserts voltage in series with the line by protecting the sensitive loads opposing to power quality issues such as voltage Swell, Sag, flickers, imbalance e.t.c., is a dynamic voltage restorer. From Fig. 8, the simplified single line diagram of DVR is shown. Even though if the input AC of the power system may distorted or not, the voltage across the receiving end be constant owed to the dynamic injection of voltage control by DVR. This DVR performs the dynamic operations via transformer i.e., consumes the reactive power from the line and stores this energy in a dc capacitor (in rectifier mode) under voltage swell conditions and similarly generates the stored energy by balancing reactive power in the line (inverter mode-controlled case) which is under voltage sag conditions. Finally with this dynamic response from the DVR the voltage across the line can be constantly maintained by balancing the reactive power which is done through the voltage source converter. With the use of filter which is series connected with VSC, the harmonic elimination can be done. DVR will be short circuited by a switch when there are no disturbances in voltage so it resembles in minimizing losses and maximize cost effectiveness. Thus the DVR eliminates voltage sag/swell/imbalance/flickers, compensates reactive power & harmonic elimination and reduces the risk of load tripping and also maintaining the quality power.

4. Classifications Of DVR depending on Power Circuit Configuration

4.1. Multifunctional DVR in Distribution Systems

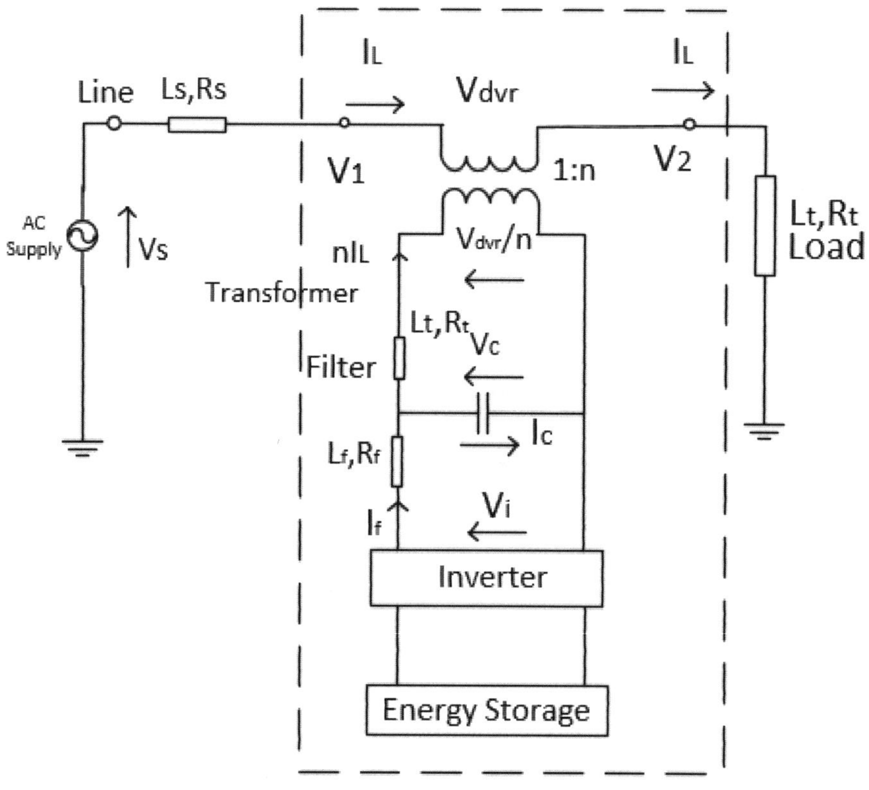

Figure. 9. DVR with Distribution system

In this Figure. 9 shown, the support of voltage level from medium to group of consumers by DVR is done. If there exists disturbance in consequent DVR's feeder, huge current flows through the system then the DVR controls the fault current without entering fault current to the line/parallel feeders by protecting load with breakers and isolators. Due to extreme fault current, the voltage across PCC may

IOP Conf. Series: Materials Science and Engineering **623** (2019) 012015 doi:10.1088/1757-899X/623/1/012015

fall and loads which are connected to feeders may also affect. During voltage compensation, DVR also commit voltage sag at PCC if control is not done properly, this leads to additional increase of fault condition. For this case, a flux charge model is best suited for limiting the fault current by the DVR in which DVR acts as virtual impedance. In this method, the DC link capacitor and battery is protected by DVR and there will not be any consumption of real power from other sources/elements in this system by DVR [3].

4.2. Interline DVR in Distribution Systems

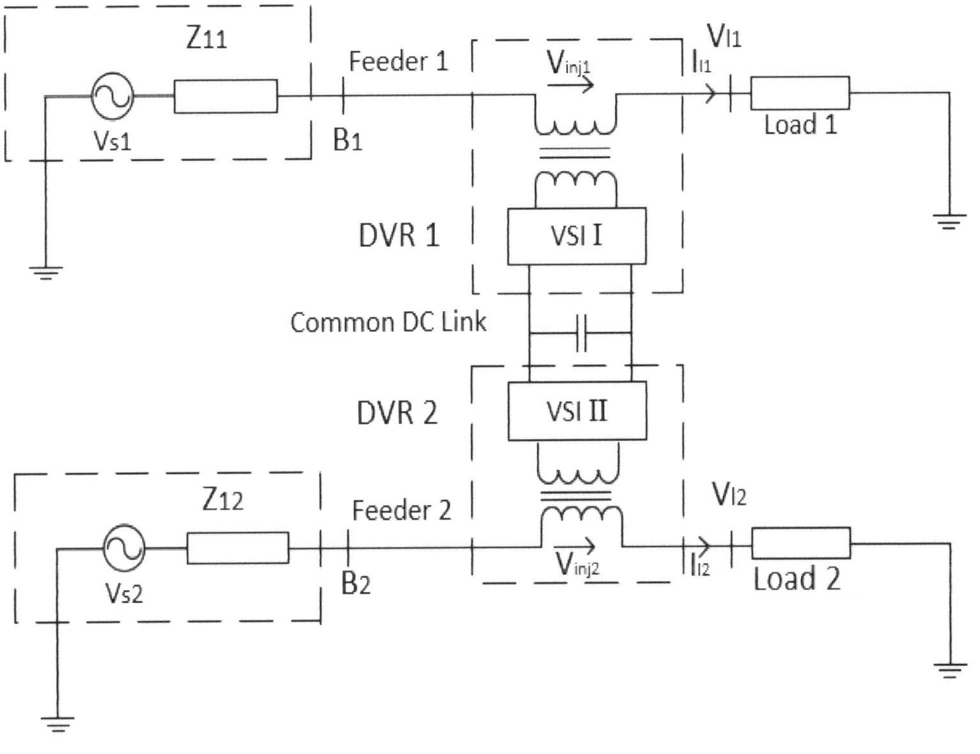

Figure. 10. Interline DVR in a 2 feeder system

Interline DVR system consists of distinct DVR's in distinct feeders, with sharing of a common dc link. Interline DVR with two line feeder system as shown in Figure 10, employs two DVR's connected to two distinct feeders which originates from two substations grid. This two feeders may consists of different or same level of voltage. If one of the DVR's compensate voltage sag, then the other DVR in the Interline DVR, which is operating at power flow mode of control to restore energy storage in the dc link which is depreciated due to real power consumption by DVR which is works on voltage sag compensation mode. Proposition of voltage sags occurs due to the power system faults, which depends on many factors such as fault current, voltage level, transformer in the Proposition track and their connection setup, etc.,. Voltage sags are likely to be in action to a larger electrical length of distance rather than in distribution system and because of these factors, two feeders of the Interline DVR system as shown in Figure. 10 are connected to two distinct substations grid, and is reasonable to consider that voltage sag in Feeder 1 can have a lower effect on Feeder 2. Consequently, the demanding generation transmission system can be considered as two independent sources for the two feeders [5], [52].

IOP Conf. Series: Materials Science and Engineering **623** (2019) 012015 doi:10.1088/1757-899X/623/1/012015

4.3. DVR in Distributed Grid with Ultra capacitor

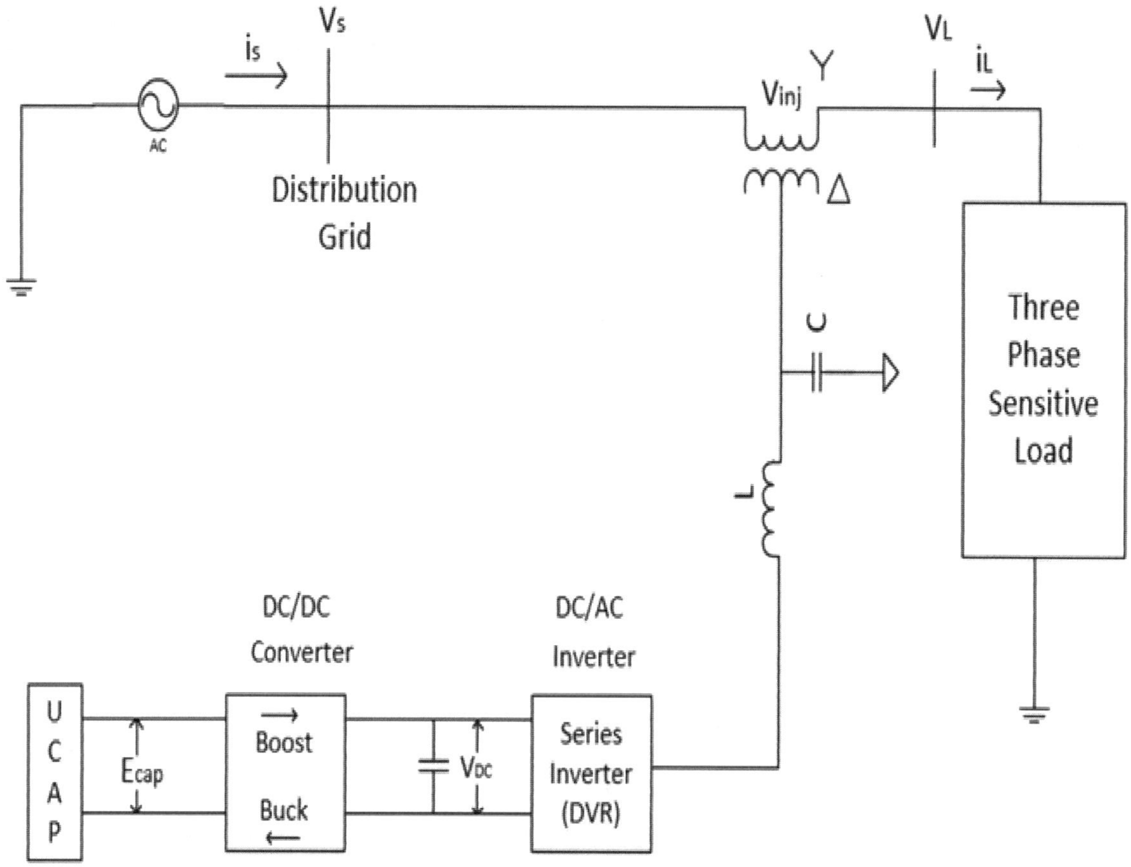

Figure. 11. DVR Single line diagram with Ultra Capacitor Energy Storage-UCAP

Out of technologies in rechargeable energy storage, active power support in milli-seconds to seconds timescale, Ultra Capacitors are best suited represented in Figure. 11. Accordingly, Ultra capacitor used integration in to DVR system is ideal, due to milli-seconds to seconds range duration of instantaneous voltage sags and voltage swells. Generally these ultra capacitors have the energy with lower density and ideal characteristics for the compensation of voltage swell and sag with high power density in which these combined functions are requires large amount of power in small time period. When compared to energy storing elements, ultra capacitors posses' large number of charge/discharge cycles even for equal size of modules which leads to ease of integration. The applications of DVR with integration of ultra capacitors can be given as.

i. For compensating voltage sag and swell in the line, the capability of generating real power to the system is done.
ii. DC-DC converter, Inverter Interface with control, and ultra capacitors preliminary demonstration can be done.
iii. To compensate voltage swell and sag in the distribution grid can be done by development of inverter and DC-DC converter.
iv. Hardware Integration and their performance validation of Ultra capacitor DVR system [13].

4.4. DVR in power system without Transformer

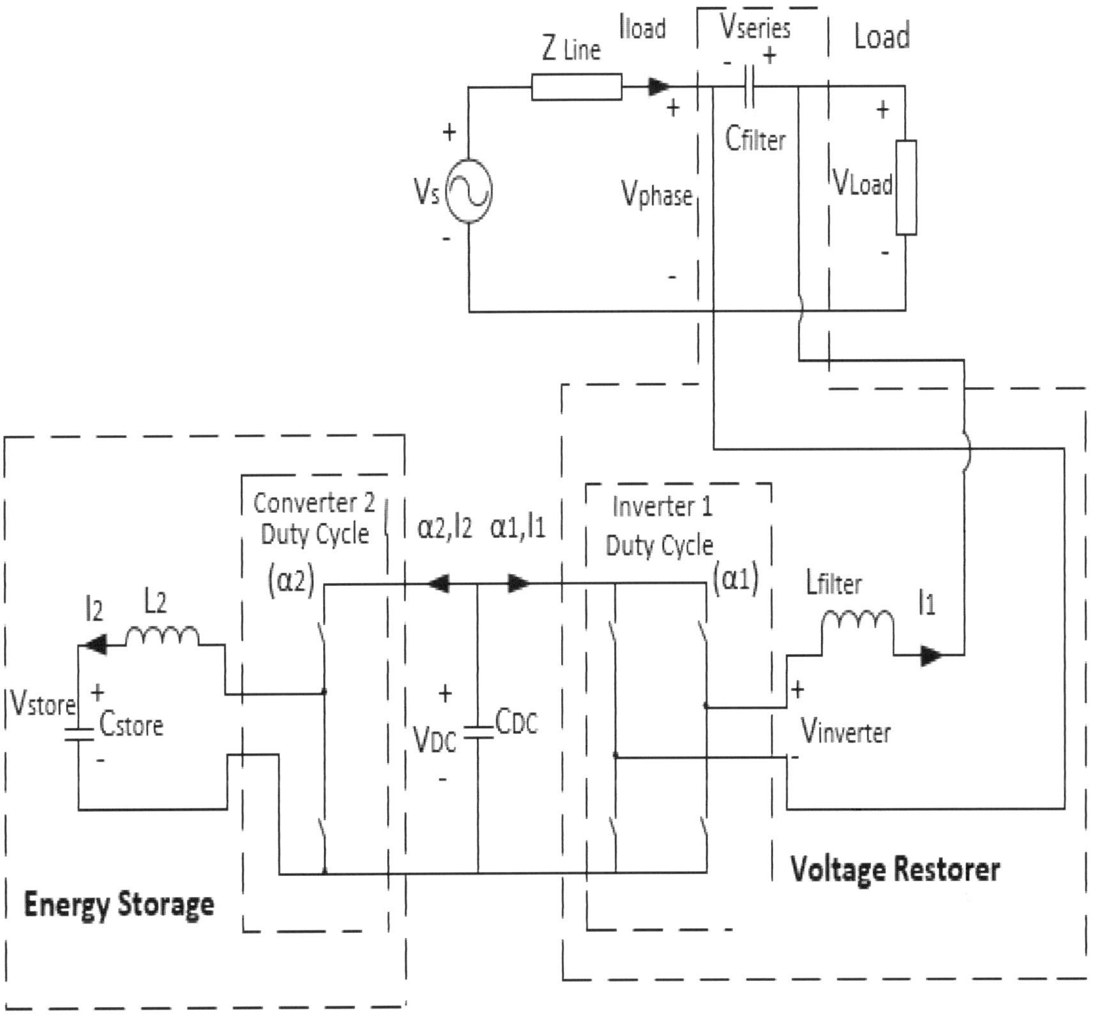

Figure. 12: Transformer less DVR

Equivalent voltage and impedance is represented in upstream source as displayed in Figure. 12. Here Load is pretended to be constant. DVR output is series inserted among the load and source and can further be segregated into the energy storage unit and voltage restorer unit which are connected by dc-link capacitor. The series injected voltage is controlled by the Restorer. In Normal operating conditions, either by a bypass switch which is connected in parallel or the voltage restorer inverter controls by maintaining 0V. Required amount of series voltage compensation is provided under sag conditions and this restorer has distinct functional modes for normal and sag conditions which can be classified as self charging and voltage restoration modes respectively. Here, the loss in phase voltage is restored by the Inverter1 which can be achieved through voltage control. When there exists healthy operating conditions, then it shifts to self charging mode [72], [77].

IOP Conf. Series: Materials Science and Engineering **623** (2019) 012015 doi:10.1088/1757-899X/623/1/012015

4.5. Grid connected DVR with Fault Current Limit

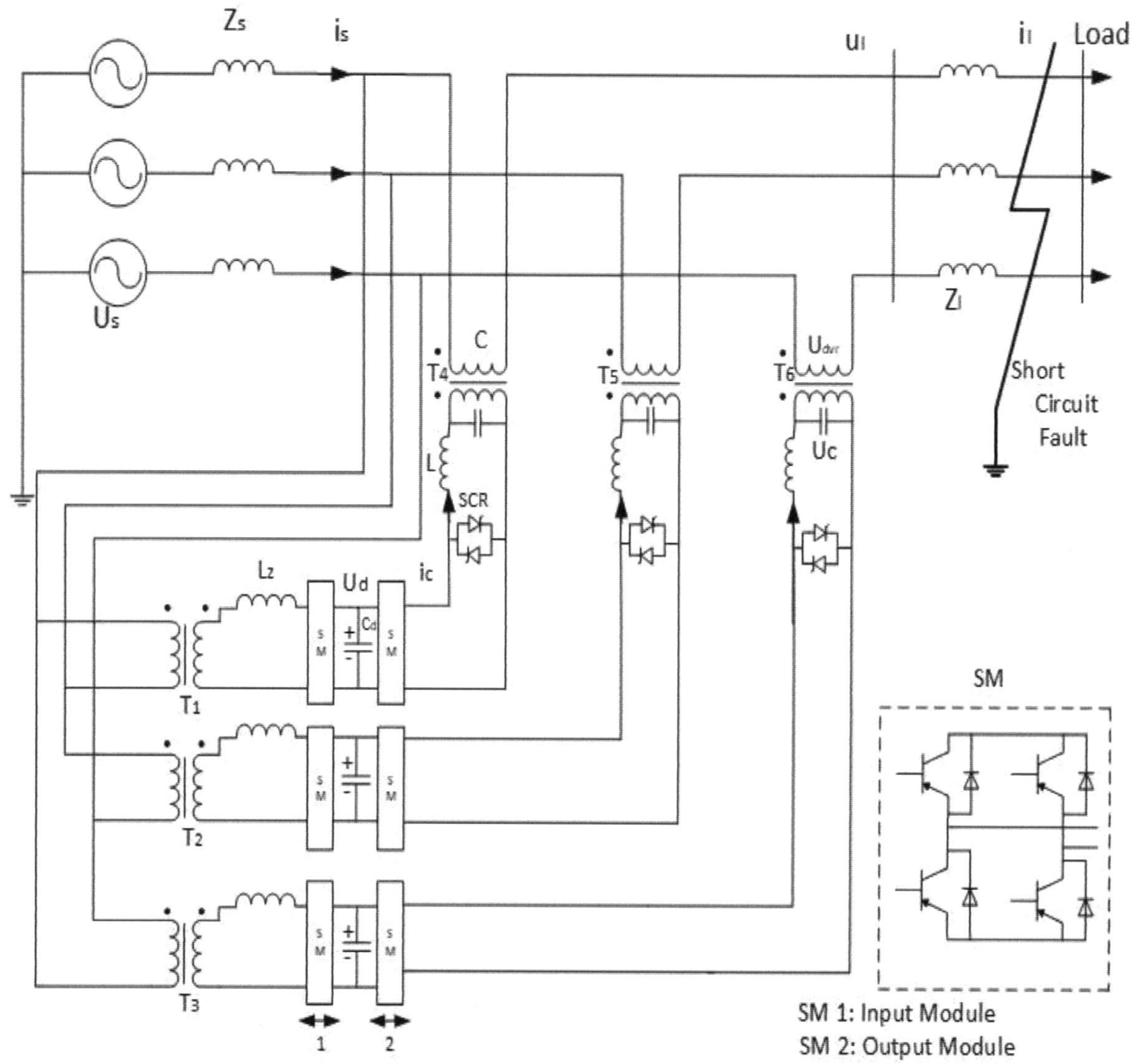

Figure. 13: Fault Current Limiting DVR topology

If, Fault Current Limiting DVR is operated in the mode of voltage compensation, the crowbar bidirectional thyristor is made inactive. Fault current limiting DVR functions as normal like voltage controlled voltage source converter as displayed in Fig. 13. By the Compensation voltage u_l (series connected with supply voltage), the control of fault current limiting DVR is achieved if any voltage unbalance/fluctuations occurs and this leads to enhancement of power quality with maintenance of voltage across load [27]. To provide the voltage across the DC- link, utilization of back to back converter input module is done. In this method, the faulty phase of inverter is non active when short circuit fault arise, crowbar bidirectional thyristor move to active mode. At this instant, on series connected transformer secondary side, insertion of reactor L is done so that the fault current is bounded. This leads to fault current limiting DVR can back to its normal mode with removal of faulty section by supporting the supply voltage with series connected transformer and back-back converter [62], [74].

4.6. DVR with Disturbance Filter

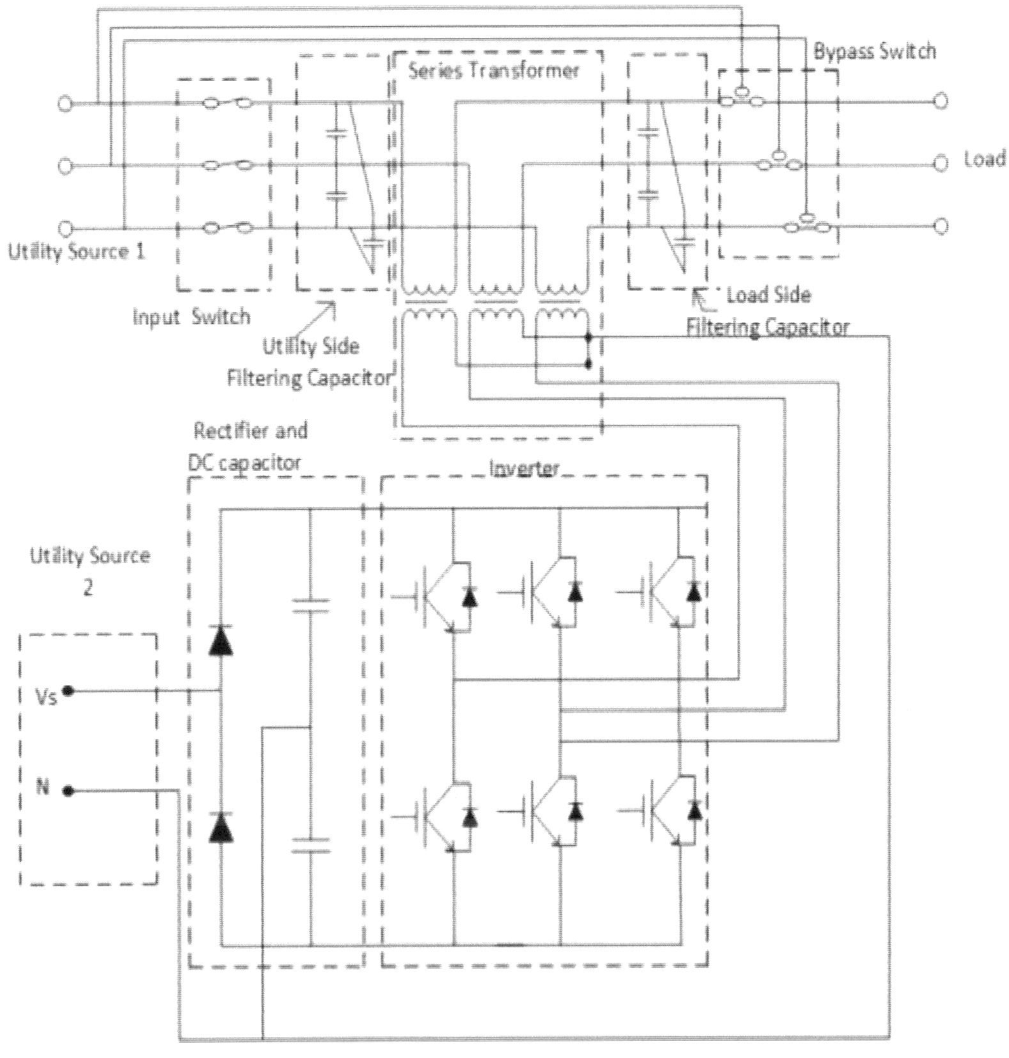

Figure. 14. DVR with Disturbance Filter

As displayed in Figure. 14 shows, the structure of the proposed DVR, where the output terminals of three phase half bridge inverter is connected via (wye) series transformer to the utility supply. To operate under unbalanced circumstances, current path of zero sequence is provided by linking series transformer neutral to ground of the inverter. Here for temporary storage, DC capacitor bank is used for storing the rectified DC power which is converted from AC through Rectifier Unit. For protecting the DVR, bypass switch is used for safe operation. The input switch can detect the fault voltage or current for the disconnection of DVR if there is any circuit fault on load. During this the bypass switch is communicated such that it can transfer load from output of DVR to utility. Thus ends up the bypass procedures. If suppose any voltage disturbance occurs, with the support of the PWM technique the output of inverter can be driven in phase AC source incoming while constant load voltage is maintained. In utility terminals and load terminals the capacitors are shunt connected for the filtering purposes which are named as load side and utility side filtering capacitor respectively. Out of these two, to inhibit the effects of voltages which are non sinusoidal from lowering the utility grid , the utility filtering capacitor is preferable. Therefore the better quality can be ensured to the demand side with such a design by the DVR [31].

4.7. Cascaded H Bridge Inverter

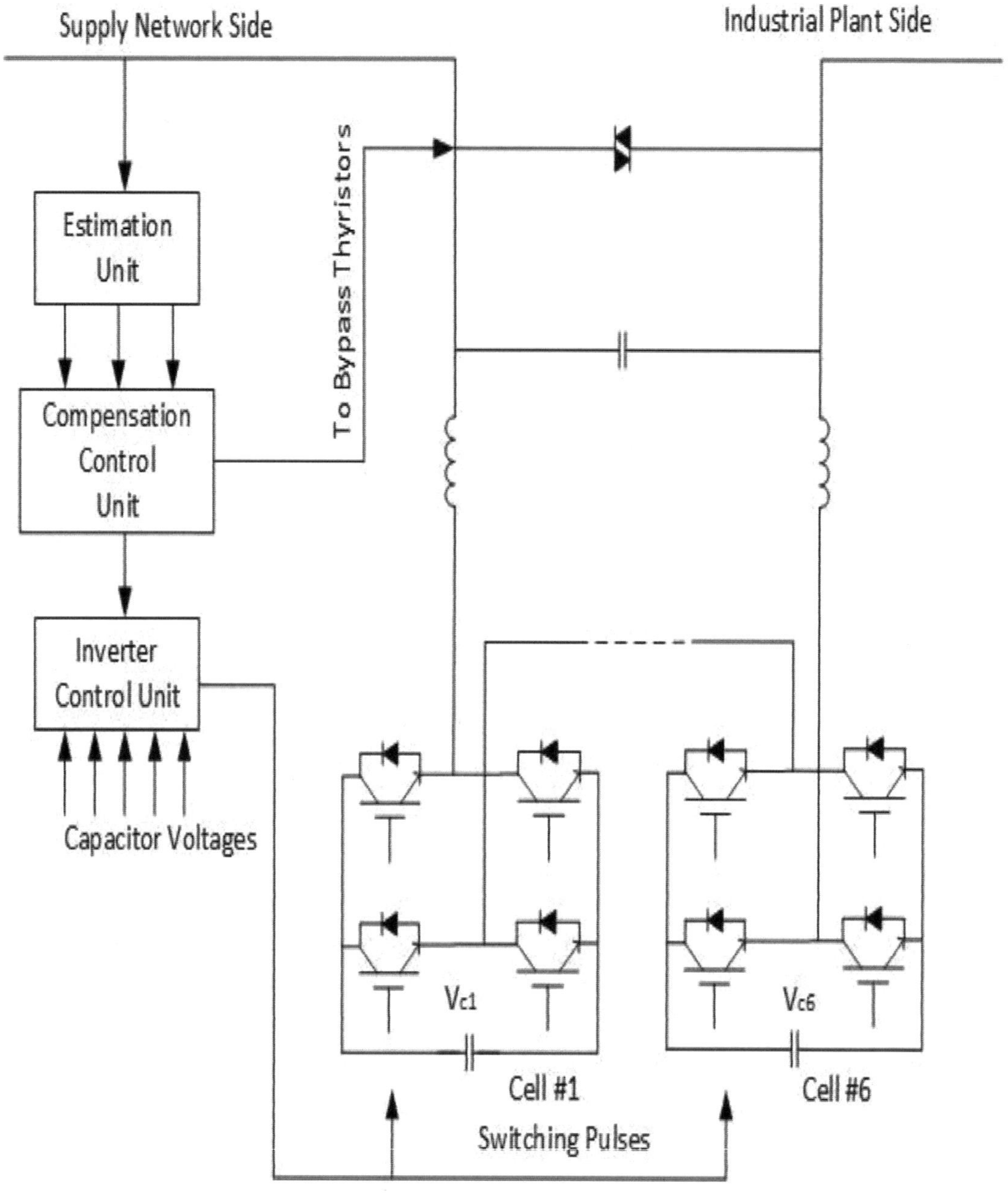

Figure. 15. Cascaded H-Bridge 7-level Inverter

Figure. 15 represents the Conventional Converter which consists series injected transformer, estimation unit, bypass thyristors, control units, control unit and output filters. In order to protect the DVR from losses and electrical faults, the two thyristors (anti parallel connected) will bypass the DVR that to when there exists normal range of voltage. As discussed, the series injected transformer operating frequency be 50/60 Hertz's with its operating voltage across secondary terminals between 20% and 25% of fundamental voltage. At the time of compensation, 10% to 20% of power in load will be passed through this series injected transformer [37].

4.8. DVR with Nonlinear and/or Unbalanced Loads:

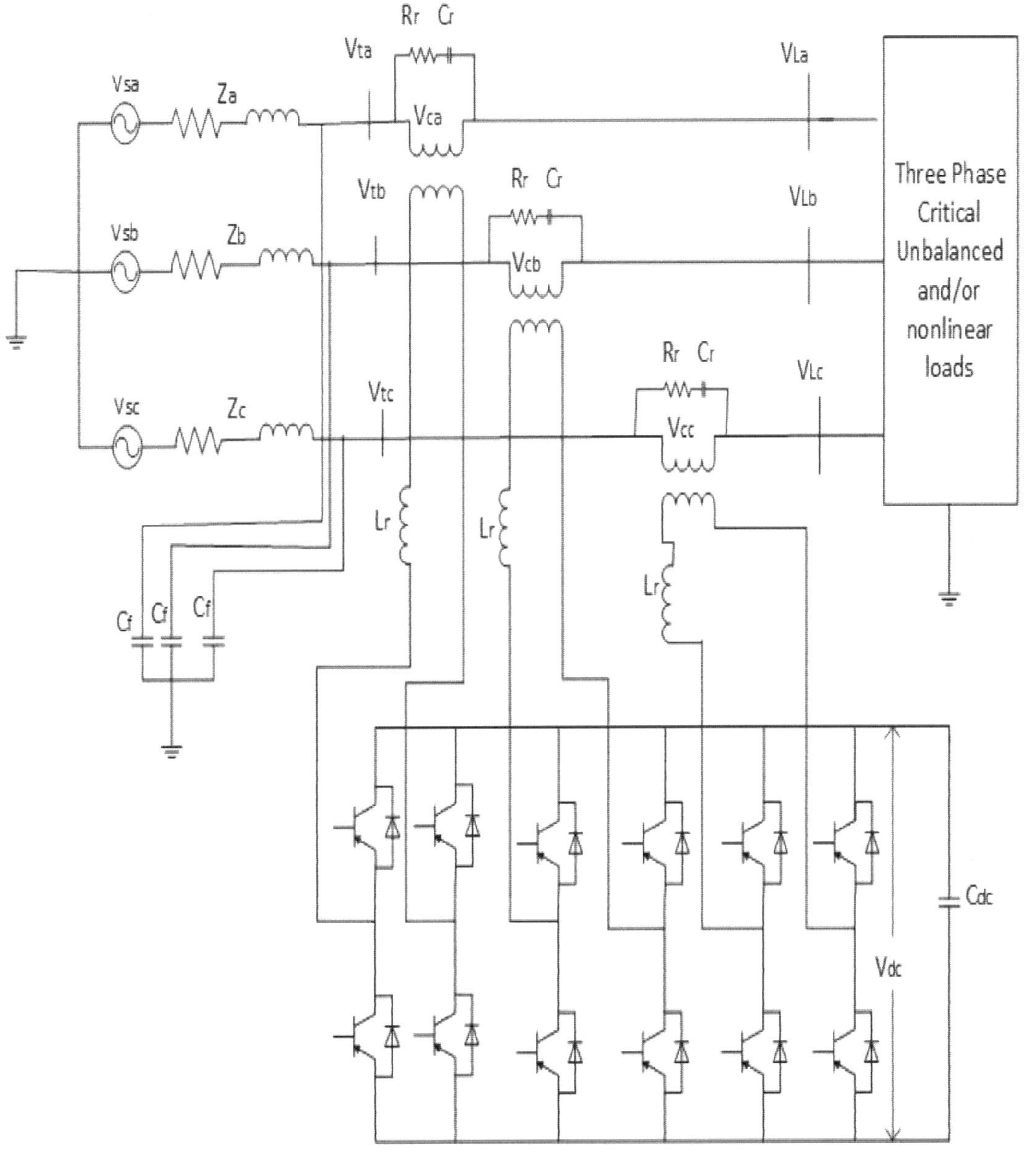

Figure. 16. power circuit of Dynamic Voltage Restorer with Nonlinear and/or Unbalanced Loads

In Fig. 16, distribution feeder of each phase are connected with 3 H-bridge VSCs through revised structure of ripple filter (Cr, Lr, Rr) & injection transformer. This not only shorten voltage required converter, but also contributes isolation among distribution feeder and converter. Shunt capacitor filter (C_f) provides low impedance path to harmonics of load currents which are of higher order, only when the current through the load is nonlinear. To trace the compensating reference voltages, revised filter structure, hysteresis band controller with constant switching frequency is used. Hysteresis band controller with constant switching frequency along with revised filter structure secures all the benefits of band controller and overthrown its difficulties by adaptive hysteresis band and revised filter structure, which is providing constant switching frequency [57], [87].

Table 2: Classification of DVR according to their compensating Variables

Compensated parameter	Type of Converter	Control/Technique	Control Strategy	Reference
Voltage Sag and also Limit the fault current	12 Pulse VSC	flux-charge model	closed-loop control system	3
Voltage Sags and real power control	2 three level VSC	PWM	Multi Loop Feedback Control System	5
voltage sags and swells	3 Phase 3 leg VSC	PWM	TMS320F28335 DSP based Controller	13
Voltage Restoration and Self Charging	Converter1 single leg and Converter 2 Two Leg	Proportional Controller	closed-loop feedback control system	16
Total time delay of the compensation voltages in DVRs	6 Leg Inverter	Feed forward and state feedback based controller	DSP Control Board TMS320VC33/15 0Mhz	17
voltage sags and swells	1 phase 2 leg VSI	Switching Frequency Band Controller	Adaptive Hysteresis Band Controller	18
Voltage Sag	3 Leg H-Bridge VSI	SVPWM	Software Phase Locked Loop	20
Voltage Injection	3 phase 3 leg VSI	PWM	Closed loop controller	21
Deep Voltage Sag and Harmonics	3 phase 3 leg VSI	Time Domain Simulation	Two Level Hysteresis Controller	23
Voltage Sag and Harmonic Voltage Distortion	3 phase 2 leg VSI	Filters	--	24
Voltage Sag	3 Phase 12 pulse VSI	PWM	Phase Locked loops	25
Voltage Sag and Voltage Restoration	1 phase 1leg	Boundary control & PLL	Closed loop controller	26
Voltage Compensation and Fault Current Limit	1 phase 3 level two 2leg VSI	Back-Back Converter	Crowbar Thyristor	27
Voltage sag and Disturbances	3 phase half-bridge Inverter	PWM	Current Controlled Inverter	31
Harmonic Distortion and Switching Losses	3-leg and 4-leg VSI's	PWM	Extended Digital Scalar and Symmetrical Component	32
Voltage sag and swell	4-leg VSC	DSP based Digital Controller	Closed Loop Control	33
Voltage Sag	9- Level Inverter	DC-DC converter	--	34

Harmonic Distortion	2 three phase Inverters 3-leg and 6-leg	PWM	Level Shifted carrier or single carrier	35
Total Harmonic Distortion	3 phase VSC	PI controller	Chaotic accelerated PSO algorithm	36
Voltage Sag and Harmonic Distortion	7 level cascade H-Bridge Inverter	DC-DC converter	Active voltage balance algorithm	37
Voltage Sag/Swell Detection	3 single phase H bridge Inverters	PWM	Fuzzy Logic Controller	41
Voltage Sag	3 phase 4 leg Inverter	--	Least Error Squares (LES)	49
Voltage sag	2 cascaded H-Bridge Inverter-7 level	Phase Shifted PWM	Closed Loop Controller	50
Voltage sag	3 Phase VSI	PWM	Phase Locked Loop	52
Weighted Total Harmonic Distortion	3-H Bridge	PWM	Space Vector	54
Voltage Sag	Single phase	PWM	DFT algorithm	55
Voltage sag	Single Phase Inverter-7 level	--	Open and closed loop control	58
Saturation in transformer	Single Phase	PWM	Space vector control	59
Voltage Regulation	Three Phase	PWM	--	60
Voltage sag and short circuit current faults	3 phase 7 level	PWM	Short circuit current limiting mode	62
Voltage Sag	Single phase High Voltage Multi level Inverter	Posicast control	Open loop and Closed loop control	63
Voltage Control	Single Phase Single Leg VSI	DSP processor	Discrete Time Model	72
Voltage Fluctuations	Three phase 3 leg VSI	PWM	double-loop control	73
Transient overshoot, settling time and steady state error	Three phase 4 leg VSI	PID Controller	Closed loop control	75
Voltage regulation	Three phase 2 three leg VSI	PWM	Phase Locked Loop	79
Voltage Sag	Three Phase VSI	DSP based Hardware	Multi Loop Control	82
Voltage protection	Three phase 3-	Synchronous	constant	87

from Unbalanced and Non-Linear Loads	Level H-bridge VSCs	Reference Frame Theory (SRF)	switching frequency hysteresis band controller	
Voltage Harmonics	Single phase and three phase	Resonant Feed back	Phase Locked Loop	90
Voltage Sag	Three Phase	PLL	multiloop control system	92
Voltage Sag	Three Phase 2 level	two degrees of freedom (2DOF)	Closed Loop system	94

5. Classification of Converter Control Techniques of Dynamic Voltage Restorer

5.1. Current limitation by Flux-Charge Model DVR

In this flux charge model as shown in Fig. 16, the protection of DVR components is done by restoring the Point of common contact voltage and fault current limitation. In Distribution feeder, DVR connected in series acts like a Virtual inductance and this can only be done by injecting appropriate voltage controlled by DVR with opposite polarity than in usual. By providing the additional communication between the circuit breaker and DVR over current tripping is possible here. Another functioning of this model is that, point of common contact voltage is protected by limiting downstream fault current. Just by adding the impedance the control of reactive power is achieved [3].

5.2. Voltage sag compensation by closed loop feedback control system DVR

In this method of closed loop feedback control, for voltage compensation, voltage loop is associated apparently to an internal loop current which is taken from current of filter capacitor of the DVR in order to avoid the drawbacks in open loop feedback control system. This scheme of control is similar to compensate voltage and real power in the system and the only change is that generation of reference signal and its dependency on control mode for real power flow which is generated according to the demand of reactive power requirements [5].

5.3. Voltage restoration by Integrated Higher Order Controller DVR.

In this method, Phase locked loop-PLL is utilised for finding rotating angle so that voltage swell and sag compensation can be done by Ultra Capacitor-DVR without any dependence on grid. The compensation of voltage is done by in phase injection of voltage with the fundamental voltage. Integrated Controller which is at higher level of is operated based on system specifications and produce inputs to DC-DC converter and inverter for controlling process [13].

5.4. Voltage sag mitigation by transformerless self-charging DVR.

In this method two modes exists i.e., voltage restoration and self-charging mode. In voltage restoration mode, the fall of phase voltage V_{phase} is controlled by Inverter 1 with control of series voltage V_{series} [16]. In order perform this action, Inverter 1 is provided with sufficient DC voltage link [59]. In the other mode i.e., self charging mode, the restoration of energy in capacitor has to be done for compensating the further any sags by the DVR when V_s returns to its normal mode of operation [72], [77].

5.5. Fault current Limit Control (FCL) DVR.

According to grid state, the fault current limiting DVR can operate in two modes i.e., Switching Strategy operation mode and detection of fault current method [27]. In this control strategy, a quick response along with the voltage compensation of low steady state error is obtained [62], [74].

5.6. Disturbance Filtering DVR

In this, the control system is operated to attempt for semiconductor switches signals and also adopted inverter with current control which comes with a quick response. The direct control of the voltage across load is performed by this method, which leads to not in calculating the series compensation voltage in prior and in a significant aspect by shortening procedures for circuit design [31].

5.7. Voltage sag/swell detection DVR
In this method V_{abc} parameter are considered for the generation of compensating voltage and activating the PWM generator by generating the pulses. In this control strategy the drawbacks of conventional methods like, transformation process, slow transient response and complexity in mathematical modelling was conquered [41].

5.8. DVR based on Synchronous Reference Frame-SRF theory
This method generates instantaneous reference voltage for compensating by controlling the self sustained SRF basic theory based DVR in order to defend the non-linear and unbalanced loads. In this positive sequence fundamental phase voltages (three phase voltages) are generated by positive sequence fundamental extractor, by suspecting only two distorted and/or unbalanced line voltages. Also, when Nonlinear loads are present the DVR alone itself suppress the voltage spikes without any use of filter capacitor which is shunt connected [57], [87].

5.9. DVR for compensating Voltage Fluctuations
In this instantaneous power theory is adopted for dynamic compensation of voltage and voltage fluctuation detection. The different compensation methods such as pre- sag, minimum energy, In-Phase, and optimized energy techniques have been adopted for generating the reference signal for voltage compensation. In this method super conducting magnetic energy storage based DVR is used for compensating Voltage as well as long tern voltage fluctuations [73].

5.10. Two Degrees of Freedom Resonant Control DVR
In this method unbalanced and balanced voltage sags are compensated by two degree Resonant Control Scheme. This control structure, is based on a scheme with two [$R_1(s)$ and $R_2(s)$] nested controllers which generates 2 degree of freedom for compensation of voltage sag. A number of parameters are defined in this methodology, which are equal to the number of poles, rising to closed loop system [94].

6. Conclusion
This paper presents the comprehensive review and discussion on DVR to enhance power quality at grid and industrial competence. Modern developments and different topologies of DVR have been discussed by focusing on their advantages and advantages. Different control strategies and functioning of compensation methods for power quality enhancement by mitigating the power quality issues using DVR has been discussed. With this review it is been concluded that DVR are beneficial in voltage Sag mitigation and other voltage related power quality issues.

7. References
[1] A. O. Ibrahim, T. H. Nguyen, D. Lee and S. Kim, *"A Fault Ride-Through Technique of DFIG Wind Turbine Systems Using Dynamic Voltage Restorers,"* in IEEE Transactions on Energy Conversion, vol. 26, no. 3, pp. 871-882, Sept. 2011.

[2] D. A. Fernandes, F. F. Costa and M. A. Vitorino, *"A Method for Averting Saturation From Series Transformers of Dynamic Voltage Restorers,"* in IEEE Transactions on Power Delivery, vol. 29, no. 5, pp. 2239-2247, Oct. 2014.

[3] F. M. Mahdianpoor, R. A. Hooshmand and M. Ataei, *"A New Approach to Multifunctional Dynamic Voltage Restorer Implementation for Emergency Control in Distribution Systems,"* in IEEE Transactions on Power Delivery, vol. 26, no. 2, pp. 882-890, April 2011.

[4] H. K. Al-Hadidi, A. M. Gole and D. A. Jacobson, *"A Novel Configuration for a Cascade Inverter-Based Dynamic Voltage Restorer With Reduced Energy Storage Requirements,"* in IEEE Transactions on Power Delivery, vol. 23, no. 2, pp. 881-888, April 2008.

[5] D. Vilathgamuwa, H. M. Wijekoon and S. S. Choi, *"A Novel Technique to Compensate Voltage Sags in Multiline Distribution System—The Interline Dynamic Voltage Restorer,"* in IEEE Transactions on Industrial Electronics, vol. 53, no. 5, pp. 1603-1611, Oct. 2006.

[6] O. S. Senturk and A. M. Hava, *"A Simple Sag Generator Using SSRs,"* in IEEE Transactions on Industry Applications, vol. 48, no. 1, pp. 172-180, Jan.-Feb. 2012.

[7] Y. Lu, G. Xiao, B. Lei, X. Wu and S. Zhu, *"A Transformerless Active Voltage Quality Regulator With the Parasitic Boost Circuit,"* in IEEE Transactions on Power Electronics, vol. 29, no. 4, pp. 1746-1756, April 2014.

[8] P. Roncero-Sanchez, E. Acha, J. E. Ortega-Calderon, V. Feliu and A. Garcia-Cerrada, *"A Versatile Control Scheme for a Dynamic Voltage Restorer for Power-Quality Improvement,"* in IEEE Transactions on Power Delivery, vol. 24, no. 1, pp. 277-284, Jan. 2009.

[9] T. Jimichi, H. Fujita and H. Akagi, *"An Approach to Eliminating DC Magnetic Flux From the Series Transformer of a Dynamic Voltage Restorer,"* in IEEE Transactions on Industry Applications, vol. 44, no. 3, pp. 809-816, May-june 2008.

[10] H. M. Wijekoon, D. M. Vilathgamuwa and S. S. Choi, *"Interline dynamic voltage restorer: an economical way to improve interline power quality,"* in IEE Proceedings - Generation, Transmission and Distribution, vol. 150, no. 5, pp. 513-520, 15 Sept. 2003.

[11] A. M. Rauf and V. Khadkikar, *"An Enhanced Voltage Sag Compensation Scheme for Dynamic Voltage Restorer,"* in IEEE Transactions on Industrial Electronics, vol. 62, no. 5, pp. 2683-2692, May 2015.

[12] S. Jothibasu and M. K. Mishra, *"An Improved Direct AC–AC Converter for Voltage Sag Mitigation,"* in IEEE Transactions on Industrial Electronics, vol. 62, no. 1, pp. 21-29, Jan. 2015.

[13] D. Somayajula and M. L. Crow, *"An Integrated Dynamic Voltage Restorer-Ultracapacitor Design for Improving Power Quality of the Distribution Grid,"* in IEEE Transactions on Sustainable Energy, vol. 6, no. 2, pp. 616-624, April 2015.

[14] A. Elserougi, A. M. Massoud, A. S. Abdel-Khalik, S. Ahmed and A. A. Hossam-Eldin, *"An Interline Dynamic Voltage Restoring and Displacement Factor Controlling Device (IVDFC),"* in IEEE Transactions on Power Electronics, vol. 29, no. 6, pp. 2737-2749, June 2014.

[15] D. Somayajula and M. L. Crow, *"An Ultracapacitor Integrated Power Conditioner for Intermittency Smoothing and Improving Power Quality of Distribution Grid,"* in IEEE Transactions on Sustainable Energy, vol. 5, no. 4, pp. 1145-1155, Oct. 2014.

[16] E. K. K. Sng, S. S. Choi and D. M. Vilathgamuwa, *"Analysis of series compensation and DC-link voltage controls of a transformerless self-charging dynamic voltage restorer,"* in IEEE Transactions on Power Delivery, vol. 19, no. 3, pp. 1511-1518, July 2004.

[17] Hyosung Kim and Seung-Ki Sul, *"Compensation voltage control in dynamic voltage restorers by use of feed forward and state feedback scheme,"* in IEEE Transactions on Power Electronics, vol. 20, no. 5, pp. 1169-1177, Sept. 2005.

[18] S. Sasitharan and M. K. Mishra, *"Constant switching frequency band controller for dynamic voltage restorer,"* in IET Power Electronics, vol. 3, no. 5, pp. 657-667, September 2010.

[19] J. G. Nielsen, M. Newman, H. Nielsen and F. Blaabjerg, "Control and testing of a dynamic voltage restorer (DVR) at medium voltage level," in IEEE Transactions on Power Electronics, vol. 19, no. 3, pp. 806-813, May 2004.

[20] V. K. Ramachandaramurthy, C. Fitzer, A. Arulampalam, C. Zhan, M. Barnes and N. Jenkins, *"Control of a battery supported dynamic voltage restorer,"* in IEE Proceedings - Generation, Transmission and Distribution, vol. 149, no. 5, pp. 533-542, Sept. 2002.

[21] G. J. Li, X. P. Zhang, S. S. Choi, T. T. Lie and Y. Z. Sun, *"Control strategy for dynamic voltage restorers to achieve minimum power injection without introducing sudden phase shift,"* in IET Generation, Transmission & Distribution, vol. 1, no. 5, pp. 847-853, September 2007.

[22] A. E. Leon, M. F. Farias, P. E. Battaiotto, J. A. Solsona and M. I. Valla, *"Control Strategy of a DVR to Improve Stability in Wind Farms Using Squirrel-Cage Induction Generators,"* in IEEE Transactions on Power Systems, vol. 26, no. 3, pp. 1609-1617, Aug. 2011.

[23] F. A. L. Jowder, *"Design and analysis of dynamic voltage restorer for deep voltage sag and harmonic compensation,"* in IET Generation, Transmission & Distribution, vol. 3, no. 6, pp. 547-560, June 2009.

[24] S. S. Choi, B. H. Li and D. M. Vilathgamuwa, *"Design and analysis of the inverter-side filter used in the dynamic voltage restorer,"* in IEEE Transactions on Power Delivery, vol. 17, no. 3, pp. 857-864, July 2002.

[25] T. Jimichi, H. Fujita and H. Akagi, *"Design and Experimentation of a Dynamic Voltage Restorer Capable of Significantly Reducing an Energy-Storage Element,"* in IEEE Transactions on Industry Applications, vol. 44, no. 3, pp. 817-825, May-june 2008.

[26] C. N. Ho, H. S. H. Chung and K. T. K. Au, *"Design and Implementation of a Fast Dynamic Control Scheme for Capacitor-Supported Dynamic Voltage Restorers,"* in IEEE Transactions on Power Electronics, vol. 23, no. 1, pp. 237-251, Jan. 2008.

[27] Z. Shuai, P. Yao, Z. J. Shen, C. Tu, F. Jiang and Y. Cheng, *"Design Considerations of a Fault Current Limiting Dynamic Voltage Restorer (FCL-DVR),"* in IEEE Transactions on Smart Grid, vol. 6, no. 1, pp. 14-25, Jan. 2015.

[28] B. H. Li, S. S. Choi and D. W. Vilathgamuwa, *"Design considerations on the line-side filter used in the dynamic voltage restorer,"* in IEE Proceedings - Generation, Transmission and Distribution, vol. 148, no. 1, pp. 1-7, Jan. 2001.

[29] A. Ghosh, A. K. Jindal and A. Joshi, *"Design of a capacitor-supported dynamic voltage restorer (DVR) for unbalanced and distorted loads,"* in IEEE Transactions on Power Delivery, vol. 19, no. 1, pp. 405-413, Jan. 2004.

[30] V. Majchrzak, G. Parent, J. Brudny, V. Costan and P. Guuinic, *"Design of a Coupling Transformer With a Virtual Air Gap for Dynamic Voltage Restorers,"* in IEEE Transactions on Magnetics, vol. 52, no. 7, pp. 1-4, July 2016, Art no. 8401104.

[31] Chi-Jen Huang, Shyh-Jier Huang and Fu-Sheng Pai, *"Design of dynamic voltage restorer with disturbance-filtering enhancement,"* in IEEE Transactions on Power Electronics, vol. 18, no. 5, pp. 1202-1210, Sept. 2003.

[32] D. A. Fernandes, F. F. Costa and E. C. dos Santos, *"Digital-Scalar PWM Approaches Applied to Four-Leg Voltage-Source Inverters,"* in IEEE Transactions on Industrial Electronics, vol. 60, no. 5, pp. 2022-2030, May 2013.

[33] S. R. Naidu and D. A. Fernandes, *"Dynamic voltage restorer based on a four-leg voltage source converter,"* in IET Generation, Transmission & Distribution, vol. 3, no. 5, pp. 437-447, May 2009.

[34] E. Babaei, M. F. Kangarlu and M. Sabahi, *"Dynamic voltage restorer based on multilevel inverter with adjustable dc-link voltage,"* in IET Power Electronics, vol. 7, no. 3, pp. 576-590, March 2014.

[35] G. A. de Almeida Carlos, E. C. dos Santos, C. B. Jacobina and J. P. R. A. Mello, *"Dynamic Voltage Restorer Based on Three-Phase Inverters Cascaded Through an Open-End Winding Transformer,"* in IEEE Transactions on Power Electronics, vol. 31, no. 1, pp. 188-199, Jan. 2016.

[36] M. R. Khalghani, M. A. Shamsi-nejad and M. H. Khooban, *"Dynamic voltage restorer control using bi-objective optimisation to improve power quality's indices,"* in IET Science, Measurement & Technology, vol. 8, no. 4, pp. 203-213, July 2014.

[37] S. Galeshi and H. Iman-Eini, *"Dynamic voltage restorer employing multilevel cascaded H-bridge inverter,"* in IET Power Electronics, vol. 9, no. 11, pp. 2196-2204, 7 9 2016.
[38] B. Wang and G. Venkataramanan, *"Dynamic Voltage Restorer Utilizing a Matrix Converter and Flywheel Energy Storage,"* in IEEE Transactions on Industry Applications, vol. 45, no. 1, pp. 222-231, Jan.-feb. 2009.
[39] R. Zhu, F. Deng, Z. Chen and M. Liserre, *"Enhanced Control of DFIG Wind Turbine Based on Stator Flux Decay Compensation,"* in IEEE Transactions on Energy Conversion, vol. 31, no. 4, pp. 1366-1376, Dec. 2016.
[40] A. M. Massoud, S. Ahmed, P. N. Enjeti and B. W. Williams, *"Evaluation of a Multilevel Cascaded-Type Dynamic Voltage Restorer Employing Discontinuous Space Vector Modulation,"* in IEEE Transactions on Industrial Electronics, vol. 57, no. 7, pp. 2398-2410, July 2010.
[41] A. Teke, K. Bayindir and M. Tumay, *"Fast sag/swell detection method for fuzzy logic controlled dynamic voltage restorer,"* in IET Generation, Transmission & Distribution, vol. 4, no. 1, pp. 1-12, January 2010.
[42] F. Badrkhani Ajaei, S. Farhangi and R. Iravani, *"Fault Current Interruption by the Dynamic Voltage Restorer,"* in IEEE Transactions on Power Delivery, vol. 28, no. 2, pp. 903-910, April 2013.
[43] C. Wessels, F. Gebhardt and F. W. Fuchs, *"Fault Ride-Through of a DFIG Wind Turbine Using a Dynamic Voltage Restorer During Symmetrical and Asymmetrical Grid Faults,"* in IEEE Transactions on Power Electronics, vol. 26, no. 3, pp. 807-815, March 2011.
[44] Changjiang Zhan, A. Arulampalam and N. Jenkins, *"Four-wire dynamic voltage restorer based on a three-dimensional voltage space vector PWM algorithm,"* in IEEE Transactions on Power Electronics, vol. 18, no. 4, pp. 1093-1102, July 2003.
[45] J. V. Milanovic and Y. Zhang, *"Global Minimization of Financial Losses Due to Voltage Sags With FACTS Based Devices,"* in IEEE Transactions on Power Delivery, vol. 25, no. 1, pp. 298-306, Jan. 2010.
[46] Y. Zhang and J. V. Milanovic, *"Global Voltage Sag Mitigation With FACTS-Based Devices,"* in IEEE Transactions on Power Delivery, vol. 25, no. 4, pp. 2842-2850, Oct. 2010.
[47] M. Barghi Latran, A. Teke and Y. Yoldaş, *"Mitigation of power quality problems using distribution static synchronous compensator: a comprehensive review,"* in IET Power Electronics, vol. 8, no. 7, pp. 1312-1328, 7 2015.
[48] S. A. Saleh, C. R. Moloney and M. A. Rahman, *"Implementation of a Dynamic Voltage Restorer System Based on Discrete Wavelet Transforms,"* in IEEE Transactions on Power Delivery, vol. 23, no. 4, pp. 2366-2375, Oct. 2008.
[49] E. Ebrahimzadeh, S. Farhangi, H. Iman-Eini, F. Badrkhani Ajaei and R. Iravani, *"Improved Phasor Estimation Method for Dynamic Voltage Restorer Applications,"* in IEEE Transactions on Power Delivery, vol. 30, no. 3, pp. 1467-1477, June 2015.
[50] M. Shahabadini and H. Iman-Eini, *"Improving the Performance of a Cascaded H-Bridge-Based Interline Dynamic Voltage Restorer,"* in IEEE Transactions on Power Delivery, vol. 31, no. 3, pp. 1160-1167, June 2016.
[51] A. M. Rauf and V. Khadkikar, *"Integrated Photovoltaic and Dynamic Voltage Restorer System Configuration,"* in IEEE Transactions on Sustainable Energy, vol. 6, no. 2, pp. 400-410, April 2015.
[52] D. M. Vilathgamuwa, H. M. Wijekoon and S. S. Choi, *"Interline dynamic voltage restorer: a novel and economical approach for multiline power quality compensation,"* in IEEE Transactions on Industry Applications, vol. 40, no. 6, pp. 1678-1685, Nov.-Dec. 2004.
[53] S. Subramanian and M. K. Mishra, *"Interphase AC–AC Topology for Voltage Sag Supporter,"* in IEEE Transactions on Power Electronics, vol. 25, no. 2, pp. 514-518, Feb. 2010.
[54] G. A. d. A. Carlos, C. B. Jacobina and E. C. dos Santos, *"Investigation on Dynamic Voltage Restorers With Two DC Links and Series Converters for Three-Phase Four-Wire Systems,"* in IEEE Transactions on Industry Applications, vol. 52, no. 2, pp. 1608-1620, March-April 2016.

[55] B. Bae, J. Lee, J. Jeong and B. Han, *"Line-Interactive Single-Phase Dynamic Voltage Restorer With Novel Sag Detection Algorithm,"* in IEEE Transactions on Power Delivery, vol. 25, no. 4, pp. 2702-2709, Oct. 2010.

[56] D. Ramirez, S. Martinez, C. A. Platero, F. Blazquez and R. M. de Castro, *"Low-Voltage Ride-Through Capability for Wind Generators Based on Dynamic Voltage Restorers,"* in IEEE Transactions on Energy Conversion, vol. 26, no. 1, pp. 195-203, March 2011.

[57] G. Chen, M. Zhu and X. Cai, *"Medium-voltage level dynamic voltage restorer compensation strategy by positive and negative sequence extractions in multiple reference frames,"* in IET Power Electronics, vol. 7, no. 7, pp. 1747-1758, July 2014.

[58] H. K. Al-Hadidi, A. M. Gole and D. A. Jacobson, *"Minimum Power Operation of Cascade Inverter-Based Dynamic Voltage Restorer,"* in IEEE Transactions on Power Delivery, vol. 23, no. 2, pp. 889-898, April 2008.

[59] C. Fitzer, A. Arulampalam, M. Barnes and R. Zurowski, *"Mitigation of saturation in dynamic voltage restorer connection transformers,"* in IEEE Transactions on Power Electronics, vol. 17, no. 6, pp. 1058-1066, Nov. 2002.

[60] E. Babaei, M. F. Kangarlu and M. Sabahi, *"Mitigation of Voltage Disturbances Using Dynamic Voltage Restorer Based on Direct Converters,"* in IEEE Transactions on Power Delivery, vol. 25, no. 4, pp. 2676-2683, Oct. 2010.

[61] J. V. Milanovic and Y. Zhang, *"Modeling of FACTS Devices for Voltage Sag Mitigation Studies in Large Power Systems,"* in IEEE Transactions on Power Delivery, vol. 25, no. 4, pp. 3044-3052, Oct. 2010.

[62] F. Jiang, C. Tu, Z. Shuai, M. Cheng, Z. Lan and F. Xiao, *"Multilevel Cascaded-Type Dynamic Voltage Restorer With Fault Current-Limiting Function,"* in IEEE Transactions on Power Delivery, vol. 31, no. 3, pp. 1261-1269, June 2016.

[63] Poh Chiang Loh, D. M. Vilathgamuwa, Seng Khai Tang and H. L. Long, *"Multilevel dynamic voltage restorer,"* in IEEE Power Electronics Letters, vol. 2, no. 4, pp. 125-130, Dec. 2004.

[64] J. D. Barros and J. F. Silva, *"Multilevel Optimal Predictive Dynamic Voltage Restorer,"* in IEEE Transactions on Industrial Electronics, vol. 57, no. 8, pp. 2747-2760, Aug. 2010.

[65] F. Jurado, *"Neural network control for dynamic voltage restorer,"* in IEEE Transactions on Industrial Electronics, vol. 51, no. 3, pp. 727-729, June 2004.

[66] B. Wang and M. Illindala, *"Operation and control of a dynamic voltage restorer using transformer coupled H-bridge converters,"* in IEEE Transactions on Power Electronics, vol. 21, no. 4, pp. 1053-1061, July 2006.

[67] S. Biricik and H. Komurcugil, *"Optimized Sliding Mode Control to Maximize Existence Region for Single-Phase Dynamic Voltage Restorers,"* in IEEE Transactions on Industrial Informatics, vol. 12, no. 4, pp. 1486-1497, Aug. 2016.

[68] R. Gupta, A. Ghosh and A. Joshi, *"Performance Comparison of VSC-Based Shunt and Series Compensators Used for Load Voltage Control in Distribution Systems,"* in IEEE Transactions on Power Delivery, vol. 26, no. 1, pp. 268-278, Jan. 2011.

[69] M. Vilathgamuwa, A. A. D. Ranjith Perera and S. S. Choi, *"Performance improvement of the dynamic voltage restorer with closed-loop load voltage and current-mode control,"* in IEEE Transactions on Power Electronics, vol. 17, no. 5, pp. 824-834, Sept. 2002.

[70] P. M. Garcia-Vite, F. Mancilla-David and J. M. Ramirez, *"Per-Sequence Vector-Switching Matrix Converter Modules for Voltage Regulation,"* in IEEE Transactions on Industrial Electronics, vol. 60, no. 12, pp. 5411-5421, Dec. 2013.

[71] B. Bae, J. Jeong, J. Lee and B. Han, *"Novel Sag Detection Method for Line-Interactive Dynamic Voltage Restorer,"* in IEEE Transactions on Power Delivery, vol. 25, no. 2, pp. 1210-1211, April 2010.

[72] C. Kumar and M. K. Mishra, *"Predictive Voltage Control of Transformerless Dynamic Voltage Restorer,"* in IEEE Transactions on Industrial Electronics, vol. 62, no. 5, pp. 2693-2697, May 2015.

[73] J. Shi et al., *"SMES Based Dynamic Voltage Restorer for Voltage Fluctuations Compensation,"* in IEEE Transactions on Applied Superconductivity, vol. 20, no. 3, pp. 1360-1364, June 2010.

[74] V. K. Ramachandaramurthy, A. Arulampalam, C. Fitzer, C. Zhan, M. Barnes and N. Jenkins, *"Supervisory control of dynamic voltage restorers,"* in IEE Proceedings - Generation, Transmission and Distribution, vol. 151, no. 4, pp. 509-516, 11 July 2004.

[75] A. Y. Goharrizi, S. H. Hosseini, M. Sabahi and G. B. Gharehpetian, *"Three-Phase HFL-DVR With Independently Controlled Phases,"* in IEEE Transactions on Power Electronics, vol. 27, no. 4, pp. 1706-1718, April 2012.

[76] S. Gao, X. Lin, S. Ye, H. Lei and Y. Kang, *"Transformer inrush mitigation for dynamic voltage restorer using direct flux linkage control,"* in IET Power Electronics, vol. 8, no. 11, pp. 2281-2289, 11 2015.

[77] B. H. Li, S. S. Choi and D. M. Vilathgamuwa, *"Transformerless dynamic voltage restorer,"* in IEE Proceedings - Generation, Transmission and Distribution, vol. 149, no. 3, pp. 263-273, May 2002.

[78] C. -. Zhan et al., *"Two electrical models of the lead-acid battery used in a dynamic voltage restorer,"* in IEE Proceedings - Generation, Transmission and Distribution, vol. 150, no. 2, pp. 175-182, March 2003.

[79] J. C. Rosas-Caro, F. Mancilla-David, J. M. Ramirez-Arredondo and A. M. Bakir, *"Two-switch three-phase ac-link dynamic voltage restorer,"* in IET Power Electronics, vol. 5, no. 9, pp. 1754-1763, November 2012.

[80] S. Alaraifi, A. Moawwad, M. S. El Moursi and V. Khadkikar, *"Voltage Booster Schemes for Fault Ride-Through Enhancement of Variable Speed Wind Turbines,"* in IEEE Transactions on Sustainable Energy, vol. 4, no. 4, pp. 1071-1081, Oct. 2013.

[81] E. Babaei and M. F. Kangarlu, *"Voltage quality improvement by a dynamic voltage restorer based on a direct three-phase converter with fictitious DC link,"* in IET Generation, Transmission & Distribution, vol. 5, no. 8, pp. 814-823, August 2011.

[82] D. M. Vilathgamuwa, A. A. D. R. Perera and S. S. Choi, *"Voltage sag compensation with energy optimized dynamic voltage restorer,"* in IEEE Transactions on Power Delivery, vol. 18, no. 3, pp. 928-936, July 2003.

[83] *"Voltage Sag Correction by Dynamic Voltage Restorer with Minimum Power Injection,"* in IEEE Power Engineering Review, vol. 21, no. 5, pp. 56-58, May 2001.

[84] C. Fitzer, M. Barnes and P. Green, *"Voltage sag detection technique for a dynamic voltage restorer,"* in IEEE Transactions on Industry Applications, vol. 40, no. 1, pp. 203-212, Jan.-Feb. 2004.

[85] C. Lam, M. Wong and Y. Han, *"Voltage Swell and Overvoltage Compensation With Unidirectional Power Flow Controlled Dynamic Voltage Restorer,"* in IEEE Transactions on Power Delivery, vol. 23, no. 4, pp. 2513-2521, Oct. 2008.

[86] A. Prasai and D. M. Divan, *"Zero-Energy Sag Correctors—Optimizing Dynamic Voltage Restorers for Industrial Applications,"* in IEEE Transactions on Industry Applications, vol. 44, no. 6, pp. 1777-1784, Nov.-dec. 2008.

[87] P. Kanjiya, B. Singh, A. Chandra and Kamal-Al-Haddad, *"“SRF theory revisited” to control self supported dynamic voltage restorer (DVR) for unbalanced and nonlinear loads,"* 2011 IEEE Industry Applications Society Annual Meeting, Orlando, FL, 2011, pp. 1-8.

[88] S. Jothibasu and M. K. Mishra, *"A Control Scheme for Storageless DVR Based on Characterization of Voltage Sags,"* in IEEE Transactions on Power Delivery, vol. 29, no. 5, pp. 2261-2269, Oct. 2014.

[89] J. G. Nielsen and F. Blaabjerg, *"A detailed comparison of system topologies for dynamic voltage restorers,"* in IEEE Transactions on Industry Applications, vol. 41, no. 5, pp. 1272-1280, Sept.-Oct. 2005.

[90] M. J. Newman, D. G. Holmes, J. G. Nielsen and F. Blaabjerg, *"A dynamic voltage restorer (DVR) with selective harmonic compensation at medium voltage level,"* in IEEE Transactions on Industry Applications, vol. 41, no. 6, pp. 1744-1753, Nov.-Dec. 2005.

[91] T. Jimichi, H. Fujita and H. Akagi, *"A Dynamic Voltage Restorer Equipped With a High-Frequency Isolated DC–DC Converter,"* in IEEE Transactions on Industry Applications, vol. 47, no. 1, pp. 169-175, Jan.-Feb. 2011.

[92] F. B. Ajaei, S. Afsharnia, A. Kahrobaeian and S. Farhangi, *"A Fast and Effective Control Scheme for the Dynamic Voltage Restorer,"* in IEEE Transactions on Power Delivery, vol. 26, no. 4, pp. 2398-2406, Oct. 2011.
[93] A. M. Gee, F. Robinson and W. Yuan, *"A Superconducting Magnetic Energy Storage-Emulator/Battery Supported Dynamic Voltage Restorer,"* in IEEE Transactions on Energy Conversion, vol. 32, no. 1, pp. 55-64, March 2017.
[94] A. P. Torres, P. Roncero-Sánchez and V. F. Batlle, *"A Two Degrees of Freedom Resonant Control Scheme for Voltage-Sag Compensation in Dynamic Voltage Restorers,"* in IEEE Transactions on Power Electronics, vol. 33, no. 6, pp. 4852-4867, June 2018.
[95] Design Approach of Grid Coupled Solar Inverter. Bandla Pavan Babu, V. Indragandhi, R. RajaSingh, Ramani Kannan. MATEC Web Conf. 225 06001 (2018).doi:10.1051/matecconf/201822506001

Inter turn Fault Analysis on Wound Rotor Induction Machine

D. Nabanita, S Sreedhar, Z. Neha, B. Onkar, P. Pratik, *R. Raja Singh
Advanced Drives Laboratory, Department of Energy and Power Electronics,
Vellore Institute of Technology, Vellore, Tamil Nadu, India

E-mail: *rrajasingh@gmail.com

Abstract. Prevention of industrial machineries against internal and external faults is mandatory for uninterrupted operation. Identifying the exact faults or anomalies in the system and predicting the machine's health continuously is an essential preventive measure. To realize the protection system, pre-analysis of various faults under different load conditions is mandatory and hence, paper analysis a critical fault "Inter turn fault" in the wound rotor induction machine (WRIM). Inter turn fault is developed experimentally in 2.2 kW WRIM with 20 percent of internal windings shorted under different load conditions (0 to 1 per unit load). Obtained results are analysed for various electrical and mechanical parameters, and the faulty effects are revealed.

Index Terms: Wound Rotor Induction motor, inter turn fault, power analyser

1. Introduction

Induction motor is a vital device for most of the industries and comprises magnetic circuit interlink between two circuits which are important parts of induction machine i.e. stator and rotor [1]. The stationary part is stator and rotational part is rotor. Through electromagnetic induction power is transferred. Preventive maintenance of electrical motors for condition monitoring of abnormal electrical and mechanical condition of the motor has been done for anomaly detection and protect the motor from failure [2]. Continuous monitoring of electrical parts of the motor helps to reduce the maintenance cost, save the energy, make the system more energy efficient. Online monitoring of induction motor helps to save the machine from major failure which is expensive for recovering. Condition monitoring technology can detect incipient fault [3]. There are various faults can be occurred in induction motor for various reasons like stator fault, rotor fault, winding faults, rotor inter turn faults, stator inter turn faults, broken rotor bar, unbalanced stator and rotor windings etc. Several fault detection methods have been introduced to identify these faults like vibration monitoring, current monitoring, temperature monitoring, chemical monitoring, acoustic emission monitoring etc. All these monitoring technique needs specialized tools, various sensors, data acquisition system. Verities of sensors can be used to analysis the critical damage of the system. These sensors measure stator voltage and currents, magnetic flux densities, output torque, speed, external temperature, rotor position and speed, internal and external temperature, case vibrations etc. A variety of motor failures can be monitored by system. The failures include opens, bearing failure, conductor shorts etc. It is seen that in a consisting manner failure monitoring is possible and different sensors can physically measure the failures.

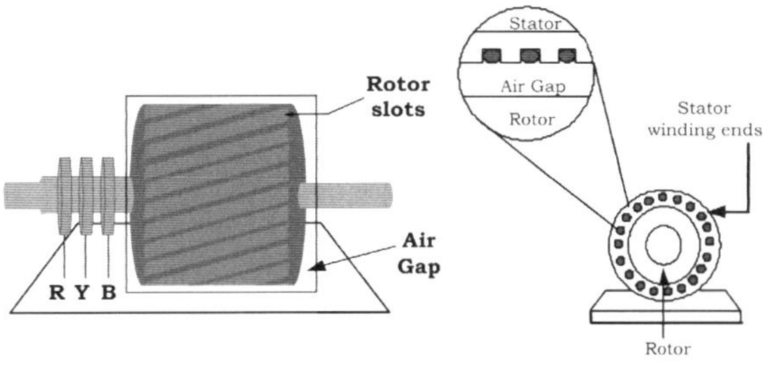

Figure 1 Schematic diagram of Induction motor

In Figure 1, the schematic diagram of wound rotor induction motor where three phases R, Y, B are shown and rotor slots are highlighted. Separately front view of stator and rotor are shown in the figure. With the help of wound rotor induction motor the experiment has been done here (Figure 1).

This paper is present the stator fault detection of WRIM and specially inter turn fault has been described here which is vital problem in industrial sector. As the main aim of the industry is to reduce operational and maintenance cost so condition monitoring of the machine is highly important to them. The failure of whole machine leads to cost of the additional equipment and maintenance cost for every time is also sometimes leads to loss for industry. So continuous monitoring of the winding is very much essential. Once in a six month the testing of winding, lubrication, insulation, resistance checking is needed. Inter turn fault causes degradation of winding, stator rotor failure etc. So overall stator winding fault has been described in this article.

2. Faults in induction motor

Among various faults Induction motor is highly affected by electrical faults and mechanical fault and mainly bearing faults. The induction motor faults are recognized as bearing fault (69%), stator winding fault (21%), rotor bar (7%), shaft/ coupling fault on the basis of IEEE standard 3004.8 the statistics have been shown in the figure. There are mainly two types of faults one is electrical fault another is mechanical fault. Under electrical fault stator, rotor fault and under mechanical fault eccentricity, bearing and broken rotor bar are seen. Please follow these instructions as carefully as possible so all articles within a conference have the same style to the title page. This paragraph follows a section title so it should not be indented.

2.1. Mechanical Fault

Mechanical faults are such types of faults which are generally occurred in internal housing of three phase induction motor. This fault is subdivided into various categories.

- *Bearing Fault*

Bearing failure is progressive but ultimately machine has a catastrophic effect for this fault. Improper forcing of bearing shaft or housing causes premature failure and physical damage of the system [5]-[10].

- *Eccentricity Fault*

Unbalanced air gap between rotor and stator of induction motor causes eccentricity and this leads for incorrect position of stator and rotor [13].

- *Broken Rotor Bar*

Squirrel cage induction motor belongs to rotor bars and shorted end ring. Any time bars will damage or partially also cracked then this broken rotor bar has been seen [11].

2.2. Electrical Fault

In induction motor very crucial fault is electrical fault which can damage the whole system within very short time period. This research is based on mainly on electrical fault detection and focuses more on inter turn fault. Induction motor cannot be break down suddenly but due to overheating or over running causes severe damage. This electrical fault has four categories.

- *Single Phasing Fault (External Fault)*

General working of three phase induction motor three phases voltage supply is must, if in case one of the phases is damaged only two phases will be available which will lead to burned or causes excessive heating of the motor.

- *Reverse Phase Sequencing Fault (External Fault)*

Reverse phase sequencing fault is occurred when any one phase of three phase voltage supply is reversed which leads to change in the direction of the motor.

- *Under Voltage and Over Voltage (External Fault)*

Due to load variations the over and under voltage phenomenon occur in the induction motor. When the circuit is overloaded this causes over voltages and under voltage fault causes due to reduced voltage supply. The rated mechanical load supplies reduced voltage and that moment under voltage problem happen [17]. Under voltage causes increased current and excess heating of machine and increased stator rotor losses. Load switching, capacitor bank switching off these are the examples of under voltage causes.

- *Unbalanced supply voltage (External Fault)*

Normal balanced operating condition positive sequence voltage is developed. Negative sequence voltage is developed when supply voltage is unbalanced. Motor behaves like a super position of two separate motors due to neglecting of non-linearity and saturation and when one is running at slip s with terminal voltage Vp and other slip is (2-s) with terminal voltage Vn [19].

- *Internal Faults*

Inter turn fault is most common electrical fault in induction motor. It is recognized as inter turn short circuit fault in stator or rotor side. The phase to phase and phase to ground fault occurs due to increase heat and short circuit condition which leads to stator circuit breakdown. Several different phenomena can be seen for short circuit faults. Mechanical stress also a reason behind short circuit fault. This fault creates a big impact in stator fault. Due to moisture flow of current from scratch to scratch causes hot spot and thereby insulation also will destroy. When stator is supplied by a PWM voltage source partial discharge can happen to destroy the insulation which causes short circuit fault. Moisture causes flow of current which leads to make hot spot and destroy the insulation. Between turns high alternating voltage causes partial discharge when PWM voltage source is supplied by stator. Various faults in induction motor and its effects has been described (Table 1).

Table 1: Various faults of Induction motor and its cause and effects

Name of the faults		Causes of faults	Percentage of faults	Effects of faults	Ref
Mechanical Faults					
	a) Bearing Fault	The external causes of bearing faults are contamination, corrosion, improper lubrication, improper installation or brine-ling. Defective bearing installation	40-50%	The rolling elements of the machine may generate vibrations and noise. Continuous break causes local fragments of the materials and causes break loss. These excessive vibrations cause physical damage of the machine.	[1], 2005
	b) Eccentricity	Static air gap between stator and rotor causes eccentricity.	10-20%	Serious damage of stator core and winding for which dynamic electricity can damage.	[5], 2016
	c) Broken rotor bar	Manufacturing defect, hotspot, thermal unbalance, mechanical stress leads to create bearing fault and metal fatigue. This is mainly happening in squirrel cage induction motor.	7-10%	Rotor bar can break and fundamental frequency increases	[8], 2011
Electrical Faults					
External Faults	**a) Single phasing faults**	In the motor if winding is open	33%	Stator can induce more current causes winding failure, for creation of vibration rotor winding can fail.	[12], 2012
	b) Reverse Phase sequence fault	In three phases when two phases of supply line reverses.	23%	Major winding failure	[8], 2011
	c) Under voltage and Over voltage	Load variation or fluctuation in the system	10-15%	Harmful effects on machine insulation	[20], 2014
	d) Unbalance supply voltage	Unbalance loading and supply voltage, unequal tap settings, unbalance primary voltage, High resistance connections	21-25%	Motor efficiency will reduce and winding will fail prematurely.	[23], 2010
Internal Faults	**Inter turn fault (stator and rotor inter turn fault)**	Unbalance supply voltage, fluctuation of voltage within a very short time repeatedly.	33-35%	High current will flow within the short circuit area causes total winding failure.	[15] [16], 2010, 2010

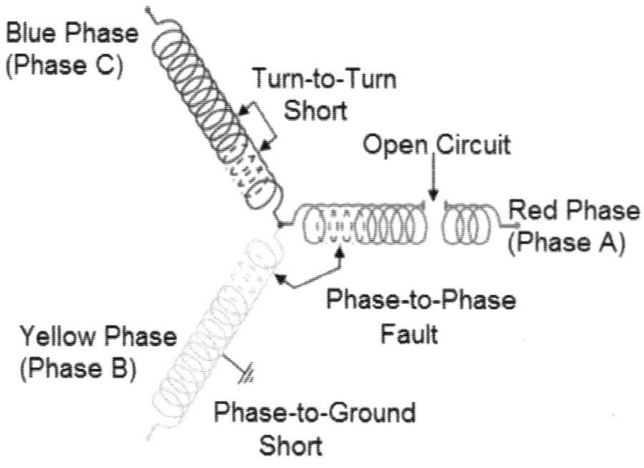

Figure 2: Different types of Stator Winding faults

An induction motor various stator winding faults are shown in Figure2.

Stator and rotor equation in the induction machine is described [6] The voltage, current, flux of stator and rotor has been expressed [1-6]

$$V_S = R_S I_S + P\lambda_S \tag{1}$$

$$0 = R_r I_r + p\lambda_S \tag{2}$$

Where

$$V_S = [V_{as1} \quad V_{as2} \quad V_{bs} \quad V_{cs}]^T \tag{3}$$

$$I_S = [I_{as} \quad (I_{as} - I_f) \quad I_{bs} \quad I_{cs}]^T \tag{4}$$

$$I_r = [I_{ar} \quad I_{br} \quad I_{cr}]^T \tag{5}$$

$$\lambda_S = [\lambda_{as1} \quad \lambda_{as2} \quad \lambda_{bs} \quad \lambda_{cs}]^T \tag{6}$$

Voltage is shown as V, current is I, λ is recognized as flux here, and stator, rotor are represented as s and r respectively, three phases are denoted as a,b,c as_1, as_2 are denoted as unfaulty and faulty part of stator.

$$V_{as2} = turn\ R_s(I_{as} - I_f + \rho\,\lambda_{as2}) = R_f I_f \tag{7}$$

$$\lambda_s = L_s I_s + L_{sr} I_r \tag{8}$$

$$\lambda_r = [L_{sr}]^T I_s + L_r I_r \tag{9}$$

These equations show the shorted part of stator winding voltage.

The resistance matrix is shown as

$$R_s = R_s\, diag[(1 - turn) \quad turn1 \quad 1] \tag{10}$$

$$R_r = R_r\, [I]_{3\times3} \tag{11}$$

The stator winding's self and mutual inductance has been shown here [12-14].

$$+\begin{bmatrix} (1-turn)^2 & turn*(1-turn) & -\frac{(1-turn)}{2} & -\frac{(1-turn)}{2} \\ turn*(1-turn) & (turn)^2 & -\frac{turn}{2} & -\frac{turn}{2} \\ -\frac{(1-turn)}{2} & -\frac{turn}{2} & 1 & -\frac{1}{2} \\ -\frac{(1-turn)}{2} & -\frac{turn}{2} & -\frac{1}{2} & 1 \end{bmatrix} \tag{12}$$

IOP Conf. Series: Materials Science and Engineering **623** (2019) 012016 doi:10.1088/1757-899X/623/1/012016

$$L_{sr} = L_{ms} \begin{matrix} (1-turn)*\cos(\theta_r) & (1-turn)*\cos(\theta_r+{}^{2\pi}/_{3}) & (1-turn)*\cos(\theta_r-{}^{2\pi}/_{3}) \\ turn*\cos(\theta_r) & turn*\cos(\theta_r+{}^{2\pi}/_{3}) & turn*\cos(\theta_r-{}^{2\pi}/_{3}) \\ \cos(\theta_r-{}^{2\pi}/_{3}) & \cos(\theta_r) & \cos(\theta_r+{}^{2\pi}/_{3}) \\ \cos(\theta_r+{}^{2\pi}/_{3}) & \cos(\theta_r-{}^{2\pi}/_{3}) & \cos(\theta_r) \end{matrix} \quad (13)$$

$$L_r = \begin{bmatrix} L_{lr}+L_{RM} & -\frac{L_{RM}}{2} & -\frac{L_{RM}}{2} \\ -\frac{L_{RM}}{2} & L_{lr}+L_{RM} & -\frac{L_{RM}}{2} \\ -\frac{L_{RM}}{2} & -\frac{L_{RM}}{2} & L_{lr}+L_{RM} \end{bmatrix} \quad (14)$$

Turn is denoted as pu for phase a, θ_r represented as rotor position, self-inductance is shown as L_S rotor self-inductance is shown as L_r Stator to rotor mutual inductance is represented as L_{Sr}.

3. Experimental procedure

The experiment is carried out with the help of wound rotor induction motor whose rotor windings are in star connection. The purpose of this research is to find out intern turn short circuit fault's effect by creating intern turn fault in the machine. The interesting current signature can be shown by stator current and harmonics are shown by rotor current. An experimental setup has been formed for three phase wound rotor machines with 5hp, 220V, 50Hz, 4.7A (stator side), 7.5 A (rotor side), 1500 RPM. In WRIM both the stator and rotor are accessible. The DC generator is coupled with WRIM for excitation purpose. One protection circuit has been connected through current sensor with the WRIM R phase to protect the circuit when fault has been created. One circuit breaker has been connected as a protection purpose to trip the circuit when it exceeds rated current of the machine. Here the process has been limited in up to 6 amp as the resistance value is higher. If 6-amp current exceeds then winding can be burnt out. But if resistance value will less current can be increased.

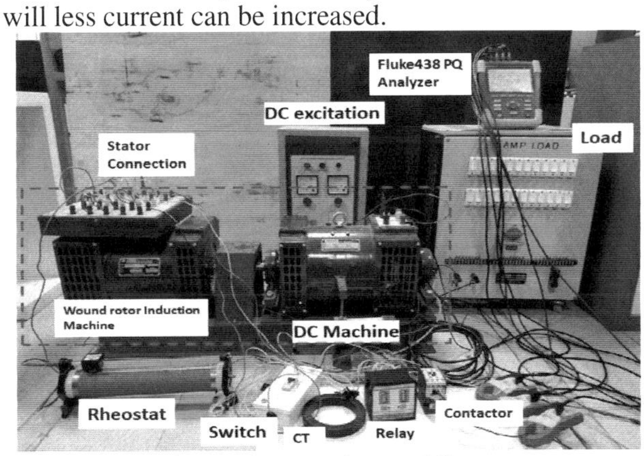

Figure. 3. Experimental Setup

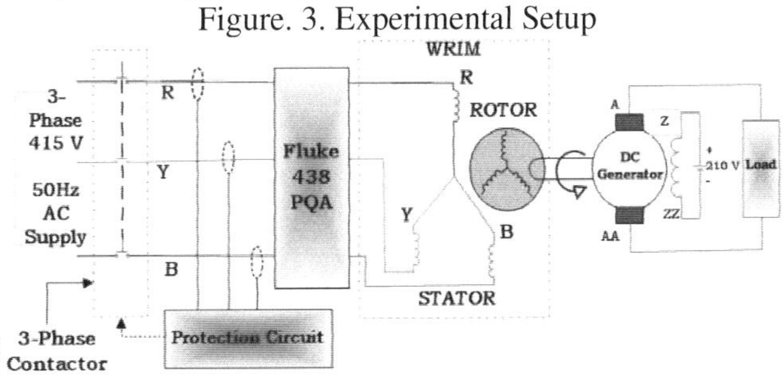

Figure. 4a. Circuit diagram of the experiment

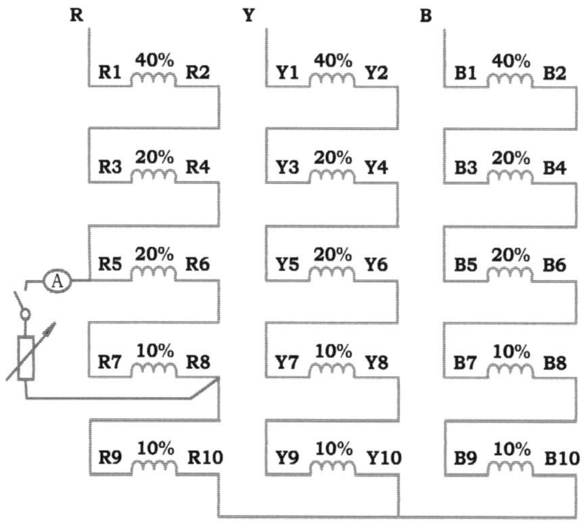

Figure .4b. Faulty winding connection

The WRIM machine three phase supply is not directly connected to the supply as in the experiment voltage is varied slowly. So, in this experiment the three phases of the induction machine stator side are connected to the supply through auto transformer. The Fluke 438 PQA is connected in between auto transformer and three phase supply of the induction machine. One protection circuit through current sensor is connected with WRIM for protection purpose. The circuit is made with relay coil, single phase switch and rheostat. One DC excitation is connected with the rotor of the induction machine. The rotor is coupled with the stator by shaft. A lamp load is connected with rotor armature. First experiment has been done with the no fault condition from 0 to full load with the 12 loads approximately. Then with same loads R6 to R7 sorted (Figure4a, Figure4b). So, 20% of the inter turn fault created to check the experiment. Now after 6 A the circuit breaker isolates the circuit as it exceeds the rated current. In the result maximum current, power is higher in the faulty condition rather than no faulty condition. Figure 4 shows the circuit diagram of the experiment and Figure3 shows the experimental hardware setup. Experimental results are shown in Figure 5.

4. Experimental results

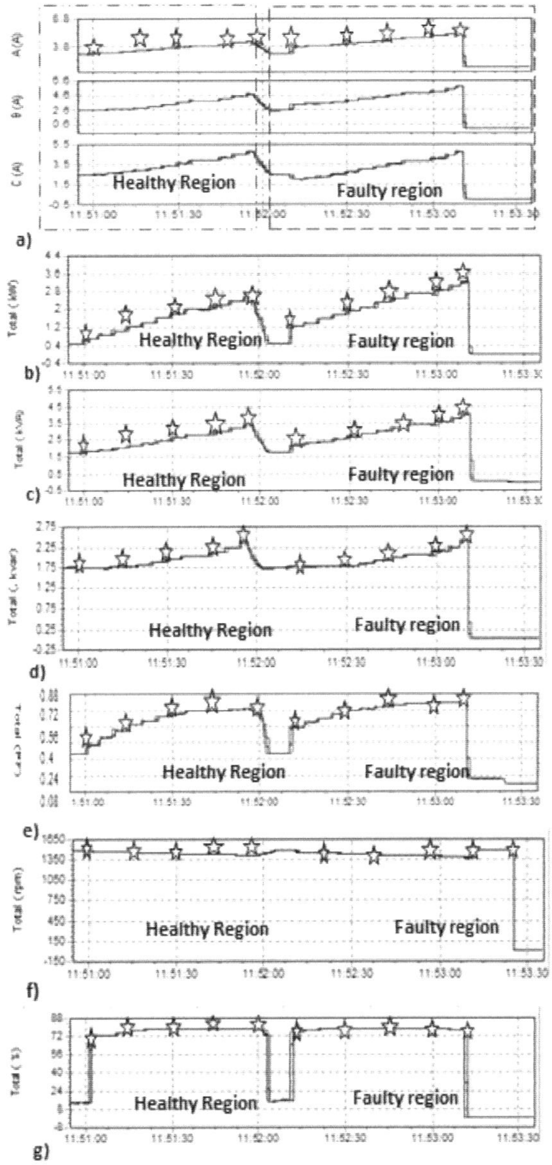

Figure 5. Experimental results

a) For 0 load healthy to faulty condition current change is 17.03%. For 0.25 load healthy to faulty region current change is 47.33%. For 0.5 load healthy to faulty region change of current is 58.77%. For 0.75 load healthy to faulty region current change is 60.68%. For 1 load healthy to faulty region current change is 61.98%.

b) For 0 load healthy to faulty condition power change is 47.03%. For 0.25 load healthy to faulty region power change is 63.15%. For 0.5 load healthy to faulty region power change is 63.15%. For 0.75 load healthy to faultyregion power change is 66.07%. For 1 load healthy to faulty region power change is 68.25%.

c) For 0 load healthy to faulty condition apparent power change is 8.3%. For 0.25 load healthy to faulty condition apparent power change is 28.03%. For 0.5 load healthy to faulty region apparent power change

is 35.70%. For 0.75 load healthy to faulty condition apparent power change is 46.34%. For 1 load healthy to faulty condition apparent power change is 48.31%.
d) For 0 load healthy to faulty condition reactive power change is 1.72%. For 0.25 load healthy to faulty condition reactive power change is 1.8%. For 0.5 load healthy to faulty region reactive power change is 2.76%. For 0.75 load healthy to faulty region reactive power change is 12.34%. For 1 load healthy to faulty region reactive power change is 14.38%.
e) For 0 load healthy to faulty region power factor change is 31.03%. For 0.25 load healthy to faulty region power factor change is 33.68%. For 0.5 load healthy to faulty region power factor change is 36.98%. For 0.75 load healthy to faulty region power factor change is 40.03%. For 1 load healthy to faulty region power factor change is 41.35%.
f) For 0 load healthy to faulty region speed change is 1.5%. For 0.25 load healthy to faulty region speed change is 2.67%. For 0.5 load healthy to faulty region speed change is 2.7%. For 0.75 load healthy to faulty region speed change is 2.7%. For 1 load healthy to faulty region speed change is 6.74%.
g) For 0 load efficiency change of healthy to faulty region is 13.80%. For 0.25 load efficiency change of healthy to faulty region is 13.96%. For 0.5 load efficiency change in healthy to faulty region is 15.69%. For 0.75 load efficiency change of healthy to faulty region is 14.01%. For 1 load healthy to faulty region efficiency change is 14.38%. Table 2 describes the overall analysis of the results.

Table 2: Analysis of the results

Parameters	**Load (p.u)**				
	0	**0.25**	**0.5**	**0.75**	**1**
Current (A)	17.03% ↑	47.33% ↑	58.77%↑	60.68% ↑	61.98% ↑
Power (kw)	47.03% ↑	63.15% ↑	48.21%↓	66.07% ↑	68.25% ↑
Apparent power (kva)	8.3% ↑	28.03% ↑	35.7%↑	46.34% ↓	48.31% ↑
Reactive power (kvar)	1.72% ↑	1.8% ↑	2.76%↑	12.34% ↓	14.38% ↑
Power factor	31.03% ↑	33.68% ↓	36.98%↑	40.03% ↓	41.35% ↑
Speed (rpm)	1.5% ↓	2.67% ↓	2.84%↓	2.7% ↓	6.74% ↓
Efficiency (%)	13.8% ↓	13.96% ↓	15.69%↓	14.01% ↓	14.38% ↓
	↑ Increase in percentage ↓ decrease in percentage				

5. Conclusion

The paper is on inter turn fault in WRIM. Different experimental techniques can be implemented for stator and rotor short circuit fault and stator rotor voltage, current, flux can be shown as different signals here. Here machine is used as motor and it can be used as generator also. Stator can provide interesting power signature module and rotor current can show new harmonics also. The use of current sensor in the rotor side detect the electrical fault very prominently in the system. It is seen from the curves that when fault occurs current is higher than the normal condition and it affects the other two phases where fault not created also. Total power, apparent power, reactive power, power factor is higher than the normal condition in faulty condition.

References

[1] Siddique, A., Yadava, G. S., & Singh, B. (2005). A review of stator fault monitoring techniques of induction motors. IEEE transactions on energy conversion, 20(1), 106-114.
[2] Gritli, Y., Zarri, L., Rossi, C., Filippetti, F., Capolino, G. A., & Casadei, D. (2012). Advanced diagnosis of electrical faults in wound-rotor induction machines, IEEE Transactions on Industrial Electronics, 60(9), 4012-4024.
[3] Sarma, N., Li, Q., Djurovic, S., Smith, A. C., & Rowland, S. M. (2016). Analysis of a wound rotor induction machine low frequency vibroacoustic emissions under stator winding fault conditions.

[4] Harrou, F., Ramahaleomiarantsoa, J. F., Nounou, M. N., & Nounou, H. N. (2016). A data-based technique for monitoring of wound rotor induction machines: A simulation study. Engineering science and technology, an international journal, 19(3), 1424-1435.

[5] Harrou, F., Ramahaleomiarantsoa, J. F., Nounou, M. N., & Nounou, H. N. (2016). A data-based technique for monitoring of wound rotor induction machines: A simulation study. Engineering science and technology, an international journal, 19(3), 1424-1435.

[6] Natarajan, R., Kohler, J. L., & Sottile, J. (1988). Condition monitoring of slip-ring induction motors, Electric Power Systems Research, 15(3), 189-195.

[7] Agrawal, D., Yadav, N., & Saini, S. (2015). Condition monitoring of slip-ring induction motor, International Journal of Innovation Research in Advanced Engineering, 2(3), 78-84.

[8] Ceban, A., Pusca, R., Romary, R., & Lecointe, J. P. (2011). Diagnosis of inter-turn short circuit fault in induction machine. Annals of the University of Craiova, Electrical Engineering series, 35, 103-110.

[9] Yektaniroumand, T., Niaz Azari, M., & Gholami, M. (2018). Optimal Rotor Fault Detection in Induction Motor Using Particle-Swarm Optimization Optimized Neural Network. International Journal of Engineering, 31(11), 1876-1882.Mini, V. P., & Ushakumari, S. (2014). Rotor fault detection and diagnosis of induction motor using fuzzy logic. AMSE JOURNALS, 87(2), 19-40.

[10] Alamyal, M., Gadoue, S. M., & Zahawi, B. (2013, August). Detection of induction machine winding faults using genetic algorithm. In 2013 9th IEEE International Symposium on Diagnostics for Electric Machines, Power Electronics and Drives (SDEMPED) (pp. 157-161). IEEE.

[11] Alwodai, A., Gu, F., & Ball, A. D. (2012). A comparison of different techniques for induction motor rotor fault diagnosis. In Journal of Physics: Conference Series (Vol. 364, No. 1, p. 012066). IOP Publishing.

[12] Baradieh, K., Al-Hamouz, Z., & Abido, M. (2018). ANN Based Broken Rotor Bar Fault Detection in LSPMS Motors. J Electr Electron Syst, 7(273), 2332-0796.

[13] Roshanfekr, R., & Jalilian, A. (2015). Analysis of rotor and stator winding inter-turn faults in WRIM using simulated MEC model and experimental results. Electric Power Systems Research, 119, 418-424.

[14] Yazidi, A., Henao, H., Capolino, G. A., Betin, F., & Capocchi, L. (2010, July). Experimental inter-turn short circuit fault characterization of wound rotor induction machines. In 2010 IEEE International Symposium on Industrial Electronics (pp. 2615-2620). IEEE.

[15] Yazidi, A., Henao, H., Capolino, G. A., Betin, F., & Capocchi, L. (2010, September). Inter-turn short circuit fault detection of wound rotor induction machines using bispectral analysis. In 2010 IEEE Energy Conversion Congress and Exposition (pp. 1760-1765). IEEE.

[16] Siddique, A., Yadava, G. S., & Singh, B. (2005). A review of stator fault monitoring techniquesof induction motors. IEEE transactions on energy conversion, 20(1), 106-114.

[17] Lin, Y., Tu, L., Liu, H., & Li, W. (2016). Fault analysis of wind turbines in China. Renewableand Sustainable Energy Reviews, 55, 482-490.

[18] Mellit, A., Tina, G. M., & Kalogirou, S. A. (2018). Fault detection and diagnosis methods for photovoltaic systems: A review. Renewable and Sustainable Energy Reviews, 91, 1-17.

[19] Djurović, S., Vilchis-Rodriguez, D. S., & Smith, A. C. (2014). Investigation of wound rotor induction machine vibration signal under stator electrical fault conditions. The Journal of Engineering, 2014(5), 248-258.

[20] Razafimahefa, D. T., Randrianarisoa, E., Sambatra, E. J. R., & Heraud, N. (2014, October). Modeling and faults detection of small power wound rotor induction machine. In 2014 International Conference and Exposition on Electrical and Power Engineering (EPE) (pp. 311-316). IEEE.

[21] Gritli, Y., Zarri, L., Mengoni, M., Rossi, C., Filippetti, F., & Casadei, D. (2013, September). Rotor fault diagnosis of wound rotor induction machine for wind energy conversion system under time - varying conditions based on optimized wavelet transform analysis. In 2013 15th European Conference on Power Electronics and Applications (EPE) (pp. 1-9). IEEE.

[22] Yazidi, A., Henao, H., Capolino, G. A., & Betin, F. (2010, September). Rotor inter-turn short circuit fault detection in wound rotor induction machines, XIX International Conference on Electrical Machines-ICEM 2010 (pp. 1-6). IEEE.

Authors profile

Srisailam Sreedhar obtained a Bachelor's degree in Electrical and Electronics Engineering from Amrita University, India in 2015 and Master's degree in Electrical Power Systems from Sree Vaidyanathan Engineering College, Andhra Pradesh, India in 2017. He is currently pursuing Ph.D. at Vellore Institute of Technology, India. He is a member of IEEE and also an Associate Member of Institute of Engineers, India (IEI). His area of interest includes Advanced Drives, Lightning protection, Renewable energy integration to the grid, Power convertor-based applications.

Nabanita Dutta received her bachelor degree from Dr Sudhir Chandra sur degree engineering college under West Bengal university of technology in 2015. She completed her M. Tech in Power Electronics and drives from Vellore Institute of Technology (2016-2018). Now she is perusing her PHD degree in Vellore Institute of Technology on fault detection of pumping system by artificial intelligence. Her area of interest is Machine learning, Artificial Intelligence, IoT, automation technology, renewable Energy, Advance Drives.

Neha Zade obtained a Bachelor's degree in Electrical Engineering from R.T.M.N University, India in 2017. She is currently pursuing Master's degree at Vellore Institute of Technology, India. Her area of interest includes Advanced Drives, Renewable energy, Power converter.

Onkar Bhalerao obtained a Bachelor's degree in Electrical Engineering from Pune University, India in 2018 He is currently pursuing Master's degree at Vellore Institute of Technology, India. His area of interest includes Advanced Drives, Renewable energy, Power converter.

Pratik Potdar obtained a Bachelor's degree in Electrical Engineering from Shivaji University, India in 2017 and pursuing Master's degree in Power Electronics and Drives at Vellore Institute of Technology, India. His area of interest Power converters, Electrical drives, Renewable Energy.

IOP Conf. Series: Materials Science and Engineering **623** (2019) 012017 doi:10.1088/1757-899X/623/1/012017

Recognition of fault and security of three phase induction motor by means of programmable logic controller

M.Pradeep[1], S.Sathishkumar[2],C.John de britto[3],A.T.Sankara Subramanian[4]

[1*]Assistant Professor, Rajalakshmi Engineering College, Chennai, Tamilnadu, India.
[2]Assistant Professor, Rajalakshmi Institute of Technology, Chennai, Tamilnadu, India. [3]Assistant Professor, Jeppiaar Engineering College, Chennai, Tamilnadu, India.
[4]Assistant Professor,K.Ramakrishnan College of Technology, Trichy, Tamilnadu, India.
Corresponding Email Id: pradeep7raj@gmail.com

Abstract. This paper presents a fault protection of three phase induction motor. Induction motor is common in many industries. Fault can be occurred due to over speed, over load, high temperature, vibration, over current and over voltage. Complete monitoring is necessary to prevent the motor from the damage. Various protection schemes are implemented to protect the motor from the fault. Here we come with the PLC (Programmable Logic Controller) and SCADA (Supervisory Control and Data Acquisition) for the detection and protection of three phase induction motor from the fault. PLC is used to control the system and the SCADA is used to control and monitor the motor. Motor status can be viewed and controlled in remote location by using the SCADA software. If any fault occurs, motor automatically turns off and gives the warning to the person working in the vicinity. Hereby we can avoid unexpected failure of the motor and prevent the whole industrial process from shutting down all of a sudden which can be dangerous to the people working in the industry. This possesses high efficiency, low cost and more safety benefits than relay based system.

Keywords – Protection, Induction Motor, PLC, SCADA

1. Introduction

The three phase induction motor is necessary for manufacturing. It is preferred because it can also be operated in hazardous conditions. The advantage of alternating current three phase induction motor is that it offers rugged construction, cost effective pricing and easy maintenance. These factors have result in the development of the manufacturing infrastructure which has led to a vast installed base of motors. AC induction motor is used in worldwide industries. Unexpected fault occurrence can damage the motor and it cause industrial issue. So continuous monitoring of induction motor has received considerable attention in recent years.
Various types of faults occurring in three phase induction motor are over current, over voltage, under current, under voltage, temperature and bearing fault. These faults can be controlled and monitored by Programmable Logic Controller (PLC) and Supervisory Control and Data Acquisition (SCADA). Programmable Logic Controller (PLC) has achieved recent development in industrial applications. Control circuit of industrial automation system is the most widely used area of programmable logic controller.

PLC offers high reliability control, ease of programming and process fault diagnosis. Input devices state is continuously monitored for an industrial application using PLC. It controls the entire system.

Supervisory Control and Data Acquisition (SCADA) is used to monitor and control the system simultaneously. While many types of motor fault detection and protection methods have been proposed, practical detection techniques for three phase induction motors are generally provided by some combination of mechanical and electrical monitoring techniques. Digitally conditioned monitoring of electrical machines has received considerable attention in recent years. In this paper various types of fault in three phase induction motor occurring in operation have been achieved with the help of PLC and SCADA. In this study, the method applied is PLC and SCADA based protection system of three phase induction motor.

2. Programmable Logic Controllers

A PLC (Programmable Logic Controllers) is used to run more complex processes. It is possible to connect more PLC to a central computer. Induction motors are used worldwide in industrial applications. These motors not only used for general purposes, but also in severe environments and hazardous locations. These motor is reliable but they are prone to some undesirable stresses which causes faults and failure of induction motor. A motor failure that is not identified in an initial stage may become severe and the induction motor may suffer severe damage like, production shutdowns. Such shutdowns are costly in terms of production time, maintenance costs, and wasted raw materials. In order to protect the induction motor recent technology is introduced in our model. Due to recent development in Programmable logic control (PLC) it can be used efficiently for fault diagnosis of induction motor.
The earlier models include Fault diagnosis of induction motor using PLC in 2013 [1] which involves Single phase Induction Motor connected to the protection system through the measuring components. The proposed PLC-controlled protective relays deals with the failures such as the over current, over voltage, overload, under voltage, under current, temperature, over speed and vibration. In the year 2014, Fault identification and protection of induction motor using PLC and SCADA [2] was developed which involves Advanced Signal Processing techniques and Stator current spectral signature analysis.

In 2015, Harmonic tracking analysis: a novel method for fault diagnosis in induction machine [3] was developed which involves usage of Fourier transform which relies on tracking the frequency signature of each type of fault in current's spectrum. In 2015, Microcontroller Based Protection and Control of Three-Phase Induction Motor [4] was developed in which implemented protection system can be programmed to suit wide range of induction motor sizes. In 2016, automated SCADA based system for identification of induction motor bearing fault [5] used in process control operation was developed .It includes analyzing of vibration amplitude using standard deviation faulty bearing. Faulty bearing can be identified from the vibration signal.
In 2017, A review and comparison of fault detection and diagnosis for squirrel cage induction motors [6] was developed. In this model, the effect of simultaneous or cascaded faults with respect to different combinations of fault severity on each individual is desired to explore.
We have proposed a system which includes the following features:

Fault detection and protection of three phase Induction Motor using PLC and SCADA. Incremental encoder is used to converts linear or rotary displacement into digital or pulse signal. Vibration sensor used to rectify the bearing fault. Temperature sensor with contact type is used, which is in physical contact with the object being sensed and use conduction to monitor change in temperature.

3. Block Diagram:

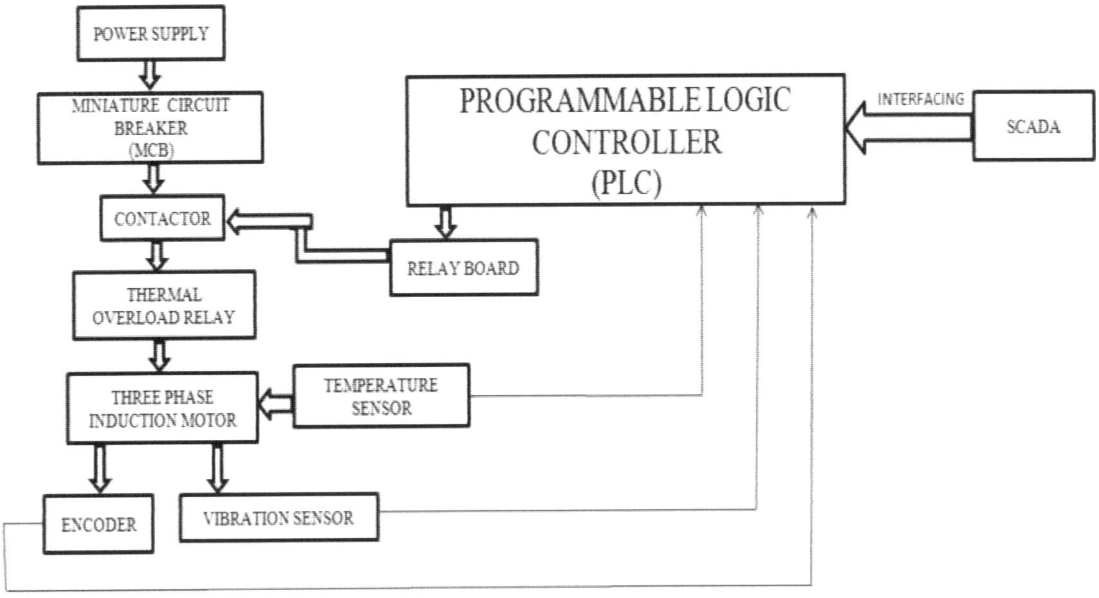

Figure.1 Block Diagram of Proposed system

In this article, we have designed the protection system for three phase induction motor. The three phase induction motor will have the rating as 415V, 3.50A and 50 Hz. The protection is done by means PLC and SCADA. The PLC used is OMRON PLC which can be utilized by CX-Programmer for CP1E. It allows faster programming by using the easy input editor software.

Quicker symbol creation because of management and memory allocation is automatic. Wonder ware In touch SCADA software is used to control and manage the process of remote area from the control room. It is the most sophisticated graphical tool which can be monitor more than 1,00,000 plants and factories. Thermal overload relay is used to control the over current, over voltage from the power supply. A contactor with 24V is used to control the 240V motor switch.

The motor is connected with the sensors like temperature sensor, vibration sensor and encoder. The temperature sensor is to control the temperature in the induction motor which has the limit of 50ambient. The vibration sensor is used to protect the motor from the bearing fault. Encoder gets the speed of the motor which is very much essential to run the motor without stopping it.

Temperature sensor, vibration sensor, encoder is fixed with PLC to operate without making issues in changing of values beyond the range. The SCADA and PLC is interfaced to control and manage the fault detection and protection of three phase induction motor without causing major damage to the motor. The solution of various faults of the phase currents, phase voltages, winding temperature and vibration of the induction motor occurring can be achieved with the help of PLC and can be monitored on PC. If any fault is observed during operation of the motor then the warning message will be displayed and then the motor will be automatically stopped.

IOP Conf. Series: Materials Science and Engineering **623** (2019) 012017 doi:10.1088/1757-899X/623/1/012017

4.Methodology:

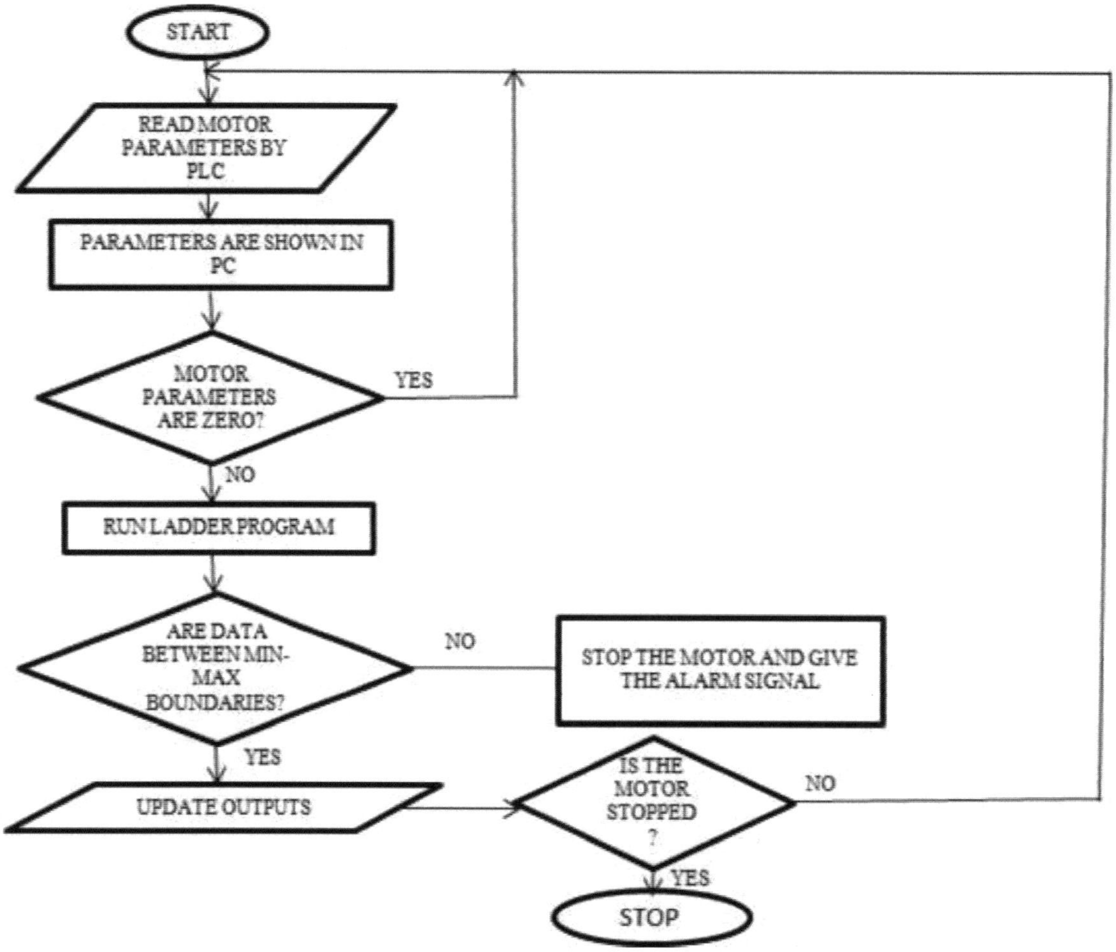

Step 1: SCADA frameworks are significant for modern associations since they help to look after productivity, process information for more astute choices, and impart framework issues to help moderate personal time.

Step 2: The fundamental SCADA design starts with programmable rationale controllers (PLCs) or remote terminal units (RTUs).

Step 3:PLCs and RTUs are microcomputers that speak with a variety of items, for example, plant machines, HMIs, sensors, and end gadgets, and afterward course the data from those articles to PCs with SCADA programming.

Step 4: The SCADA programming forms, appropriates, and shows the information, helping administrators and different workers break down the information and settle on significant choices. For instance, the SCADA framework rapidly tells an administrator that a cluster of item is demonstrating a high rate of mistakes.

Step 5: The administrator delays the task and perspectives the SCADA framework information through a HMI to decide the reason for the issue. The administrator audits the information and finds that Machine 4 was breaking down.

IOP Conf. Series: Materials Science and Engineering **623** (2019) 012017 doi:10.1088/1757-899X/623/1/012017

Step 6: The SCADA framework's capacity to tell the administrator of an issue encourages him to determine it and counteract further loss of item.

1. Software Tools :

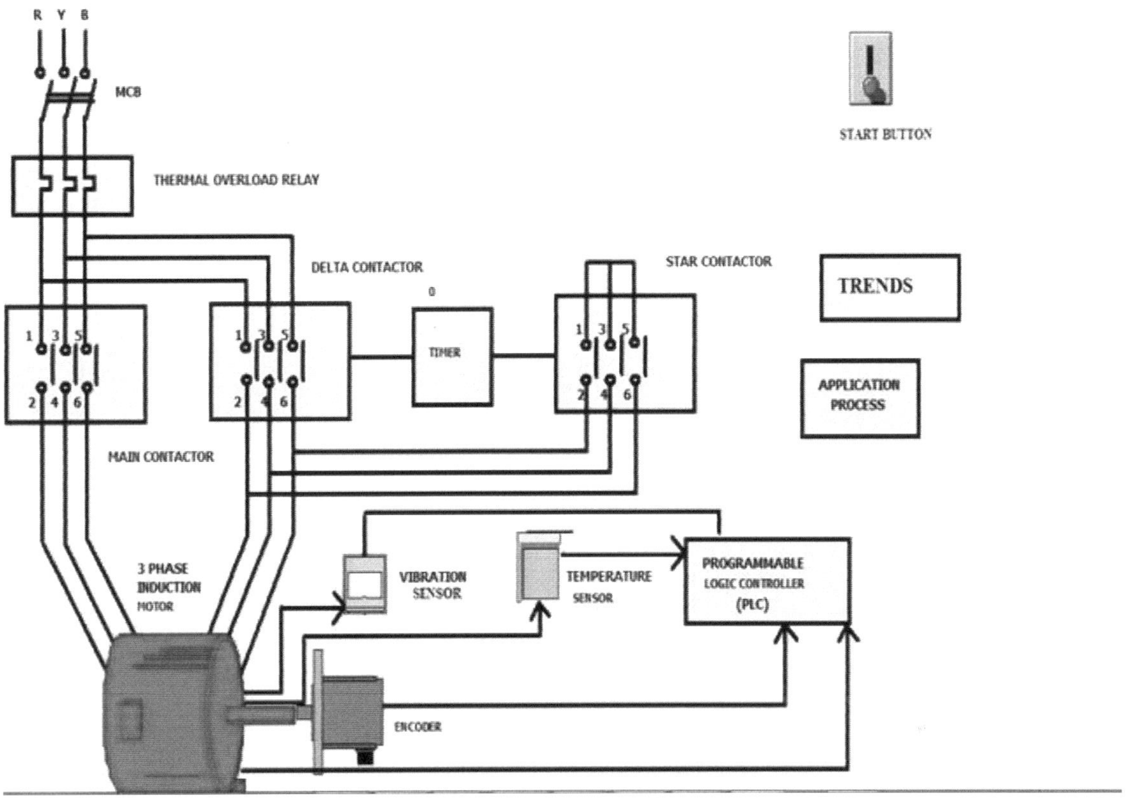

The snapshot shows the SCADA design for the fault detection and protection of three phase induction motor. Here comes the thermal overload relay, main contactor, delta and star contactor, encoder, vibration sensor and temperature sensor. When the power supply is given to the motor, initially the motor runs in star connection then it runs in delta connection because of the use of star delta starter. Starter is always essential in a motor.

Ia = V-Eb/Ra

(I =V/R, Ohm Law) Suppose

A 5 HP motor with 415V having the armature resistance as 0.25ohms and normal full load current is 50 amperes.

Direct connection to the motor without a starter the result will be

Ia = 415-0/0.25

= 1660 A

This high current will destroy the winding because it is 33.2 times high with normal full load current. So this is the reason to install a starter with a motor.

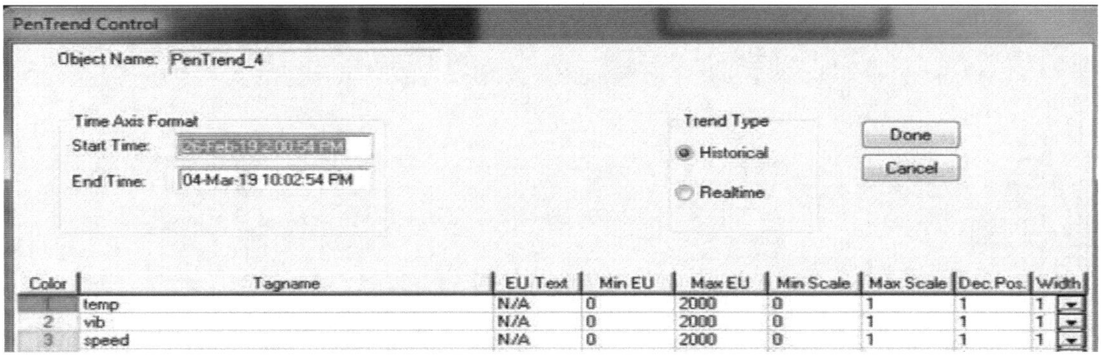

Figure 2 : Pen Trend Control

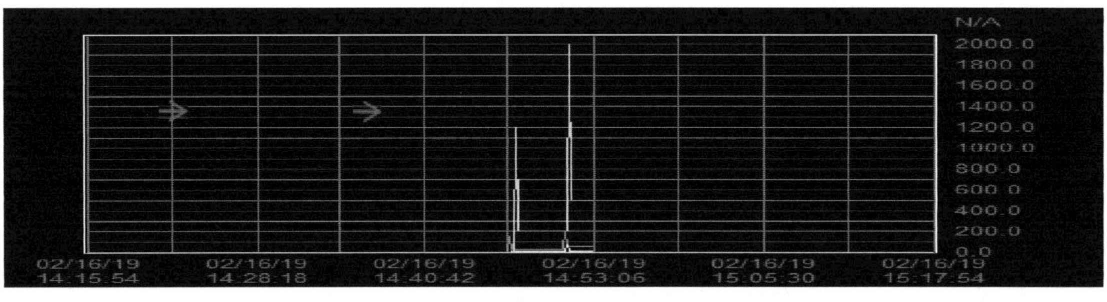

Figure 3 : Fault condition of the Motor

Figure 4.2 show the time period to control and monitor the process. As mentioned by date and time it will show the graph in the name of trend and the process update. Figure 4.3 is the fault condition of the motor and during this condition the motor will stop and the alarm is heard and the detail of alarm is shown in figure4.4

Date	Time	State	Class	Type	Priority	Name	Group
23 Feb	21:23		EVENT	SYSTEM	999	$Operator	$Systen
23 Feb	21:23		EVENT	SYSTEM	999	$HistoricalLogging	$Systen
23 Feb	21:23		EVENT	SYSTEM	999	$OperatorDomain	$Systen
23 Feb	21:23		EVENT	SYSTEM	999	$OperatorName	$Systen
23 Feb	21:23		EVENT	SYSTEM	999	$AccessLevel	$Systen
23 Feb	21:23		EVENT	SYSTEM	999	$LogicRunning	$Systen
23 Feb	21:32		EVENT	SYSTEM	999	$NewAlarm	$Systen
23 Feb	21:32	UNACK	VALUE	HI	1	temp	$Systen
23 Feb	21:32	ACK_RTN	VALUE	HI	1	temp	$Systen
23 Feb	21:32		EVENT	SYSTEM	999	$NewAlarm	$Systen
23 Feb	21:32	UNACK	VALUE	HI	1	vib	$Systen
23 Feb	21:32		EVENT	SYSTEM	999	$NewAlarm	$Systen
23 Feb	21:32	ACK_RTN	VALUE	HI	1	vib	$Systen

Update Successful **Default Query**

Figure 4: Alarm status of the each stepping process

Figure 5 : PLC Ladder Diagram

This shows the simple PLC ladder diagram for the fault detection process.

5.Conclusion:

The solution of various faults of the phase currents, phase voltages, winding temperature and vibration of the induction motor occurring can be achieved with the help of PLC and can be monitored on PC. If any fault is observed during operation of the motor then the warning message will be displayed and then the motor will be automatically stopped.

6.References

[1] Ravi Masand, DeepikaJadwani, Prof.S.P.Shukla, " Fault diagnosis of induction motor using PLC ", International Journal of Advanced Research in Electrical, Electronics and Instrumentation Engineering, vol. 2, issue 12, December 2013.

[2] Sujith John Mathew and B.Hemalatha, "Fault identification and protection of induction motor using PLC and SCADA", International Journal of Advanced Research in Eletrical, Electronics and Instrumentation Engineering, vol. 3, issue 4, April2014.

[3] Angel Sapena-Bano, Manuel Pineda- Sanchez and Javier Martinez-Roman, "Harmonic order tracking analysis: A Novel Method for Fault Diagnosis in Induction Machines", IEEE transactions on energy conversion, vol. 30, issue 3, September 2015.

[4] Ms.MadhuriBalasahebZambre, Mr.Vivek and Ms.Bhupali, "Microcontroller Based Protection and Control of Three-Phase Induction Motor", International Journal on recent and innovation trends incomputing, 2018.

[5] S.Mithra and C.Koley, "An Automated SCADA based system for identification of Induction Motor Bearing Fault used in Process Control Operation", 2016 2nd International Conference on Control, Instrumentation, Energy andCommunication(CIEC).

[6] Manoj M. Kharade and N.R. Kulkarni, "Automation in speed control and overcurrent protection of three phase induction motor", i-manager's Journal on Instrumentation and control Engineering, vol. 4, no. 2, April2016.

[7] Yiqi Liu and Ali M. Bazzi, "A review and comparison of fault detection and diagnosis for squirrel cage induction motors", ISA transaction,vol. 70, September2017.

[8] Pradeep. M, Priyanka. R, Sathiskumar. S, Edwin. M, "Improving voltage quality at EV charging stations by using smart – grid systems" , International Journal of Pure and Applied Mathematics, Volume 119 No. 15 2018, 613-620, ISSN: 1314-3395.

[9] Ms.PriyaV.Kale, Mr.AmitM.Kale and Mr.NikhilR.Kamdi, "Protection of Induction Motor

Using Classical Method", International Research Journal of Engineering and Technology", vol. 4, issue: 01, January2017.

[10] M. Pradeep, M. Senthil Kumar, S. Sathiskumar, and S. Hakkim Raja, "Interleave Isolated Boost Converter as a Front End Converter for Solar/Fuel Cell Application to Attain Maximum Voltage in MATLAB", Indian Journal of Science and Technology, Vol 9(16), April 2016, ISSN (Print) : 0974-6846.

[11] R. Suguna, S. Jalaja, M. Pradeep, R. Senthil Kumar, S. SrikrishnaKumar and K. R. Sugavanam, "Transient Stability Improvement using Shunt and Series Compensators", Indian Journal of Science and Technology, Vol 9(11), March 2016, ISSN (Print) : 0974-6846.

[12] Akshaya. S. R, R. NiranjanKumar ,D.Vinodini, M.Pradeep, "Implementation of High Step-up Interleaved Converter with VSI", International Journal of Pure and Applied Mathematics, Volume 119 No. 15 2018, 3779-3784.

[13] M.Pradeep, N.Udhayaraj, P.Sathyanathan, "Low Cost Power Quality Analysis For Domestic And Industrial Applications", International Journal of Mechanical and Production Engineering Research and Development (IJMPERD) ISSN (P): 2249-6890, Vol. 8, Special Issue 3, Dec 2018, 1549-1558 ©.

[14] N. Udhayaraj, M. Pradeep, Anup Kumar &S.AmoseDinakaran, "Application Of Genetic Algorithm For India's Electricity Consumption Estimation", International Journal of Mechanical and Production Engineering Research and Development (IJMPERD) ISSN (P): 2249-6890, Vol. 8, Special Issue 3, Dec 2018, 1565-1573 ©.

[15] AN. Udhayaraj, P. Sathyanathan, M. Pradeep, Bharani, P. Mani, R. Sakthivel, "Analysis and Implementation of Single Phase Seven Level Inverter for Utility Supplies application", Journal of Chemical and Pharmaceutical Sciences,Special Issue 8: June 2017, ISSN: 0974-2115.

IOP Conf. Series: Materials Science and Engineering **623** (2019) 012018 doi:10.1088/1757-899X/623/1/012018

An advanced control technique for a Grid-Tied Hybrid Renewable energy system with Reciprocated sinusoidal PWM methodology

M D Udayakumar[1], P Sabarish[2], V Indragandhi[3], A Rajkumar[4]

[1,2]Assistant Professor, Department of EEE, K.Ramakrishnan College of Technology, Trichy, Tamilnadu, India.

[3]Associate Professor, School of Electrical Engineering, Vellore Institute of Technology, Vellore.

[4]Professor&Head, Department of EEE, K.Ramakrishnan College of Technology, Trichy, Tamilnadu, India.

Abstract. This research paper deals with the algorithm based on the extraction of maximum powerknown as Maximum Power Point Tracking (MPPT) fromGrid-tied renewable energy system. Thesporadic nature of solarinsolationand the variable wind speed create the necessity to find out the optimal power that makes certainof obtaining of maximum energy from the renewable resources. The optimization can be done by implementing the MPPT mechanism in the controller design. A new parabolic prediction method is suggested for tracking the MPP continuously and Reciprocated sine PWM technique is applied for minimizing the unwanted harmonic contents in the delivered output power. The simulation and experimental results show the efficiencyand outcome of the proposed methodologies with regard tothe control ofthe output power delivered as well as the reduction in harmonics from the system.

1. Introduction

The present energy scenario shows the hugeeradication of fossil fuels likecoal, petroleum oil andnatural gas which takes more decades to shape its original form [1], [3]. This major consumption of fossil fuels highly threatens the future generation from running out of these energy resources. Also thetremendous usage of fossil fuels leads to many environmental related problems [2]. So it has become much important to switch and utilize the renewable energy resources like wind and solar for electric power generation [5],[6].

The wind power stands out to be one of the most promising renewable energy sources of electrical power in the near decades. Most of the countries keep promoting the wind power technology in various ways [10]. In next two decades,the use of wind turbines for electric power generation will hold the highest share among all other renewable schemes [8]. A modern wind turbine starts producing the electrical power at wind speeds around 5 m/s. As wind speed increases, the power output increases linearly until it equals the rated power of the wind turbine, so that it reaches its maximum power output level [7],[10].

Solar Photovoltaic(PV) power is one of the fastest growing renewable energy technologies around the world. The sun provides the energy in the form of solar radiation [13],[14]. The intensity of this radiation received from the sun is called irradiance and also referred to as insolation. A PV cell captures the solar radiation and converts into the electrical energy [11].Due to its modular character, reliability, reduced level of noise and harmful emission, the PV cell becomes more attractive among all other renewable energy resources[12].

The integration of electrical power generated from the renewable energy sources into electricity grid is influenced by various factors such as the technology for integration, availability of resources, variation predictability,capacity of each generator and moreover the variation with time of power

IOP Conf. Series: Materials Science and Engineering **623** (2019) 012018 doi:10.1088/1757-899X/623/1/012018

generated [15]. So it is very much important to integrate the renewable energy generation options with suitable control and storage methodologies [17].

2. Comparison of MPPT Techniques

The tracking of Maximum Power Point is much essential in modern photovoltaic systems [18]. A solar panel is able to transform about 35% of the solar insolation which receives on its surfaceinto electrical energy [11]. Numerous literatures show the effectiveness of various MPPT techniques used for producing maximum power. From the maximum power transfer theorem, one can better understand the concept of extraction of maximum power using MPPT techniques [16]. According to that theorem, for obtaining the maximum power output, the load impedance should match with the circuit thevenin's impedance. It can be accomplished by introducing a boost converter, which maximizes the output power by increasing the output voltage for the same current rating [17]. If the duty ratio of the converter is adjusted to an appropriate value, load impedance can be made equal to source impedance. We can discuss various MPPT techniques as highlighted in the literature part.

Perturb & Observe method, one of the well-known MPPT techniques, operates by changing the duty ratio of the converter based on the perturbation i.e. increment or decrement in voltage or current parameterof the PV array in order to find outthe power [14]. Then the estimated power is observed to increase or decrease the duty ratiocorresponding to its previous measured value. The measured voltage and current of the PV array require two sensors for power measurement.

And also, by this method, tracking the maximum power is very difficult under rapidly changingclimatic conditions. To manage the above situation, Incremental conductance method can be used for tracking the MPP [15]. By this method, MPP can be easily tracked for varying irradiance and atmospheric conditions with higher accuracy. Even though this method is faster than P&O method, it increases the complexity.

To reduce the complexity while tracking the MPP, Fuzzy logic & neural network based techniques can be proposed. Even though these techniques perform well, the cost for practical implementation is much higher [21] [22].

3. Proposed Methodology

This research work deals with a downward parabola based MPPT technique in the converter side and Reciprocated sine PWM technique in the inverter side for performance optimization ofthe given system. The optimal Maximum Power Point of solar panel is determined using the working points & the parabolic curve correspondingly based on progressive renewal process [2]. Also, this derived parabolabased systematic procedure has been suggested for fine tuning and the wrong tracking direction possibilities are decreased efficiently.

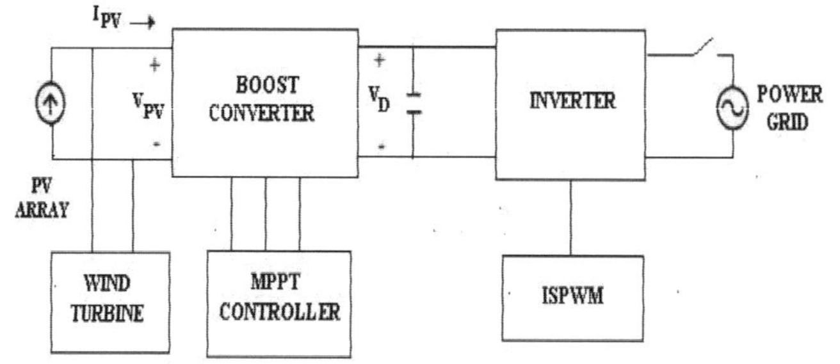

Figure 1. Functional blocks of the proposed system

Figure 1 shows the configuration of suggested hybrid system, comprising of a PV array, wind turbine, an inverter and a boost converter. A dc link which helps to establish the proper voltage level in order to connect the hybrid system with the utility network after the conversion made by an inverter.

3.1Converter side

The proposed MPPT procedure initially assigns three duty ratio values as D_1, D2 and D_3. After the measurement of V and I in PV array, we can obtain the three points as (D_1, P_1), (D_2, P_2) and (D_3, P_3) collectively known as a set of (P, D) measurement. With the help of these three points, we can draw the parabolic curve as shown in figure 2.

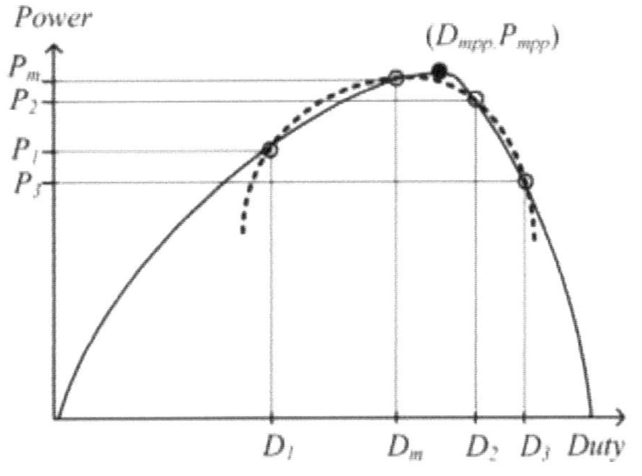

Figure 2. Parabolic curve for proposed MPPT Procedure

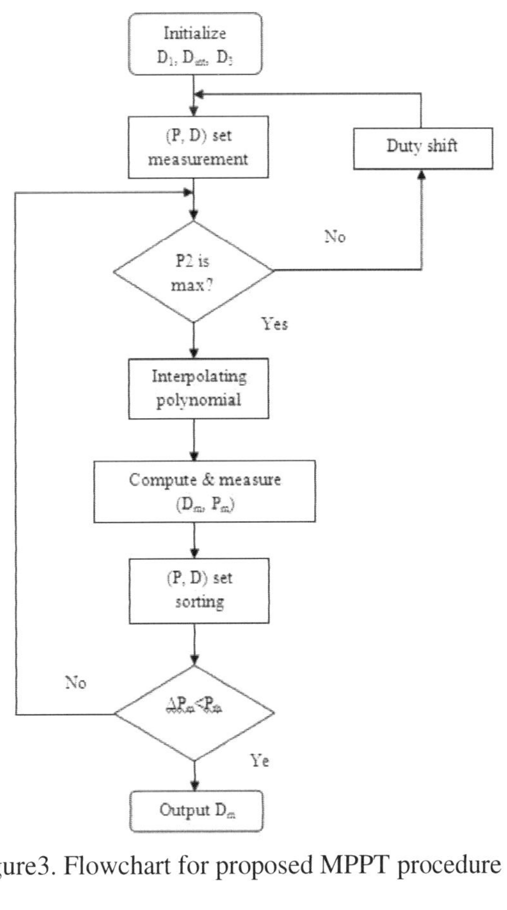

Figure3. Flowchart for proposed MPPT procedure

IOP Conf. Series: Materials Science and Engineering **623** (2019) 012018 doi:10.1088/1757-899X/623/1/012018

A parabolic function Q(x) can be represented for estimating the objective function f(x) and is given by

$$Q(x) = Ax^2 + Bx + C \quad \text{------(1)}$$

$$Q(x) = f(x_0)\frac{(x-x_1)(x-x_2)}{\Delta x_{01} \bullet \Delta x_{02}} + f(x_1)\frac{(x-x_0)(x-x_2)}{\Delta x_{10} \bullet \Delta x_{12}} + f(x_2)\frac{(x-x_0)(x-x_1)}{\Delta x_{20} \bullet \Delta x_{21}} \quad \text{------ (2)}$$

where,

$$\Delta x_{ij} = x_i - x_j \; i, j = 0,1,2 \quad \text{------ (3)}$$

Also we can predict a local maximum value denoted as D_m. So, one new point can be obtained as (D_m, P_m). After measuring the new point from the parabolic curve, it is necessary to sort the(P, D) set to release the small output power among them. This procedure is examined through two successive iterations with the estimated value of maximum power. Here the purpose of iteration procedure is to determine a convergent result. In the end, if there is any discrepancy in measured power, then we can compare the ΔP_mvalue with P_{th}(the point to define the convergence) the procedure will withdraw the lowest power point. Untilobtaining the optimum duty ratio, this procedure is repeated. The DC link voltage is varied regarding to the duty ratio variation (D_{int}) on internal switch. Thus the PV array voltage (V_{PV}) is controlled via tuning and it tends to produce maximum output power at various working conditions.

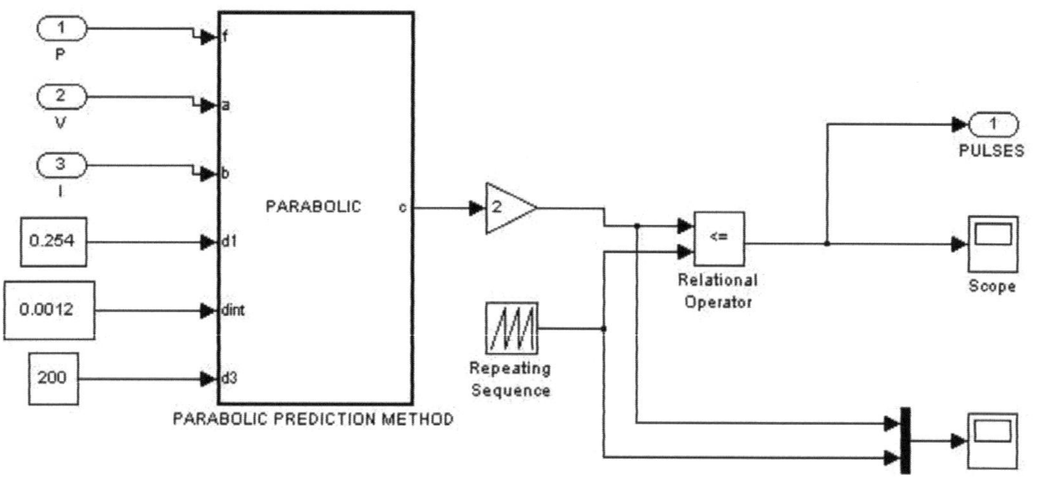

Figure 4. Simulation diagram for proposed MPPT technique

3.2. Inverter side

Here, the Reciprocated sine PWM method is used to produce gate pulses for inverter circuit. The conventional sinusoidal reference signal & the reciprocatedsine waves are utilized as two signals for

IOP Conf. Series: Materials Science and Engineering **623** (2019) 012018 doi:10.1088/1757-899X/623/1/012018

this proposed technique. For a given modulation index to enhance the output voltage, the control scheme uses a reciprocated sine carrier with high frequency range. The modulating signal amplitude is continuously measured and it is matched withthe sine carrier signal. Then the measured value is greater than the sine carrier, the gate pulses are generated. The higher fundamental output voltage with better spectral quantity achieved by this proposed technique with the minimization of total harmonic distortion (THD). Also to overcome the drawbacks of over modulation (reason for lower order harmonics in output voltage), this reciprocated sine PWM technique is proposed.

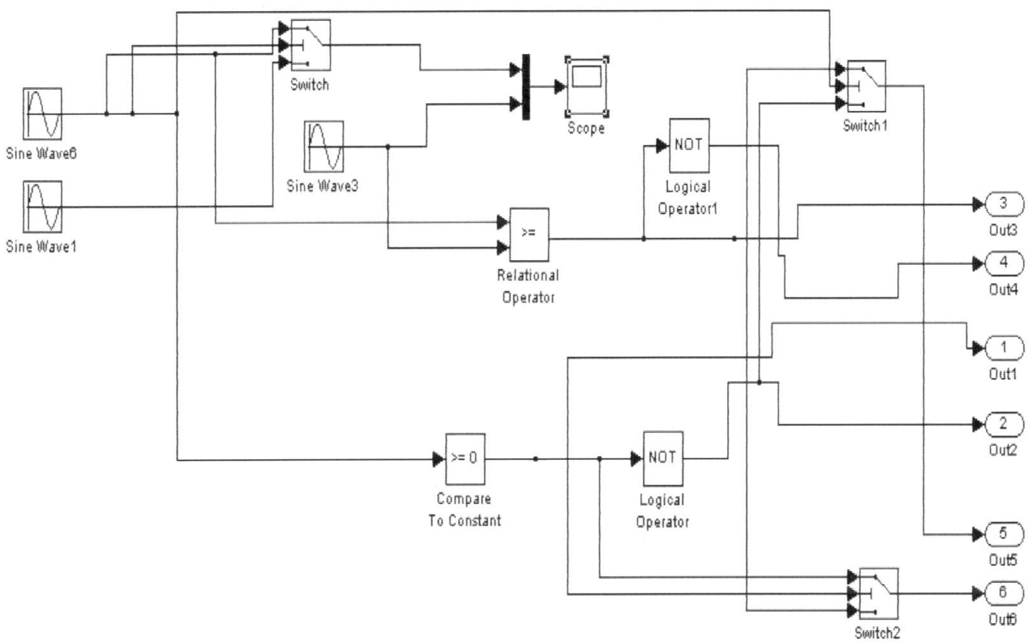

Figure 5. Simulation diagram for proposed PWM technique

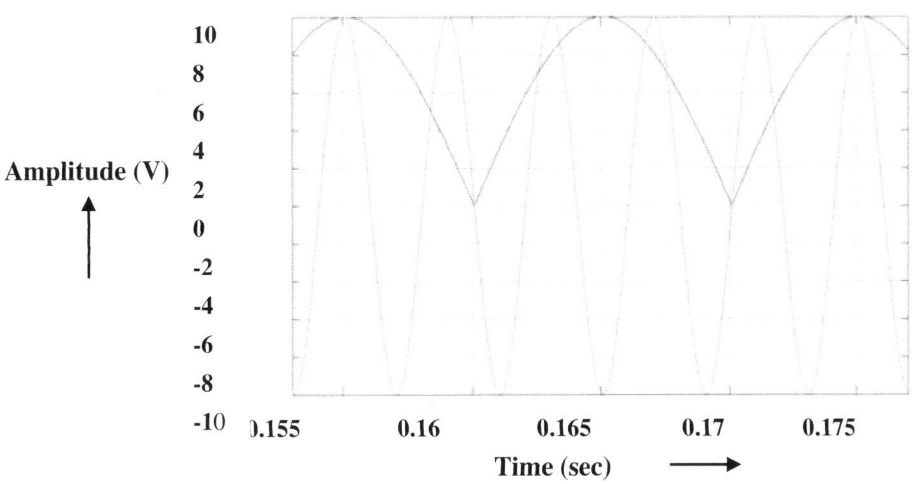

Figure 6. Carrier and reference signals

4. Simulation Results

The proposed methodology is implemented for a hybrid grid-tied renewable energy system which comprising of a wind and a solar energy source that connected to the grid network. The simulation

IOP Conf. Series: Materials Science and Engineering **623** (2019) 012018 doi:10.1088/1757-899X/623/1/012018

result shows the effectiveness of proposed techniques for both MPPT tracking and Pulse width modulating. The simulation result also shows the inverter output which is almost a sine wave with 230V magnitude and nominal 50Hz frequency with low THD.

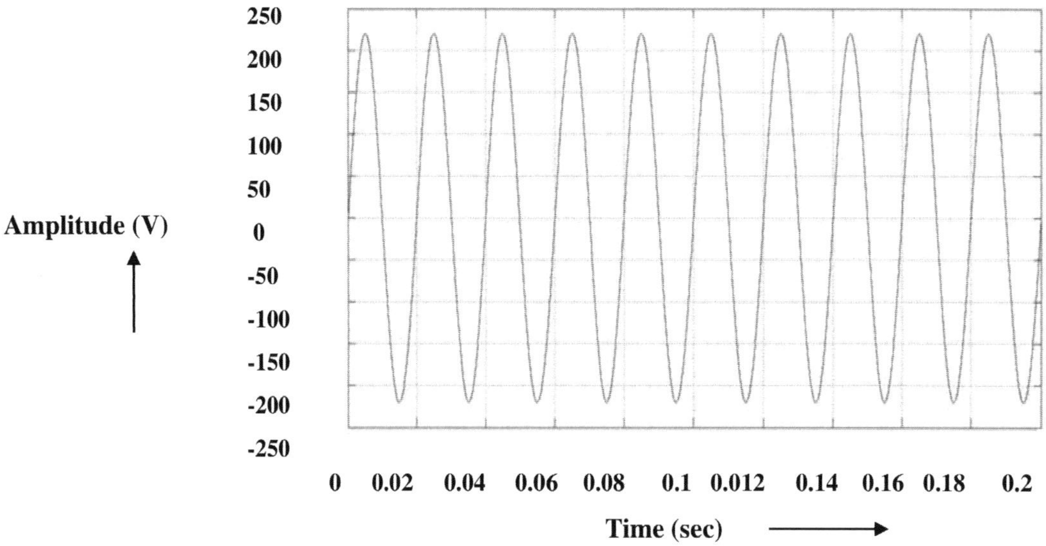

Figure 7. Inverter Output

4.1Design Specifications

Windmill Module	
Speed of Generator	200 RPM
Pitch angle	1 degree
Speed of wind	12 metre/sec
Solar Module	
Irradiance	5 W/m^2
I_{SC}	70 Amp
V_{OC}	40 Volt
Temperature	15 degCelsius
Boost Converter	
L_1	5 mH
C_1	1000μF
F_s	20 KHz
Parasitic Capacitor (C_{PV})	75 nF

Table 1. Simulation Design parameters for various components

5. Experimental Results

This hardware kit as shown in fig. 8 mainly comprising of a module of hybrid PV/Wind, a controller module, a converter &an inverter module. The voltage generatedby the PV/Wind module is provided to the boost converter. The Reciprocated Sine PWM technique has been implemented and pre-loaded to PIC16F877A controller. The generated PWM pulses from the controller will be given to the inverter circuit and it controls its ac output with low THD. The experimental results have been observed by a Digital storage Oscilloscope.

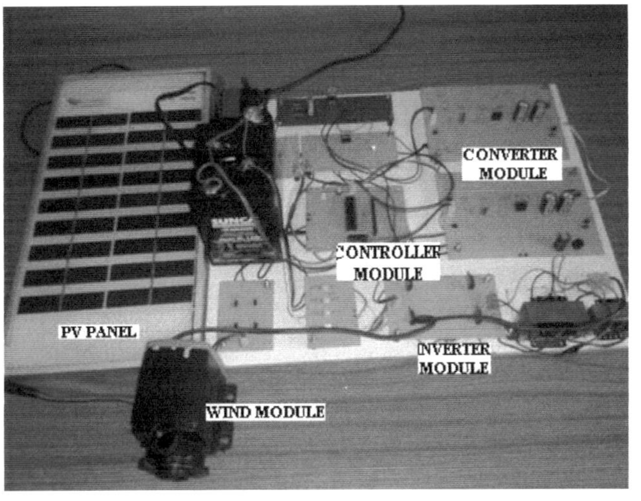

Figure 8.Hardware kit

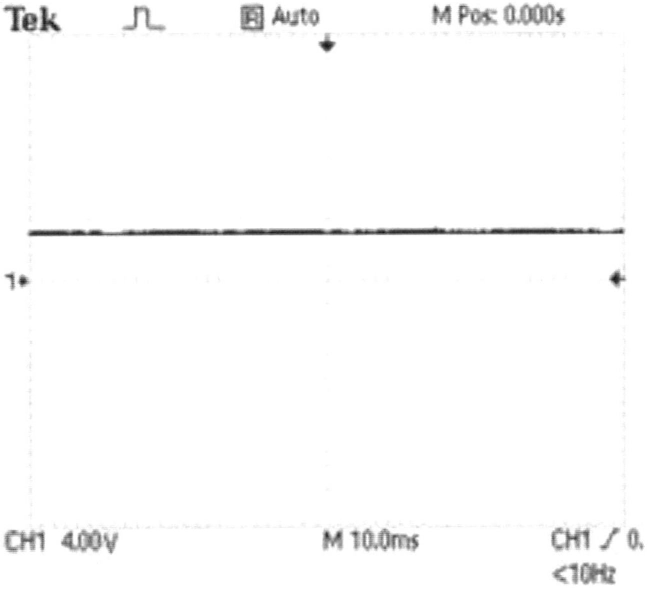

Figure 9. Input voltage

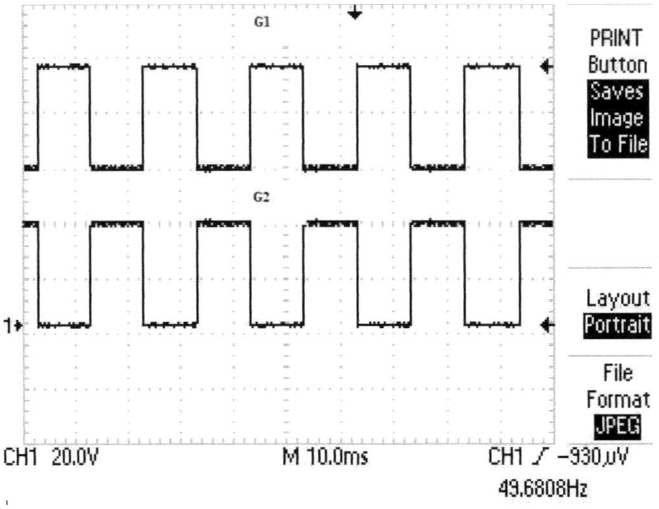

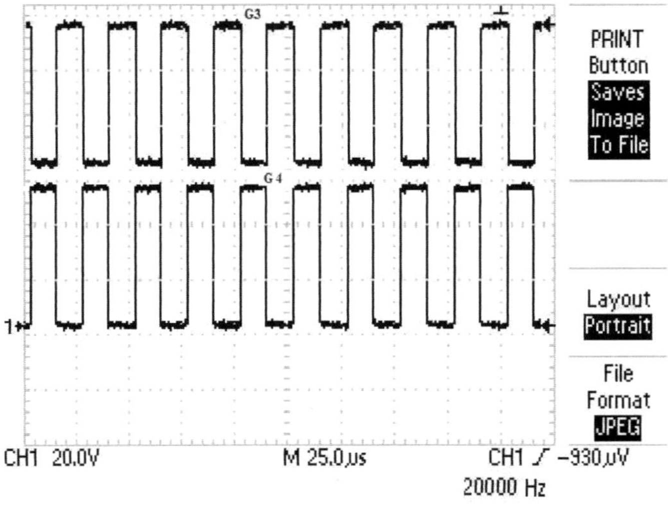

Figure 10. Gate pulse output

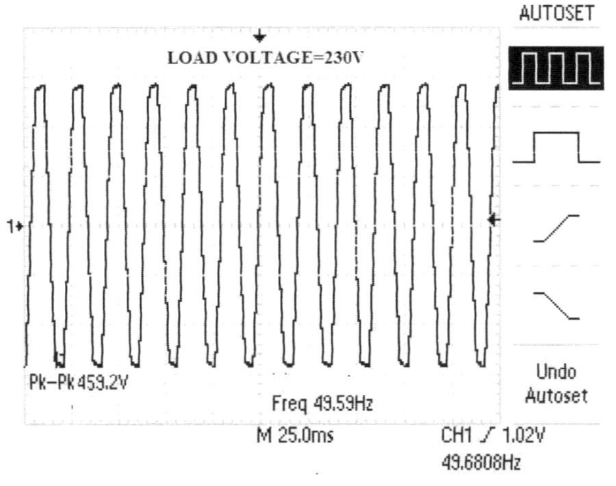

Figure 11. Inverter output

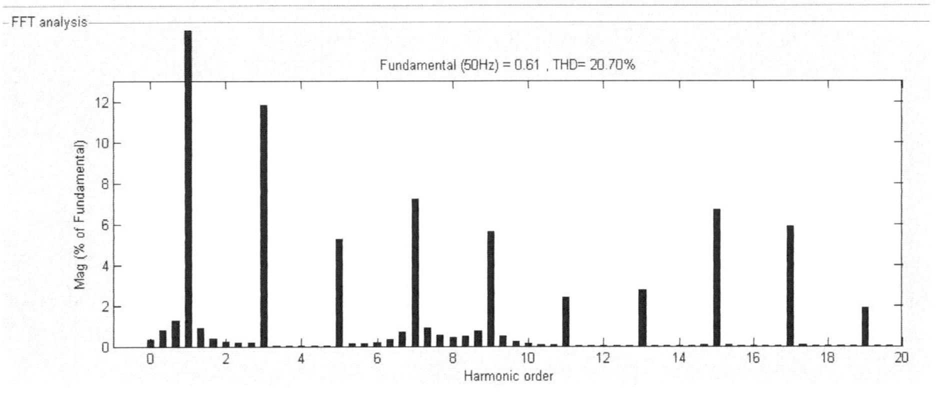

Figure 12. Harmonic spectrum without RSPWM

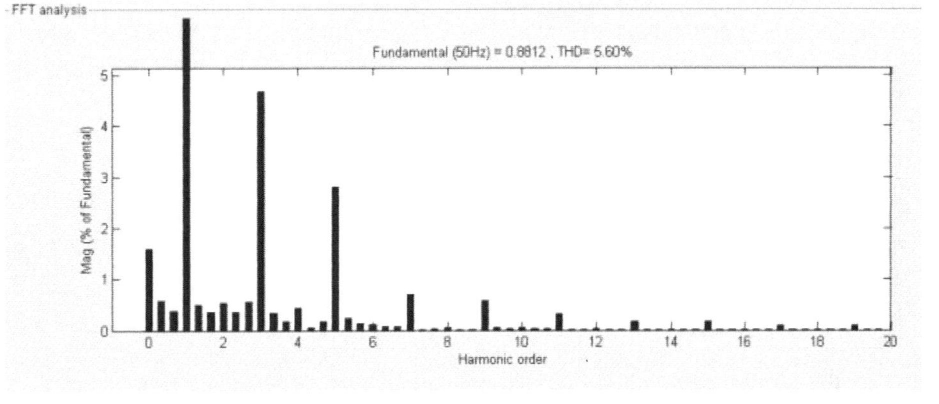

Figure 13.Harmonic spectrum with RSPWM

S.No	Description	Inverter output (V)	THD
1	Without RSPWM technique	215	20.70%
2	With RSPWM technique	230	5.60%

Table 2. Comparative analysis

6. Conclusion

This paper proposes the design and implementation of grid-tied hybrid power generation system. The maximum power from PV/wind combinationcan be extracted by using a well-defined and well-designed MPPT algorithm. The simulation and experimental studies show that power and voltage are at controllable state even under variation of environmental conditions. This proposed methodologyresults in high efficiency, better transient and stability are attained even at presence of disturbances.

7. References

[1] AriyaSangwongwanich, Yongheng Yang and FredeBlaabjerg"High-Performance Constant Power Generation in Grid-Connected PV Systems" IEEE Trans. Power Electron., vol. 31, no. 3, pp. 1822–1825, March 2016.

[2] Liming Liu, Hui Li, YaosuoXue and WenxinLiu"Reactive Power Compensation and Optimization Strategy for Grid-Interactive Cascaded Photovoltaic Systems" vol. 30, no. 6, pp. 2918–2922, June 2015.

[3] Urayai, Caston, and Gehan A. J. Amaratunga."Single sensor boost converter-based maximum powerpoint tracking algorithms", 2011 Twenty-Sixth Annual IEEE Applied Power Electronics Conference and Exposition (APEC), 2011.

[4] Pai, Fu-Sheng, Ru-Min Chao, Shin Hong Ko, and Tai-Sheng Lee. "Performance Evaluation of Parabolic Prediction to Maximum Power PointTracking for PV Array", IEEE Transactions on Sustainable Energy, 2010.

[5] Chiu, Chian-Song. "T-S Fuzzy Maximum Power Point Tracking Control of Solar Power GenerationSystems", IEEE Transactions on Energy Conversion, 2010.

[6] H.PatelandV.Agarwal,"MPPTschemeforaPV-fedsingle-phase single-stage grid connectedinverter operating in CCM with only one currentsensor,"IEEETrans.EnergyConvers.,vol.24,no1,pp.256–263,Mar.2009.

[7] F. Liu, S. Duan, F. Liu, B. Liu, and Y. Kang, "A variable step size INC MPPT method for PV systems," IEEE Trans. Ind. Electron., vol. 55,no. 7, pp. 2622–2628, Jul. 2008.

[8] SuthaPadmanabhan, KannanKaliyappan, "High Performance MPPT Based on Variable Speed Generator Driven by Wind Power Generation in Battery Applications", Journal of Electrical EnggTechnology Vol. 9, No. 1: 205-213, 2014

[9] N. Fermia, D. Granozio, G. Petrone, G. Spagnuolo, and M. Vitelli, "Predictive & adaptive MPPT perturb and observe method," IEEE Trans. Aerosp. Electron. Syst., vol. 43, no. 3, pp. 934–949, Jul. 2007.

[10] T. Esram and P. L. Chapman, "Comparison of photovoltaic array maximum power point tracking techniques," IEEE Trans. Energy Convers.,vol. 22, no. 2, pp. 439–449, Jun. 2007.

[11] Pau1l, M. Mano Raja, R. Mahalakshmi, M. Karuppasamypandiyan, A. Bhuvanesh, and R. Jai Ganesh. "Classification and Detection of Faults in Grid Connected Photovoltaic System."

[12] Jaiganesh, R., et al. "Smart Grid System for Water Pumping and Domestic Application Using Arduino Controller." International Journal for Modern Trends in Science and Technology 3.05 (2017): 385-390.

[13] Ramkumar, R., et al. "CUK-SEPIC fused MPPT converter using Inverted Sine PWM Technique for a Grid Connected Hybrid Renewable Energy System." International Journal of Scientific Research in Science, Engineering and Technology (IJSRSET), Print ISSN (2017): 2395-1990.

[14] SUBRAMANIAN, SANKAR. "A PV system based high step up converter with a Voltage Multiplier module using Fuzzy logic controller." International Journal of Applied Engineering Research 10 (2015): 7441-7446.

[15] Murugesan, S., and C. Kalavalli. "FPGA Based Multilevel Inverter With Reduce Number of Switches For Photovoltaic System." International Journal of Scientific Research in Science, Engineering and Technology (IJSRSET), Print ISSN(2017): 2395-1990.

[16] Kalavalli, C., and K. ParkaviKathirvelu. "R. Balasubramanian Department of Electrical & Electronics Engineering, SASTRA UNIVERSITY, I Single Phase Bidirectional PWM Converter for Microgrid System I." International Journal of Engineering and Technology (TJET) 5.3 (2013).

[17] Raguraman, L., and P. Sabarish. "Integrated Bridgeless PWM Based Power Converters." International Journal of Advanced Research in Engineering & Technology (IJARET) 4.5 (2013): 17-23.

[18] Subramanian, AT Sankara, P. Sabarish, and R. Jai Ganesh. "An Improved Voltage follower Canonical Switching Cell Converter with PFC for VSI Fed BLDC Motor." Journal of Science and Technology (JST) 2.10 (2017): 01-11.

[19] Sabarish, P., et al. "Performance Analysis of PV-Based Boost Converter using PI Controller with PSO Algorithm." Journal of Science and Technology (JST) 2.10 (2017): 17-24.

[20] Subramanian, AT Sankara, P. Sabarish, and A. Nazar Ali. "A power factor correction based canonical switching cell converter for VSI fed BLDC motor by using voltage follower technique." Electrical, Instrumentation and Communication Engineering (ICEICE), 2017 IEEE International Conference on. IEEE, 2017.

[21] Kanimozhi, P., S. Ramalingam, S. Murugesan, and N. Ramya. "Input Vector Observing Simultaneous BIST Architecture Utilizing SRAM Cells." (2017).

[22] Ali, A. Nazar, R. Jeyabharath, and M. D. Udayakumar. "Cascaded Multilevel Inverters for Reduce Harmonic Distortions in Solar PV Applications." Asian Journal of Research in Social Sciences and Humanities 6.11 (2016): 703-715.

[23] Karthikeyan, B. JEBASALMA. "RESONANT PWM ZVZCS DC TO DC CONVERTERS FOR RENEWABLE ENERGY APPLICATIONS." International Journal of Power Control and Computation (IJPCSC) Vol 6: 82-89.

[24] A.Rajkumar and B.Karthikeyan., Fuzzy Logic Controller for Fast Transient Response DC -DC Converter. Adv. in Nat. Appl. Sci., 8(21): 65-71, 2014

[25] A Rajkumar, B Karthikeyan, S Senthilkumar, Reliability Analysis of Fuzzy Logic Controlled Fast Transient Response DC- DC Converter. International Journal of Applied Engineering Research 10 (85), 728-730.

[26] T.AjithBosco Raj, R.Ramesh, J.R.Maglin, "Grid Connected Solar PV system with SEPICConverter compared with parallel Boost Converter based MPPT", International Journal of Photoenergy, Vol. 2014, Article ID 128327

[27] Ko, Shih-Hung, and Ru-Min Chao. "A Quadratic Maximization MPPT Method Applied to PV System on a Moving Vehicle", 2012 Asia- Pacific Power and Energy Engineering Conference (APPEEC), pp. 1-4, 2012.

Conserve to conquer

K Sundar

Abstract. That the accelerated consumption resulting in exponential depletion of fossil fuels, it is imperative to conserve energy at every possible avenue not only to ensure availability of fuels for use in the days to come but also to ably tackle the climate change which is order of the day. While many initiatives for optimum energy use emanate in various manufacturing companies and also are put into practice, emphasis on reduction of specific energy consumption has become mandatory for high power intensive industrial units. Plethora of micro level energy saving activities at various spheres of life can yield phenomenal results to benefit the country to conserve reserves as well as consequent economic growth. Well buttressed by technical efforts, certain administrative measures too are vital to achieve the said goal. Hence the need for presentation of energy conservation in industry and beyond.

Keywords : Conservation, Energy, Fuels.

1. Introduction

Energy is the fulcrum of all activities across the Globe which is available in disproportionate quantity for most users in general and in the power-intensive industrial units in particular. While generation of power is on the rise at one end, its optimum use through efficient process is also exercised constantly at the other end so as to strike a workable balance between demand and supply.

Thinkers and doers in any industry can and should explore all avenues to conserve energy for one simple verity that energy saved is energy generated without additional cost, time and efforts. Conservation of resources should be a bottom up activity for the lone reason that it is the workforce that operates plant and machinery day in and day out. Hence, they are the ones who should be aware of the ways and means of the frugal and effective use of energy at all levels. The momentum thus created by the bottom line should be well propped up by the middle management staff to perpetuate energy conservation activity. People at higher echelons are the ones to use eagle's eye to monitor and encourage all the productive suggestions as well as motivate people to come up with more and more ideas for energy conservation in a synergized pattern.

This paper covers all needed points in a broader sense as to how there had been steady and steep reduction in specific energy consumptions of various products in one of India's leading heavy chemicals manufacturers located at southern part of Tamil Nadu where the author served for over two decades. Based on the feedback on the recent lecture programs delivered by the author at few engineering colleges, the need to include energy management module in curriculum is also elaborated in this paper. Needless to mention, aligning to the title of the paper, certain administrative measures to encourage energy conservation too have been covered in the article.

2. About the industry

This chemical industrial complex has been in existence close to six decades. With modest commencement of its operation by manufacture of caustic soda way back in 1959, gradually the capacity was expanded over the years. Simultaneously, plant capacities of the byproducts viz. chlorine

IOP Conf. Series: Materials Science and Engineering **623** (2019) 012019 doi:10.1088/1757-899X/623/1/012019

and its derivatives were also enhanced. The mercury cells installed during inception were replaced by latest membrane cell technology during 2007, much ahead of stipulated deadline laid by the Government for such conversion. The switch over to sophisticated and modified manufacturing process has not only eliminated use of mercury but reduction in energy consumption substantially. Doubling the capacity of Synthetic Rutile plant in the complex (which was installed during 1970 for utilization of hydrochloric acid generated in chloralkali plant) during 2004 by few mere modifications in the manufacturing process and adding balancing equipment did result in remarkable reduction in specific energy consumption of the said product . A stand-alone PVC manufacturing plant too is in operation in the premises. Synthetic Iron Oxide plant and CPVC plant are very recent additions making the unit totally unique. Now, the entire steam and power needs of the complex is met by coal based cogeneration plant . This paper vividly covers various energy conservation measures adopted in the chemical complex during the last few decades.

There are three major avenues for achieving energy conservation in process industry viz.

- Low investment – workplace related initiatives by workmen – bottom line
- Moderate investment activities – Supervisory related monitoring as well as improvements aimed propellants involving staff – middle line
- High investment major projects handled by the management – conceptual ideas involving top executives – top line

Even a wee bit of efforts taken by individuals at any of the above three levels would result in tangible improvements and savings.

3. Initiatives with less investment

Quality Circles, which are small groups comprising workmen / shop floor supervisors surcharged with skilled awareness on energy saving must be formed. The teams with great amount of synergy do identify the areas where energy-saving can be implemented at their respective work place. The team takes cognizant of various spheres of day-to-day functioning such as safety, energy, housekeeping and maintenance issues . Even a minuscule effort on right direction can result in substantial benefit sooner or later and hence the teams are to be constantly motivated to sustain their functioning. The Quality Circles in the said industry have done about 150 minor projects over the last 10 years. It is germane to cite an interesting case study undertaken by one of the teams. Salt is the raw material for manufacture of caustic soda and it is harvested from company owned salt pans spread over 1000 acres of land. The conventional solar evaporation is the process adopted and over the years various improvements like laying of LDPE films in the salt pans to avoid percolation of brine into the ground, replacement of old pumps by energy efficient pumps to pump sea brine were carried out .

The pumps are installed at various locations in the above said 1000 acres to handle / feed the brine at different concentrations from condensers to crystallizers. The sea brine whose temperature increases during day time will evaporate during night when there will be heavy wind across the salt fields and hence salt production will be at its peak during night hours as well. When such high wind is available, if there be any pump failure at any stage in the brine flow, salt production will be directly affected.

Few years ago, frequent tripping of pumps was experienced and when such problem occurs at night, not only there would be loss of salt production, it will be extremely difficult to locate / identify the exact pump station in the vast land area and even if located, analyzing further causes of tripping during night hours was a concern since power supply will be available only for the pump house and nearby surroundings will be dark.

On such occasions, whenever inspection was made during next day morning, it was observed tripping was mostly due to "bird hit". Migrated large birds accidentally get caught in between the power lines when they take off from the electric poles during night thus tripping the electric motor.

IOP Conf. Series: Materials Science and Engineering **623** (2019) 012019 doi:10.1088/1757-899X/623/1/012019

The aforesaid problem was taken up by a Quality Circle team. While birds' movement cannot be prevented, the team members worked out possibilities of avoiding tripping as well as saving birds getting electrocuted. They came up with an idea of fixing a cross like structure atop all the electric poles where birds can land and take off safely above the live wires. The condemned wooden materials salvaged from the packing boxes at the stores were used and hence NIL cost for the project. Initially, as trial, the wooden cross were fixed at few poles (which were more prone to tripping caused by birds based on the past data) and the effort yielded fruits. Absolutely there were no tripping thereafter, but some days later when similar problem was observed, the team decided to install the wooden cross at all the electric poles in the salt to eliminate the issue once for all. The salt production simply got spiraled.

Through simple provision of a wooden structure at no cost, nuisance of tripping was totally eliminated benefiting with reduction in specific power consumption in respect of salt production. Those involved in the above project had a sigh of relief and elation in saving the migrated birds in addition to enhanced salt production. The team was encouraged to participate in State and National level Quality Circle Conventions which motivated them to go for more such innovative yet simple no-cost projects to save energy besides making other teams venture into critical analysis at their respective workplaces on various issues on the same modality.

Few other noteworthy conservation activities by Quality Circles are

- In-house invented Auto Sampler for mineral processing plant which resulted in dramatic reduction in energy use by process optimization.
- Recycling of expensive Demineralized water back into process.
- Product recovery in mineral processing plant from channels / pits.
- Use of gravity flow to avoid pumping system.

Kaizen Scheme encourages shop floor staff and workmen to come out with various suggestions to improve safety and productivity. Every month, employees who had contributed in this realms are rewarded and recognized. This system, apart from improved productivity and minimized breakdowns, enhances the morale of the workforce, as, human asset with unassuaged encouragement, energy and enthusiasm is the precious asset.

Energy conservation fortnight celebrations are yet another opportunity thrown open to employees through which even non technical personnel too are encouraged to contribute their mite. The suggestions are studied and critically evaluated prior to implementation at shop floor. During valedictory function of such celebrations, vendors are also provided opportunity to display their energy saving products and also explain the features, so that various options for usage would be studied and implementation taken up wherever feasible.

4. Moderate investment activities

Cost improvement programs are organized twice a year. Proposals from various teams (Civil, Mechanical, Electrical, Process, Safety etc.,) are presented highlighting present conditions, suggestions for modifications, improvements etc. A healthy rapport and subsequent interactions between the presenting teams and audience pave way for better understanding of the subject / issue in depth through collective mental cogitation and cognitive brain storming. With staff from all departments taking part in the meeting, pros and cons could easily be deliberated with absolute open mindedness.

Use of renewable energy sources for various applications is also debated in this program. One such outcome was "Solatube" use of sunlight for office lighting – installed in one section and studied for performance and energy savings. The Lux levels met the requirements even during cloudy weathers and implementation at few more buildings was taken up.

Few of activities undertaken in cost improvement programs which resulted in considerable energy reduction are;

- Enhanced batch size in reactors
- Flash steam recovery system
- Re engineering of pumps
- Harnessing of solar energy for certain process applications is also on the anvil.

Buyer – Seller meet is yet another event organized every year wherein issues related to product quality that in turn affects process and ultimately increases energy use are discussed with vendors. Best practices are exchanged for mutual benefit in the meet. Plant visit by vendors to discuss in detail about the requirements and observe the functioning of the materials being supplied by them at site is the key objective of the meet. Such visits by product suppliers facilitate the end user on the right application under right parameters, operation & maintenance – including preventive maintenance and throw more light on the technical knowledge which would result in optimum usage of products for sure.

Energy audits at regular intervals as also when felt needed by techies, experts and doyens in the domain of energy are organized and quick wins are addressed on priority.

Condition monitoring of critical equipment by competent technical personnel optimizes energy use apart from minimizing breakdowns.

5. High investment efforts

In the aforesaid energy conservation initiatives, wherever large investments are estimated, suitable erudite and experienced consultants from outside are engaged for detailed study to ensure precise suggestion with perfect time frame and cost implications along with exact ROI (Return On Investment).

Plant scale trials are undertaken wherever possible to generate surfeit of data which can help us implement after establishing energy savings as envisaged. One such implemented system is introduction of suitable equipment in the process flow to avoid mismatches between the upstream batch operation and downstream continuous operation. Installation of the said equipment (known as "table feeder") in addition to prodigious fuel oil reduction in the downstream rotary kiln, has also avoided requirement of additional kilns at huge capital cost. Though certain standard table feeders are available in the market, this application is very unique to handle acidic, wet, beneficiated, mineral sand and hence necessitated visit of vendor to site to understand the need and to design & incorporate suitable MOC (Material of Construction) to withstand severe duty conditions as well as adverse environ, if any.

Modernization of plants with latest energy efficient equipment, automation of process to banish human errors, retrofitting to minimize capital cost, investment on metering equipment, identifying alternate fuels for process etc., are few of the initiatives were taken up with considerable investments and successfully implemented. Conversion from mercury cell to membrane cells for caustic soda manufacture was one of the very high investments that the management took up during 2008. This conversion, as one could understand, especially those in the sphere of manufacture of caustic soda, required new plant altogether with necessary modern state of art control systems, brine purification system along with other accessories & auxiliaries .

The author , during his visit to Mozambique to source raw materials for the Synthetic Rutile plant , on return , could implement few modern techniques in mineral sampling systems and ore purification methods prevailing at the vendor's facilities which resulted in minimizing loss of products in waste stream , means enhanced production and hence reduction in specific energy consumption of the product.

IOP Conf. Series: Materials Science and Engineering **623** (2019) 012019 doi:10.1088/1757-899X/623/1/012019

6. Energy management in the curriculum

The author, who is now a freelancer, visits few technical educational institutions to deliver lecture programs to the students in various engineering disciplines on desired technical captions. Mostly the topics covered are either on the need for industry – institution interactions or applicability of theoretical knowledge and the ways to put into practice at workplace or energy conservation & use of renewable energy.

It is the strong opinion of the author, that, discrimination of branches in engineering like Civil, Mechanical, Electrical, Electronics, Chemical and Instrumentation etc. is valid until a student completes the bachelor course. The transition from college campus to industry demands conversion of knowledge into skill as a " full-fledged engineer" in his/her core area and his lectures pin point this important transformation on the part of the students.

- For instance, in a process industry radiation losses largely depend on the healthiness of coatings and insulations. Generally heat resistant painting, refractory linings, hot & cold insulations are dealt by ***Civil*** department.
- Selection and maintenance of pumps, reactors, utilities etc., lie largely with ***Mechanical*** department. Needless to mention steam, compressed air and efficiencies of equipment account for significant energy use which are monitored by utility department.
- Supply of power to the equipment is handled by ***Electrical*** department. Power factor corrections and limiting harmonics in supply lines are major steps for energy conservation. Where electrolysis is involved (like manufacture of caustic soda), every millivolt drop in the bus bars at cell house contributes for increased power consumption.
- Plant operation is managed by ***Chemical*** department. Optimization of batch size and enhancing throughput is the key function to achieve energy reduction. For caustic soda manufacture , achieving ultra purified brine for electrolysis (impurities in the order of ppb) is the main factor that directly impacts power consumption.
- For measuring of critical process parameters , the role by ***Instrumentation*** department is vital. Control room operation and installation of accurate meters to measure power and steam are the prerequisites which were rare phenomenon 30 years back is now a regular feature in any plant. The accuracy level of such meters decide the exact energy reduction.

Thus , every aspect of the above said engineering activity has strong influence in energy use . Hence , it would be advantageous if certain basic course / subject on energy management for all engineering disciplines is included in the curriculum, it will be of help to better contribute when a student joins the industry after graduation.

It is reiterated that engineering departments can't function effectively in isolation in any industry and when such inputs are provided in the bachelor's degree course, will naturally accelerate the improvements at workplace. The policy makers who frame the syllabus for technical courses are requested to kindly consider this aspect and explore possible implementation in the days to come. India has immense potential to harness solar energy and if such course / subject is covered along with basic aspects of solar power, benefits in the long run can be expected.

As an approved competent Energy Auditor, the author always eyes on energy saving and highlight this factor among students as to how they should unanimously and collectively interact with all concerned to benefit the organization they belong to in respect of energy saving!

7. Suggested administrative measures

It had been the practice to provide conveyance allowance / reimbursement at the offices to those who use their cars / two wheelers based on their positions. Now a totally upside down approach is required!!!. Yes, those who possess vehicle but still use public transport are to be encouraged with suitable "allowance for not using the car" While doing so, two main aspects to be taken care of. Firstly, it is to be ensured that the right person gets the benefit with suitable system by recording declared non usage of own vehicle. Secondly, the tax implications on payment of such incentives are to be suitably addressed. Otherwise, by getting a benefit for no usage of own vehicle, a person may lose the benefit by way of additional income tax for such extra earnings. Since, this conservation measure ultimately will reduce import of fuel benefitting the country, there should not be tax burden on the individual. To be precise, compared to loss of plausible revenue by way of income tax to the Government exchequer on account of exemption of IT for the above extra income, import duties on account of fuel would absorb the loss.

Most of the people deem it prestigious to own a vehicle either two-wheeler or four-wheeler (of course they are proclaimed to be fuel efficient) and avoid public transport system. A few others discard public transport facilities for more than one reason as Government transport facility deny their comfort. Therefore, when the Government focuses its attention on the reformation and rejuvenation of its entire transport network, the public would slowly drift towards using public transport provided sufficient awareness is also created on fuel conservation. Of course, the role of Indian Railways in the recent past in making Indian public to use its services by improving the train commuting facility across the country is commendable.

8. Conclusion

As described in this paper, the three-tier system in any industry viz. Initiatives with less or no investment, moderate investment activities and high investment efforts with specific time frame and ROI will ensure energy savings on a consistent basis. ISO 5001 system can further streamline energy management and will help us bench mark energy consumption and efforts towards continual improvements.

Energy saving efforts are perennial, perpetual, eternal and everlasting considering the depleting natural resources at one end and increasing cost of fuel at the other end. I now prefer to take it on to the student community who are going to serve various industries. Let us adopt at once the Switch Off Something (SOS) concept before the entire country sounds SOS (Save Our Souls) for want of natural energy resources.

That is the main reason of my suggesting the people at the helm of affairs in the realms of education to add conservation of energy in their syllabi so that young India would get energy conservation imbibed so strongly to implement the same at their respective work spots.

To conclude I energetically fuel the urge in the minds of the readers of this paper to go all out to conserve energy with unassuaged enthusiasm. .

9. References

[1] *energy manager – a quarterly magazine of the society of energy engineers and managers* | India April – June | 2015 | Vol :: **08** | No :: **02** ISSN0974 – 0996 pp 43-47

Author profile

The author Mr. K.Sundar was graduated from College of Engineering , Guindy , Madras , India ,in the year 1983 in Mechanical Engineering branch. He additionally qualified in Electrical Engineering from The Institution of Engineers (Kolkatta) during the year 1988. He holds Post Graduate Diploma in Business Administration from Annamalai University during 1988. He is BEE (Bureau of Energy Efficiency) certified energy auditor since 2008. He served in M/s. Dalmia Magnesite Corporation, a unit of Dalmia Cements Ltd from 1983 to 1994 and at M/s. DCW Limited from 1994 to 2017. During his 34 years' experience in industry , he shouldered responsibilities in the areas of maintenance , projects , technical services , process optimization , ISO systems , indigenization and energy conservation and while retiring held the position of Senior General Manager . He is winner of National award as Best Energy Manager by SEEM (Society of Energy Engineers and Managers) . He is Fellow life member of The Institution of Engineers (India) and SEEM and life member of Quality Circle Forum of India and Indian Society of Training and Development. He has authored for two guides viz. Basic Mechanical Engineering and Basic Electrical Engineering covering Anna University syllabus and the readers have expressed usefulness in their curriculum as well as at workplace. He is evaluator of energy manager/ auditor examination papers for the last ten years. At present , as a freelancer, he delivers lecture programs at few engineering colleges and also engaged in energy audits in chloralkali industry. The author's mail id is sundar6850@gmail.com.

IOP Conf. Series: Materials Science and Engineering **623** (2019) 012020 doi:10.1088/1757-899X/623/1/012020

An improved P&O MPPT control algorithm for increasing power extraction efficiency of solar PV module

D Dinesh Kumar[1], Challa Babu[1*], K Jyotheeswara Reddy[2], Kumar K[3], N K Kumar[4]

[1] Department of electrical and electronics engineering, Audisankara college of engineering and technology, Gudur, A.P, INDIA
[2] Department of electrical and electronics engineering, Sree Vidyanikethan Engineering College, Tirupathi, A.P, INDIA
[3] Department of electrical and electronics engineering, SV College of Engineering, Tirupahi, A.P, INDIA
[4] SELECT, Vellore Institute of Technology, Vellore, TN, INDIA

Abstract. The extraction of solar energy is done by using power point tracking techniques. The solar PV cells available have low efficiency in the order of 15-20%. The extraction efficiency depends on solar insolation, cell temperature, angle of incidence and the load current. These factors resulting a non linear V-I characteristics for a PV cell. A Perturb & Observation MPPT is developed to operate the PV cell at a reference voltage corresponding to maximum power point in the characteristic curve. This algorithm is incorporated to improve the performance and thereby increasing the efficiency of PV cell. A 225W solar PV module at different cell temperatures and solar insolation is simulated in MATLAB environment. The simulated results shows the deviation of output power of solar module at maximum power point with ideal PV power is about 12% for different cell temperature and solar insolation.

1. Introduction

Solar energy is the most emerging alternate source for electricity generation. It generates environmental friendly energy. The reform made by the nations on global warming the solar power generation is increasing day by day. India's installation capacity is about 29.42GW as of May 2019. The government has set its target to reach 40GW by 2020 [1].

The energy available in the solar spectrum can be utilized directly for generation of electricity and indirectly in the form of thermal to produce steam to drive the steam turbines [2, 3]. Solar photovoltaic cells are made up of semiconductors will generate electrons when the light energy packets known as photons are absorbed. The efficiency of conversion of solar energy to electric energy depends on the factors such as solar irradiance or solar insolation level available at particular time, cell temperature and the load current. The available solar PV cell in the market has an efficiency of 10-25%. This brings the necessary to improve the performance by incorporating power tracking techniques thereby ensuring better utilization of solar energy and decreasing the cost of investment.

Many MPPT methodologies are suggested in the literature but they depend on the factors like dynamic irradiance, low convergence speed and hardware implementation. The MPPT methods include open voltage, constant voltage, short current, incremental conductance(IC) and perturb & observe (P&O) methods [4].The design simplicity and good performance of P&O MPPT method has made to use widely in solar Photovoltaic applications[5,6].

IOP Conf. Series: Materials Science and Engineering **623** (2019) 012020 doi:10.1088/1757-899X/623/1/012020

This paper focuses on the extraction efficiency of solar PV array under different irradiance and temperature conditions. A 225W solar PV module acting as source to an inverter for isolated electric system is simulated in MATLAB environment. The performance is analyzed by using P&O MPPT algorithm. The simulates results will validate its efficiency of P&O MPPT algorithm in terms of percentage deviation with ideal PV panel output power.

2. Photovoltaic model

The commonly known PV cell is configured as a semiconductor p-n junction made of silicon. The photons in sunlight hit the solar panel and are absorbed by the semiconducting materials. This phenomenon will make the electrons to loose from their atoms when excited. The materials used in the solar cell allow these electrons to move only in single direction resulting direct current (DC) electricity. The electrically equivalent model is useful to understand the characteristics of solar cell. The solar cell is modelled as photo generated current source (I_{ph}) in shunt with a diode (D). In practical a series (R_S) and shunt (R_{Sh}) resistances are added. The circuit model of a solar cell is given in figure 1. These cells are combined together in series and in parallel to form PV modules and PV arrays.

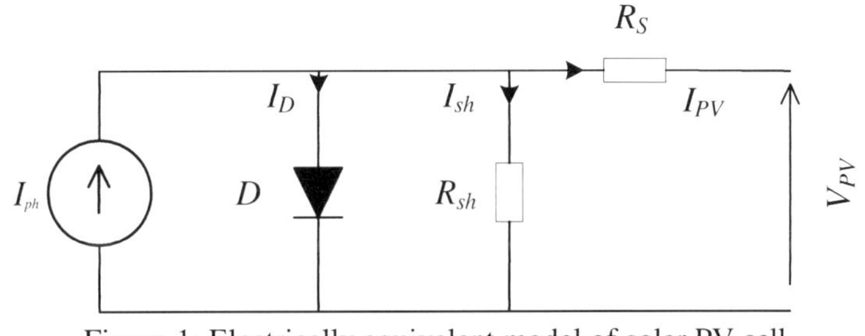

Figure 1: Electrically equivalent model of solar PV cell

From the equivalent circuit V_{pv} is output voltage and I_{pv} is the output current of PV cell. Applying Kirchhoff's current law for the current generated by the PV cell.

$$I_{pv} = I_{ph} - I_D - I_{sh} \tag{1}$$

I_D is diode current given as

$$I_D = I_0\{ exp\frac{V_j}{nV_T} - 1\} \tag{2}$$

and

$$V_T = \frac{KT}{q} \tag{3}$$

Where

V_j is the voltage across the diode and shunt resistor

I_0 is reverse saturation current

n is ideality factor
T is absolute temperature
V_T is thermal voltage at 25^{0}C approximately given as 0.025V
K is Boltzmann's constant
q is elementary charge

Therefore the current generated by the PV cell from equation (1) is given as

$$I_{pv} = I_{ph} - I_0\{exp\frac{V_j}{nV_T} - 1\} - \frac{V+IR_S}{R_{Sh}} \quad (4)$$

The above equation (4) gives a non linear characteristic curve for a solar PV cell. When the solar cell is open circuited $I_{PV}=0$, and the voltage across the diode and shunt resistance is termed as open circuit voltage V_{OC}. Similarly, when the solar cell is short circuited, the voltage V_{PV} becomes zero and $I_{SC}\approx I_{ph}$ which is termed as short circuit current.

A 225W solar array with three modules consisting of 34 cells in series is considered at different solar irradiance and cell temperatures to obtain non linear characteristics. The parameters and test conditions of the solar module at solar irradiance 1200W/m^2 are listed in the below table 1.

Table 1. Specification of Solar module.

Parameters	Notation	Value
Maximum rated power	P_{MP}	225W
Rated Voltage at maximum power	V_{MP}	45V
Rated current at maximum power	I_{MP}	5A
Open circuit voltage	V_{OC}	52V
Short circuit current	I_{SC}	5.502A
Solar cells in series	N_s	110
Solar cells in parallel	N_p	1

The solar PV model with the given specifications is simulated to obtain its I-V and P-V characteristics at 20^{0}C for different irradiances of 1200W/m^2, 800W/m^2 and 600W/m^2.

3. P&O MPPT control technique

In order to improve the solar PV module efficiency at different irradiance conditions a P&O MPPT was developed. This algorithm finds the reference voltage corresponding to maximum power point in the characteristic curve. The control block diagram is shown in the below figure 2.

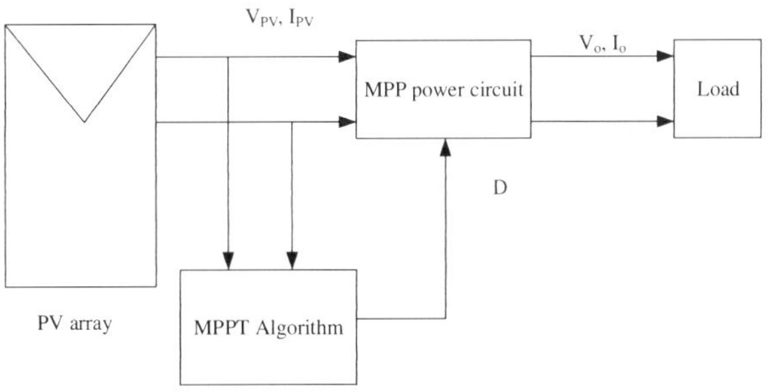

Figure 2: MPPT Control block diagram

The power circuit comprises a DC-DC boost converter. The load impedance is matched by varying converter duty cycle (D). The P&O MPPT algorithm is utilized for duty cycle to obtain the reference voltage corresponding to maximum power point. The flow chart corresponding to P&O MPPT algorithm is shown in figure 3. First the voltage and current are measured to obtain the output power. The obtained power is verified with voltage. If the voltage of PV panel is increased the output power is also increased in the same way duty cycle is also increased else the duty cycle is decreased by one step of ΔD. This is continued till the maximum point is tracked and the output voltage is maintained at that point.

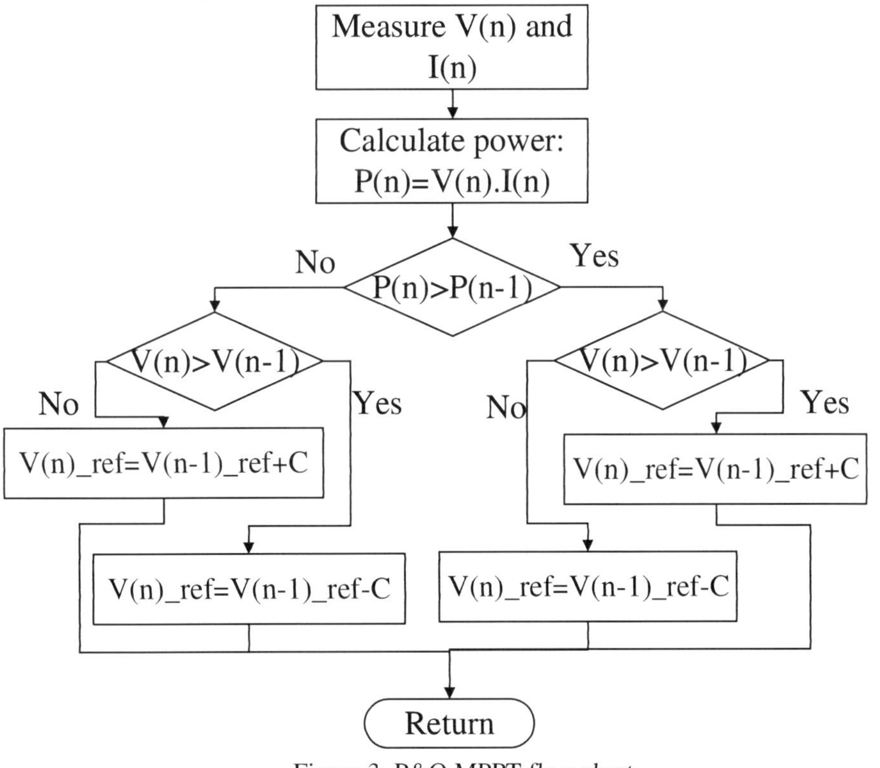

Figure 3: P&O MPPT flow chart

The duty cycle will maintain the DC-DC boost converter output voltage as constant. The output voltage of solar PV panel is fed to the input voltage of converter which is given as

$$V_{in} = V_{PV} = V_o(1 - D) \quad (5)$$

Where V_o is output voltage corresponding to the duty cycle D

For any small change in solar irradiation and temperature resulting in change of output voltage and output power of a solar PV panel. These variations are given as

$$\Delta V_{PV} = (\Delta D)V_O \quad (6)$$

Where ΔV_{PV} is change in output voltage and ΔD is change in duty cycle which is chosen as 0.0005.

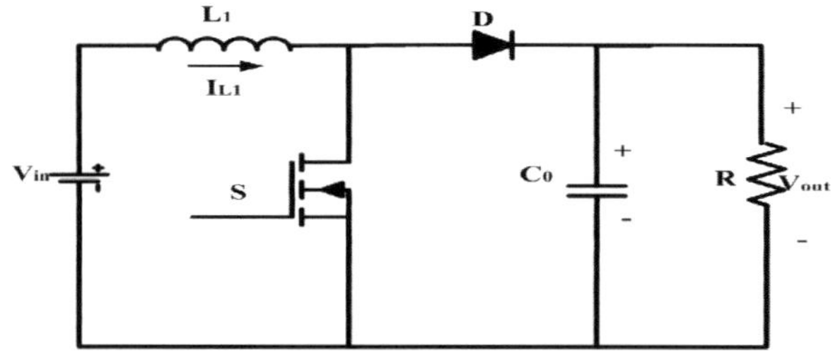

Figure 4: DC-DC boost converter

A DC-DC boost converter is having an advantage of high reliability and less number of components. Hence it is incorporated to reach maximum power point at power stage. It consisting of MOSFET for which the gate pulses are generated using PWM technique, a diode, an inductor (L_1) and a capacitor (C_0) as shown in figure (4). The boost converter specifications for simulink model are listed in the table 2.

At steady state the input to output ration of the converter is given as

$$\frac{V_{in}}{V_{out}} = 1 - D \quad (7)$$

The inductor and capacitor expression are given with respect to duty ratio

$$L = \frac{V_{in}D}{f_s \Delta I_L} \quad (8)$$

$$C = \frac{I_O D}{f_s \Delta V_{out}} \quad (9)$$

Where ΔV_{out} and ΔI_L are ripple voltage and currents respectively.

Table 2. Specification of DC-DC boost converter.

Parameters	Notation	Value
Input voltage	V_{in}	48V
Output voltage at D=0.5	V_{out}	95V
Resistive load	R_L	38.6Ω
Inductor	L_1	550μH
capacitor	C_0	110μF
Switching frequency	f_s	5KHz

4. Simulation results and analysis

The simulink model of solar PV array with P&O control algorithm shown in figure (3) is developed in MATLAB environment to understand its performance characteristics at different irradiation and cell temperatures. The MATLAB is shown in figure (5) was developed for the specifications listed in table 1 and table 2. The results taken for solar output power at 1200W/m^2, 800W/m^2 and 600W/m^2 operating at 20^{0}C, 45^{0}C and 70^{0}C. Duty cycle and PV power outputs are shown in figure (6) and figure (7). The I-V and P-V curves are shown in figure (8) and figure (9). It is observed that as the duty cycle has been decreased and the output power is decreased with the decrease in solar irradiance from 1200W/m^2 to 600W/m^2 with the increment in cell temperature from 20^{0}C to 70^{0}C. The percentage deviation of P&O MPPT solar PV array with PV maximum power output at different irradiations and cell temperature are presented in the table 3 & 4 given below.

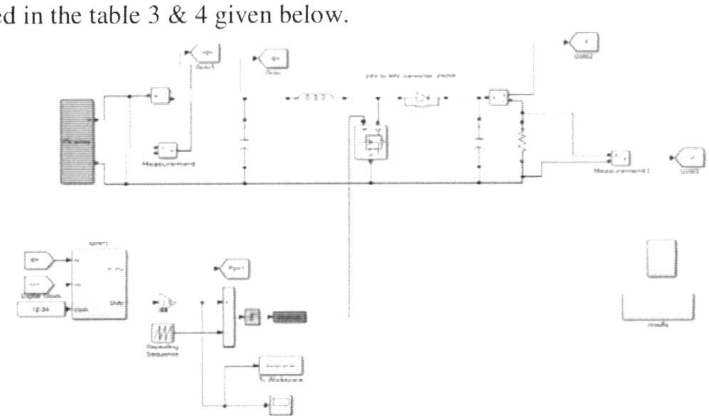

Figure 5: Simulink model of solar PV module with P&O MPPT method.

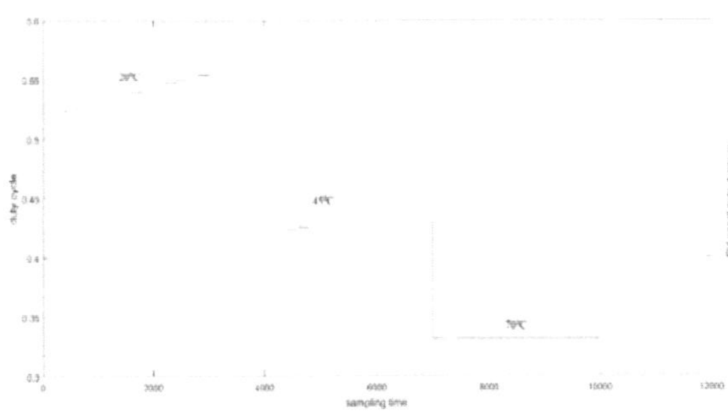

Figure 6: Duty cycle for different T and constant G

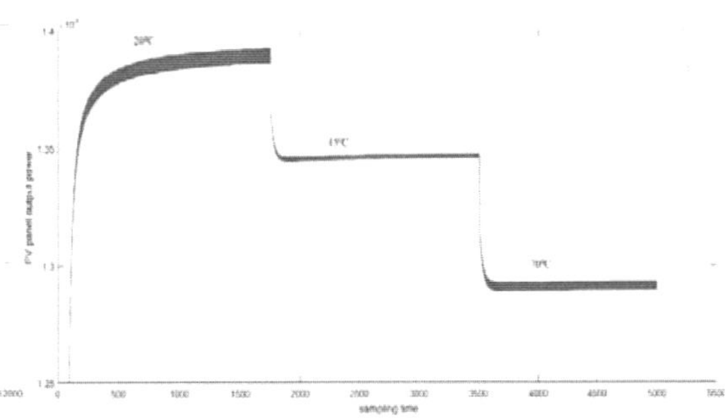

Figure 7: PV power for different T and constant G

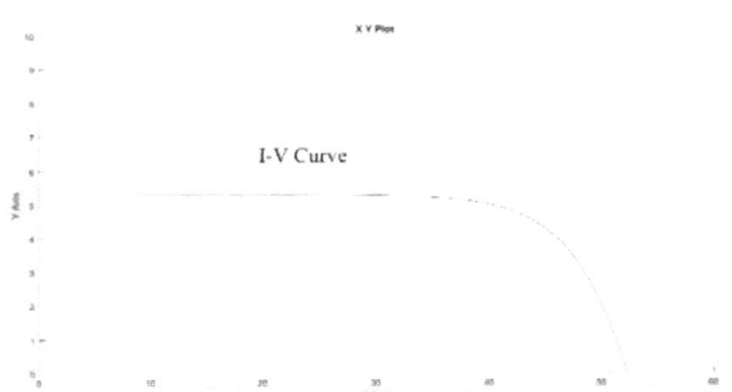

Figure 8: I-V characteristics of PV panel

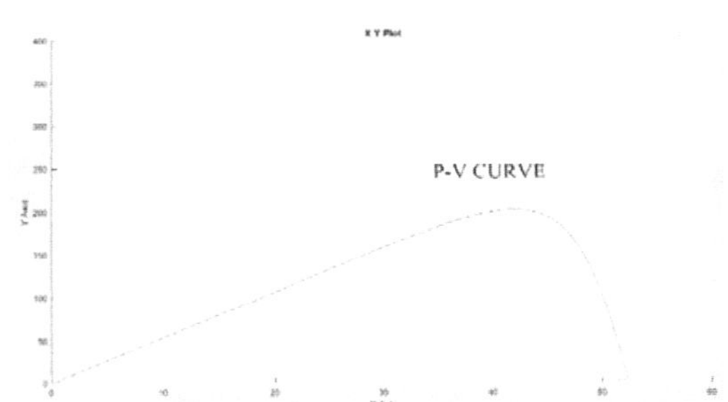

Figure 9: P-V characteristics of PV panel

Table 3. comparison at constant cell temperature of 20^{0}C

Solar irradiance (W/m^2)	PV Maximum output power(W)	Output power of PV with P&O MPPT(W)	Percentage deviation (%)
1200	221.42	214.31	3.211
800	173.86	155.49	10.56
600	126.83	114.67	9.58

Table 4. comparison at constant solar irradiance of 1200 W/m^2

cell temperature (^{0}C)	PV Maximum output power(W)	Output power of PV with P&O MPPT(W)	Percentage deviation (%)
20	221.42	214.31	3.211
45	186.43	175.23	6.007
70	150.64	138.94	7.766

5. Conclusion

An improved P&O MPPT algorithm is incorporated with the solar PV array was developed to increase its power extraction efficiency at different solar irradiance 1200W/m^2, 800W/m^2 and 600W/m^2 operating at cell temperatures of 20^{0}C, 45^{0}C and 70^{0}C. The technique used for obtaining maximum power effectively which is very nearer to PV Maximum output power. The comparison results will validate that percentage deviation of output power of PV with P&O MPPT with ideal PV output power is about 10% when operated at different irradiance and temperature conditions.

References

[1] "*Solar Power Potential in the Country-2019"*, National institute of solar energy, MNRE, Government of India. https://mnre.gov.in/national-institute-solar-energy.

[2] Babu, C., & Ponnambalam, P. (2017). The role of thermoelectric generators in the hybrid PV/T systems: A review. *Energy conversion and management*, *151*, 368-385.

[3] Babu, C., & Ponnambalam, P. (2018). The theoretical performance evaluation of hybrid PV-TEG system. *Energy conversion and management*, *173*, 450-460.

[4] D. Freeman, "Introduction to photovoltaic systems maximum power point tracking" *Texas Instruments Rapport. d'application SLVA446- November* vol. 2010, 2010.

[5] T. Esram and P. L. Chapman "Comparison of photovoltaic array maximum power point tracking techniques," *IEEE Transactions on Energy Conversion EC,* vol. 22 p. 439, 2007.

[6] A. Cupertino, J. de Resende H. Pereira, and S. S. Júnior, "A Grid-Connected Photovoltaic System with a Maximum Power Point Tracker using Passivity-Based Control applied in a Boost Converter," *the solar system* vol. 4, p. 6, 2012.

DESIGN AND SIMULATION OF PHASE SHIFTED DC-DC FULL BRIDGE CONVERTER

Prarthiv Evan Poudyal[1], V Vigneshwar[1], N K Kumar[1], V Indragandhi[1], A Nazar Ali[2]

[1]School of Electrical Engineering, Vellore Institute of Technology, Vellore
[2]Associate Professor, Department of EEE, Rajalakshmi College of Technology, Chennai

Abstract -Electric and hybrid electric vehicle (EV/HEV) architectures require a small DC-DC converterto replace any conventional vehicle's alternator. The DC-DC converter, also described as the vehicle Auxiliary Power Module (APM), provides power flow between the vehicle high voltage (HV) and low voltage (LV) DC bus. This abstract represents the phase shifted full bridge DC-DC converter scheme to design, simulate DC/DC Converter for a power rating of 80W intended for low power/ auxiliary applications in electric vehicles The proposed converter is very beneficial as it can be used for 4 quadrant operation much adaptive to the unknown parameters in the PV system and these unknown parameters are estimated through the adaptation laws in the algorithm which guarantees maximum power extraction possible from the power converter for a PV system. The converter supplies AC current as per requirement. The overall stability of the converter output is analyzed by simulation. The results are compared with other existing converter results for improving power quality even further. The proposed results indicate the robustness of the proposed scheme.

Keywords: DC-DC Converter, Phase Shift Full Bridge, Hybrid vehicles.

1 INTRODUCTION

Electric vehicles (EVs) powered by batteries are the key to reducing global warming and rising fuel cost.[1,2] A recent survey reported that EVs must be a good for most Americans' driving and commute patterns, and 40% of the respondents indicated strong interests in purchasing EVs.

IOP Conf. Series: Materials Science and Engineering **623** (2019) 012021 doi:10.1088/1757-899X/623/1/012021

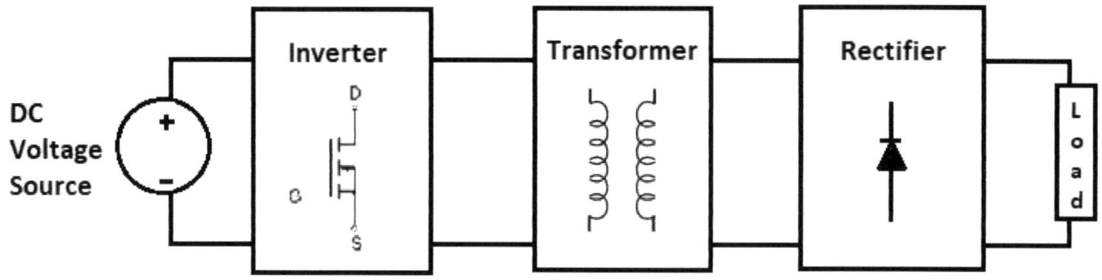

Figure 1 Isolated Topology

[2]Despite the increasing popularity of EVs, they have not yet dominated the internal combustion vehicle market for three reasons. First, EVs are expensive due mainly to the large number of battery cells required to power them. [3]There have been various ongoing researches about reducing EV energy consumption in the usage-phase that focuses on the vehicle side through, for an instance, optimization of the powertrain system, upgrade to motor control strategies, and improvements in power density of the battery. [5,7] Most of the modern-day electric cars struggle with range that they can provide to the user on a full charge of these batteries

The possible ways to tackle this can be two:

(1) Increase the capacity of the battery

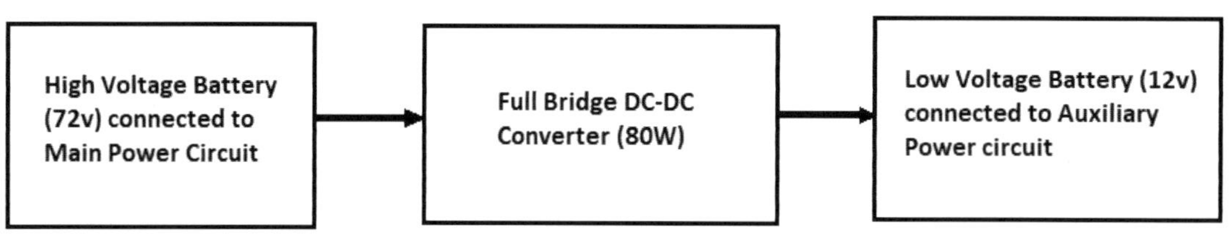

(2) Decrease the loads and improve the converter efficiency within the car.

Figure.2 Block Diagram

[4]The second way, Decrease the loads and improve the converter efficiency within the car. Main Power Circuit (Motor and Drive Circuit). Auxiliary Power Circuit (Air Conditioning Unit, Audio System, GPS and others)

2. Non-isolated converters

The non-isolated type converters are generally used wherever the voltage needs to be either stepped up or down by a relatively small ratio (less than 4:1). [7,8,9] And whenever there is no problem with the output and input having no dielectric isolation. There are five main types of converter in this non-isolated group, usually called the buck, boost, buck-boost, [11,12] Cuk and charge-pump converters. The buck converter is used for stepping down the voltage to the desired value, while the boost converter is used for stepping up the voltage to the desired value. The buck-boost and Cuk converters can be used for either stepping-down or stepping-up as per applications requirement. [13,14] The charge-pump converter is used for either voltage step-up or voltage inversion, but only in relatively low power applications.

3. Related Works

- **Mahindra e2o plus P8 Variant**

  Specifications:

  - High Voltage Battery = 72V
  - Low Voltage Battery = 12V
  - DC-DC Converter Topology = Full Bridge Phase Shifted
  - Power Rated (P_o) = 80W

- **Toyota Prius 2001 Hybrid Variant**

  Specifications:

  - High Voltage Battery = 300V
  - Low Voltage Battery = 42V
  - DC-DC Converter Topology = Full Bridge Isolated Boost
  - Power Rated (P_o) = 80W

4. SIMULATION&SYSTEM MODELLING

Designing of Transformer. Filter capacitor & Load resistance

$$Transformer\ Turns\ Ratio\ (N) = \frac{Output\ Voltage(V_o)}{Input\ Voltage\ (V_i)}$$

Here,

$$Output\ Voltage\ (V_o) = 12V,\ Input\ Voltage(V_i) = 72V$$

$$Transformer\ Turns\ Ratio\ (N) = \frac{V_o}{V_i}$$

$$Transformer\ Turns\ Ratio\ (N) = \frac{12}{72}$$

$$i.e,\ N = 1/6$$

$$Load\ Resistance\ (R) = \frac{[Output\ Voltage\ (V_o)]^2}{Ouput\ Power\ (P_o)}$$

Here,

$$Output\ Voltage\ (V_o) = 12V, Ouput\ Power\ (P_o) = 80W$$

IOP Conf. Series: Materials Science and Engineering **623** (2019) 012021 doi:10.1088/1757-899X/623/1/012021

$$Load\ Resistance\ (R) = \frac{[12]^2}{80}$$

$$i.e, R = 1.8\Omega$$

$$Filter\ Capacitor\ (CF) = \frac{1}{4fR}\left[1 + \frac{1}{\sqrt{2} \times \overline{RF}}\right]$$

Here,

$$Ripple\ Factor\ (\overline{RF}) = \frac{Ripple\ Voltage\ (V_r)}{Average\ Output\ Voltage\ (V_o)}$$

Here,

$$Frequency(f) = 100kHz,\ Load\ Resistance(R) = 1.8\Omega,$$

$$Ripple\ Voltage(V_r) = 0.12V,\ Average\ Output\ Voltage\ (V_o) = 12V$$

$$Ripple\ Factor\ (\overline{RF}) = \frac{0.12}{12} = 0.01$$

$$Filter\ Capacitor\ (CF) = \frac{1}{4fR}\left[1 + \frac{1}{\sqrt{2} \times \overline{RF}}\right]$$

$$= \frac{1}{4 \times 100 \times 10^3 \times 1.8}\left[1 + \frac{1}{\sqrt{2} \times 0.01}\right]$$

$$i.e,\ CF = 100\mu F$$

5. SPECIFICATION FOR MATLAB MODELLING.

Input Voltage (V_i)= 72V

Output Voltage (V_o) = 12V

Power Rated (P_o) = 80W

Transformer Turns Ratio ($P:S_1:S_2$) = 6:1:1

Duty Cycle = 50%

Topology = Full Bridge Phase Shift Type

Output Filter Capacitor (CF) = 100μF

Load Resistance (R) = 1.8Ω

5.1 Steady State Results:

Input Voltage (V_i)= 72V

Input Current (I_i) = 1.21A

Input Power (P_i) = 87.12W

Output Voltage (V_o)= 12V

Output Current (I_o) = 6.69

Output Power (P_o) = 80.28W

Hence, we simulate a DC/DC Converter for a power rating of 80W intended for low power applications in hybrid electric vehicles.

6. OPERATION

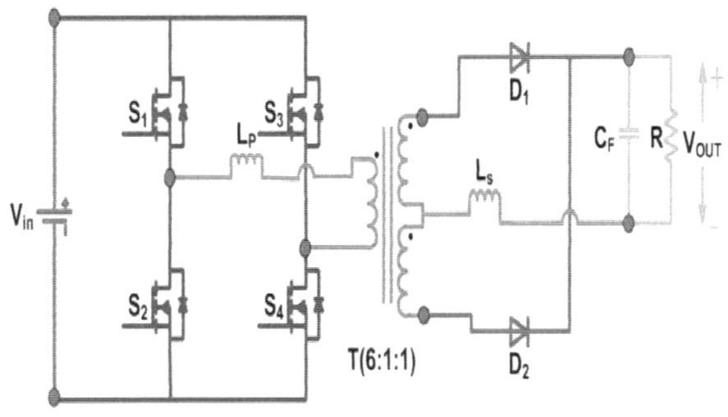

Figure.3 Circuit Diagram

A full bridge isolated phase shifted DC-DC converter is used. Full bridge refers to turning ON the switch by pairing on two of the four switches (MOSFET) crossing diagonally to transfer electric power from primary side to secondary side. Efficiency is improved with the phase shift type using soft switching. We assume the breakdown voltage required for MOSFET with the full bridge converter to be:

When S_1 & S_4 are turned ON, ON-resistance of MOSFET is low and as a result, it can be regarded as a short circuit between drain and the source. Therefore, the voltage V_{in} is applied between the drain and the source of MOSFET, S_2 & S_3 are in the OFF state. When S_2 & S_3 are turned ON, V_{in} is also applied to S_1 & S_4 that are in the OFF state. This is the measure of the breakdown voltage required for MOSFET. In fact, a design with sufficient margin is required with surge voltage and derating taken into consideration. Voltages of $+V_{in}$ & $-V_{in}$ are applied to the primary side of the transformer alternately by switching the switches. This enables an effect equivalent to that of applying the power supply voltage twice (V_{in} * 2) which allows supporting high power application. A device with high speed and low ON resistance is required as a switching element. POWER MOSFET enables this demand. There is some loss in MOSFET because of the flow of drain current Id under the application of voltage between drain and the source. During switching transient period of MOSFET, drain current I_d flows with voltage applied V_{ds}, causing a loss. Hence, soft-switching is used, switching when the current or voltage is zero, it is a counter measure for the switching loss. A method to switch during zero voltage is called ZVS and that with zero current is called ZCS. However, ZVS is more popular in the full bridge type power supply circuit using MOSFETs.

6.1 PHASE SHIFTED TYPE FULL BRIDGE MODE 1 OPERATION

The timing to control each MOSFET differs. Timing is divided into 8 segments. In addition, a dead time is applied while S_1 & S_2, and S_3 & S_4 (vertically connected switches) are NOT turned ON simultaneously. Every MOSFET is turned OFF in the hard switch type, not every MOSFET is turned OFF in the phase shift type. From t_0 to t_1, When S_1 & S_4 are turned ON, V_{in} is applied positively to the dot end of the primary transformer. At this time, the dot end of the secondary side becomes positive and is turned ON by forward biasing the diode D_1. In addition, the capacitor C_o is charged via the choke coil L_s. t_1 to t_2, at the beginning of this period the voltage between the drain and the source is zero thus it starts with soft switching. It doesn't transit to OFF instantly. The freewheel current caused by the leakage inductance L_p continues to flow even during transition period, the charge voltage finally reaches equivalent to V_{in}.

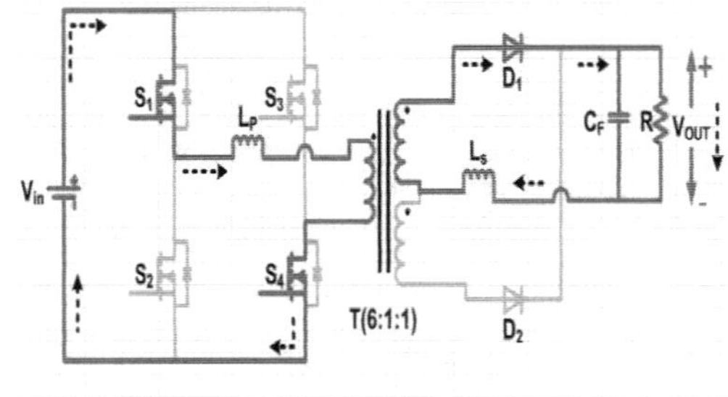

Figure 4. Mode 1 Operation

6.2 BRIDGE MODE 2 OPERATION

When S_2 & S_3 are turned ON, the voltage between the drain and source of Q_2 is zero, and during transition from OFF to ON, the freewheel current of L_p continues to flow, soft switching is completed. The non-dot side of both primary and secondary sides of the transformer becomes positive. D_2 is turned ON to charge C_o via choke coil L_s. electric power is transferred from primary to secondary side. Phase-shift type realizes soft switching by optimizing ON/OFF from S_1 through S_4.

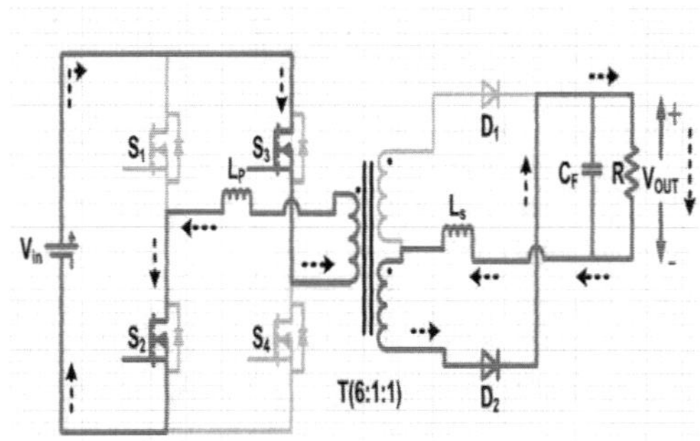

Figure 5. Mode 2 Operation

7. SIMULATION RESULTS

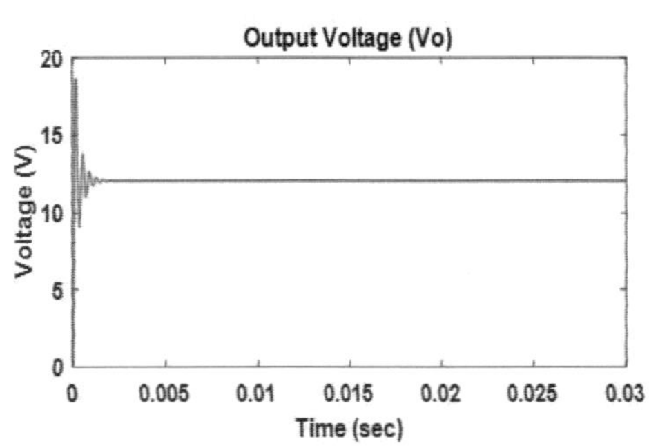

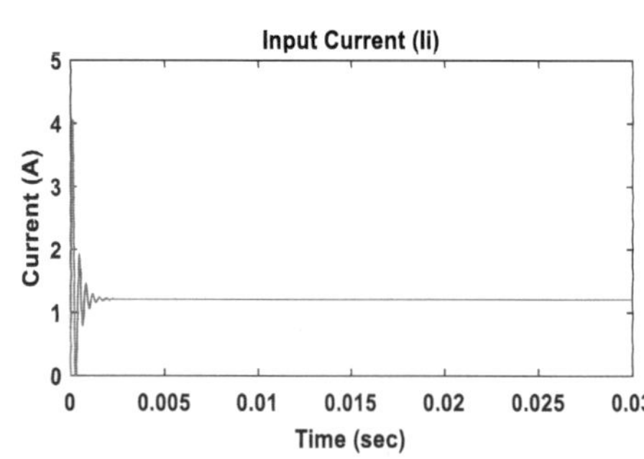

IOP Conf. Series: Materials Science and Engineering **623** (2019) 012021 doi:10.1088/1757-899X/623/1/012021

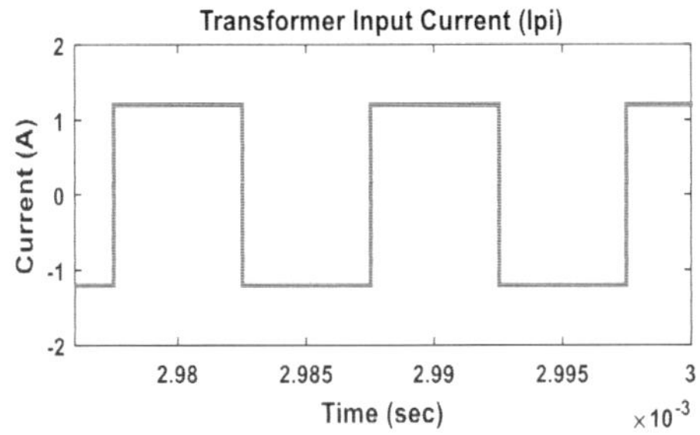

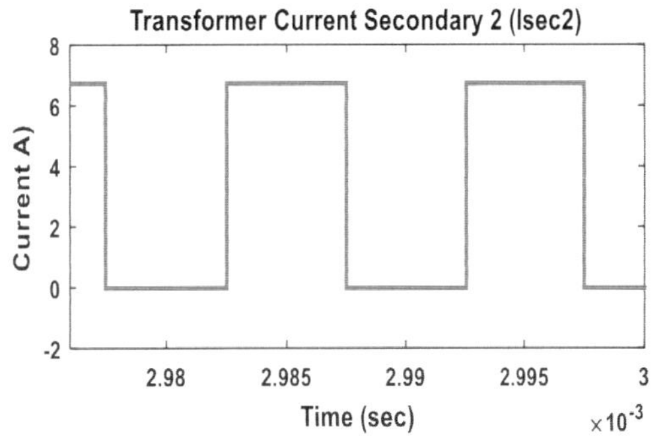

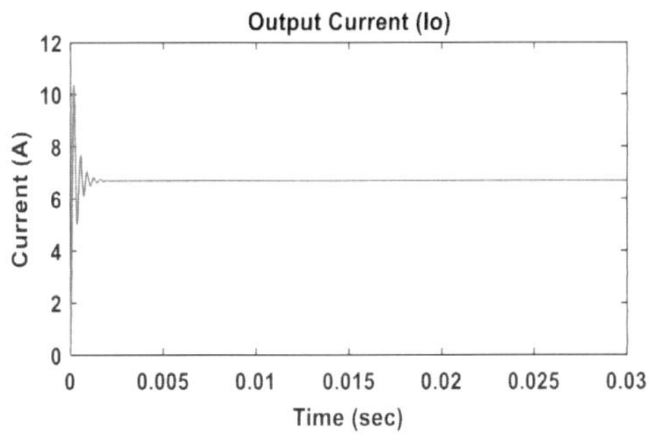

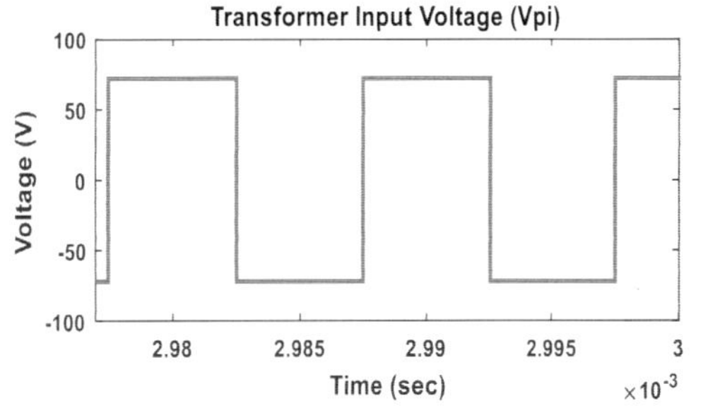

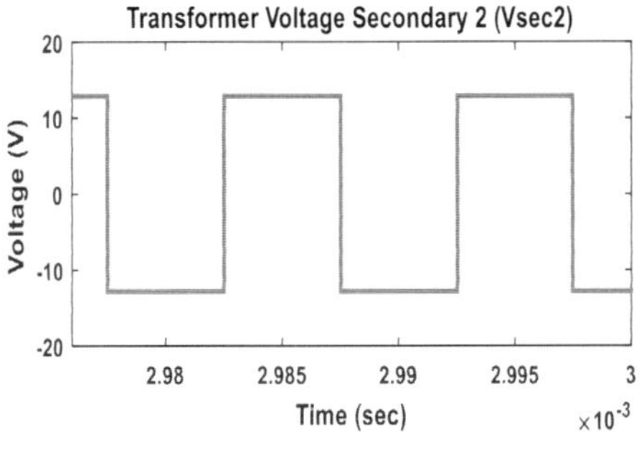

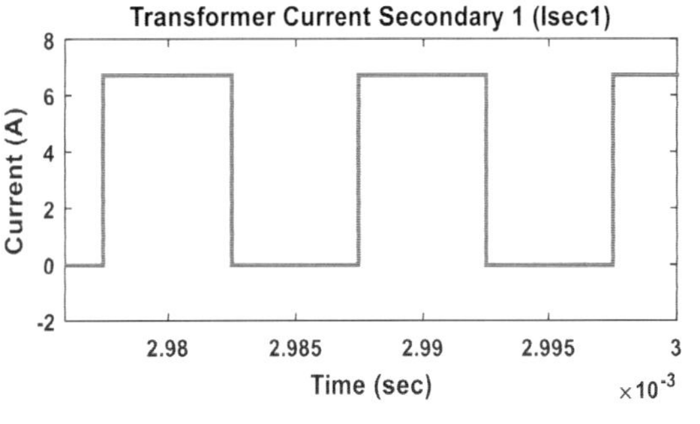

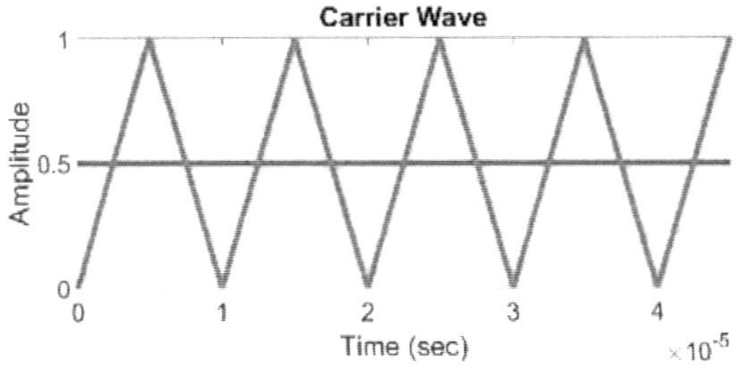

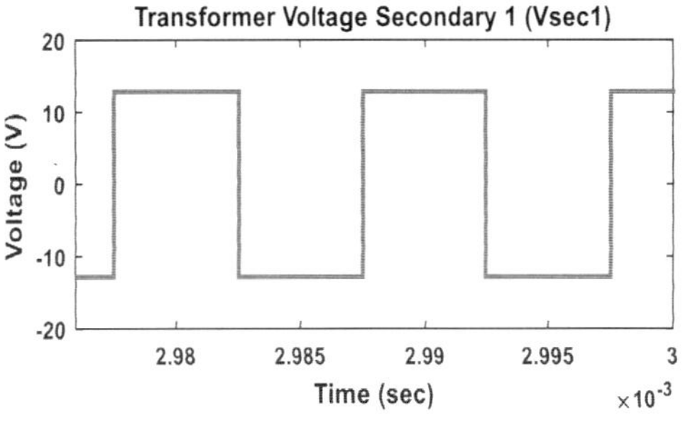

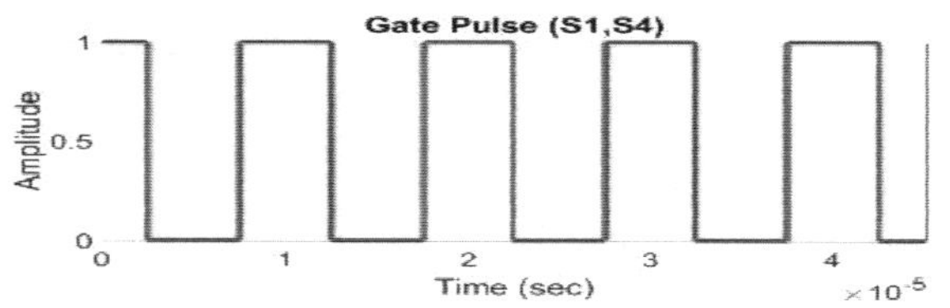

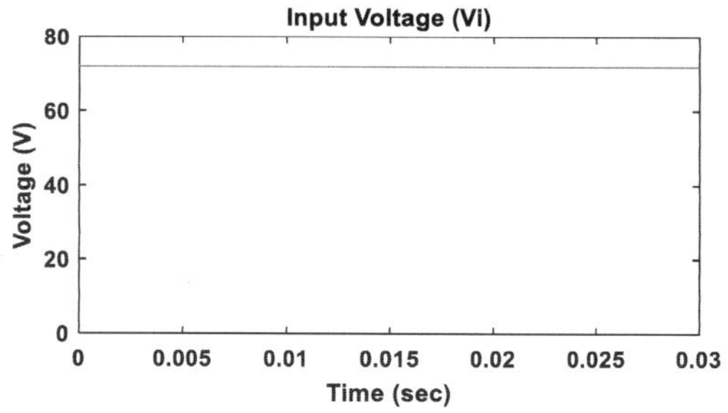

8. CONCLUSION

This paper proposes a new active soft switching circuit for the PSFB DC-DC converter. The converter's analysis is presented with the mathematical treatment required. The converter's equivalent steady-state circuit model and dynamic model are presented. The converter's design guidelines will be discussed. A design example is presented in accordance with the design guidelines. The designed converter is simulated and for the specifications a prototype is developed. In order to validate the analysis, the circuit simulations and experimental waveforms are presented. Experimental results are consistent with the theory and simulations. Converter efficiency has been improved overall. This model of proposed are validated experimentally.

9. REFERENCES

[1] K. Shi, D. Zhang, Z. Zhou, M. Zhang, D. Zhang and Y. Gu, "A Novel Phase-Shift Dual Full-Bridge Converter With Full Soft-Switching Range and Wide Conversion Range," IEEE Transactions on Power Electronics, Vol. 31, No. 11, November 2016

[2] Mallikand A. Khaligh, "Variable-Switching-Frequency State-Feedback Control of a Phase-Shifted Full-Bridge DC/DC Converter," IEEE Transactions on Power Electronics, Vol. 32, No. 8, August 2017

[3] O. Lee, "Hybrid DC-DC Converter with Phase-Shift or Frequency Modulation for NEV Battery Charger," IEEE Transactions on Industrial Electronics, Vol. 63, No. 2, February 2016

[4] S. Kim, H. Cha, H. F. Ahmed, B. Choi, H. Kim, " Isolated Double Step-down DC-DC Converter with Improved ZVS Range and No Transformer Saturation Problem ," IEEE Transactions on Industrial Electronics, Vol. 32, No. 3, March 2017

[5] Y.Shen, X.Sun, W.Li, X.Wu, and B.Wang,"A Modified Dual Active Bridge Converter With Hybrid Phase-Shift Control for Wide Input Voltage Range," IEEE Transactions on Power Electronics, Vol. 31, No. 10, October 2016

[6] J. Mason and P. K. Jain, "New Phase Shift Modulated ZVS Full Bridge DC/DC Converter with Minimized Auxiliary Current for Medium Power Fuel Cell Application," in Proc. Power Electronics Specialists Conference, 2005, pp. 244-249, June 2005.

[7] Y. Jang and M. M. Jovanovic, "A New Family of Full-Bridge ZVS Converter," IEEE Transactions on Power Electronics, vol. 19, no. 3, May 2004.

[8] Z. Chen, B. Ji, F. Ji, and 1. Shi, "Analysis and Design Considerations of An Improved ZVS Full-Bridge DC-DC Converter," in Proc. Applied Power Electronics Conference and Exposition, 2010, pp. 1471-1476.

[9] B. Yang, 1. 1. Duarte, W. Li, K. Yin, X. He, and Y. Deng, "Phase Shifted Full Bridge Converter Featuring ZVS over the Full Load Range," in Proc. 36th Annual Conference on IEEE industrial Electronics Society, 2010, pp. 644-649.

[10] X. Zhou, M. Donati, 1. Amoroso, and F. C. Lee, "Improved Light-Load Efficiency for Synchronous Rectifier Voltage Regulator Module," IEEE Transactions on Power Electronics, vol. 15, no. 5, May

2000.
[11] Texas Instruments, "Green Phase-Shifted Full-Bridge Controller with Synchronous Rectification," Datasheet SLUSAI6, Mar. 2010.
[12] R. Ramachandran and M. Nymand, "Design and analysis of an ultra-high efficiency phase shifted full bridge GaN converter," in Proc. IEEE Applied Power Electronics Conference and Exposition, Charlotte, NC, Mar. 2015, pp. 2011-2016.
[13] B. Y. Chen and Y. S. Lai, "Novel Dual Mode Operation of Phase-Shifted Full Bridge Converter to Improve Efficiency under Light Load Condition," in Proc. IEEE Energy Conversion Congress and Exposition, San Jose, CA, Sept. 2009, pp. 1367-1374.
[14] J.W. Kim, D.Y. Kim, C.E. Kim, M.Y. Kim et. al. "Switching control method for light load efficiency improvement in phase shifted full bridge converter," in Proc. IEEE Energy Conversion Congress and Exposition Asia, Melbourne, Jun. 2013, pp. 165-169.

COMPUTATIONAL STUDY OF AERO-SPIKE ON BLUNT BODY

M. Edwin[1],Sujay Iswarian Isaac[2], M. Pradeep[3] and L.Nagarajan[4]
[1]Department of Aeronautical Engineering, Tagore Engineering College Chennai, Tamilnadu
[2]Department of Aeronautical Engineering, G.K.M. College of Engineering and Technology, Tamilnadu
[3]Department of EEE, Rajalakshmi Engineering College, Chennai, Tamilnadu.
[4]Department of EEE, K.Ramakrishnan College of Technology,Trichy,Tamilnadu.

Abstract. An Aero-Spike involved to the stagnation point of a Semi-circular Blunt frame modifications its movement field and reduces the sweptback graft. The blunt body with aero-spike with different L/D relations (1, 2) and aero-spike shapes of truncated hemisphere, extended hemisphere, Simple spike and flat faced cone at Mach number 2 and at different angles of attack (0, 5, 10. 15) were computationally analysed using ANSYS Fluent. A recirculation zone forms in front of the blunt body and its gives more streamlined shape to the blunt body. Larger recirculation zone reduces the coefficient of drag efficiently, thus increasing length of the spike results in increase of recirculation sector and reduction in drag. By increasing angles of attack the coefficient of drag also increased. At 0 deg position of spell single disk with L/D ratio of 2 reduces the co efficient of drag efficiently. Keywords— Aero-Spike, Aerodynamic drag, coefficient of drag.

1. Introduction

A grind dipping aero-spike is a ploy used to diminish the sweptback graft of frank frames at supersonic hustles. The aero-spike generates a removed shudder done the figure ahead. Among the shudder and the front frame a recirculation area ensues which acts like a further modernised frame and hence sinking the graft. This concept was castoff on the UGM-96 Trident 1 and it is valued to have improved the collection by 550 Km. The Trident aero-spike comprises of a smooth spherical bowl riding on a extensible successful which is installed presently next the shell pauses complete the obvious of the liquid next take-off from sea. The habit of aero-spike acceptable a copious franker nose contour, providing better heart size for shipment and momentum minus collective the graft. For to growth in kind over the Poseidon C-3 shell Trident 1 C-4 was dynamic.

To fit within the present sea unveiling cylinders the third step motor taken to be riding in the core of the post-boost vanguard per the return vehicles organised nearby the motor. At the equal despite the circumstance (middle 1970s) aero-spike was advanced in KBM for 9M39 surface-to-air rocket of 9K38 Igla MANPADS (in order to weaken heating of ultraviolet homing chaser fairing and cut wave strain), open-handed the label to the uncut co-ordination Spike. Further development in this impression has ensued in the Air-spike. This is grown by tireless energy, since the electric arc incinerate or a beaten laser, likely forward from the frame. This has the improvement concluded a operational aero-spike that the air mass is lower than that ahead a shock as long as increased drag saving.

We study the Aero-spike concept for a blunt body to reduce the gravity drag on the blunted build. The drag reduces the thrust force and hence the range of the bullet, so we incorporate an Aero-spike on the nose of a blunt body. Also we check which Aero-disk shape gives us efficient drag reduction. That's why we fix different shapes for an Aero-disk, and change the length of the Aero-spike and check which length reduces the drag efficiently. Also we can try different combinations of aero-disk shapes.

Main neutral of the study is to diminish the burden drag on the blunt body at supersonic flow condition. We can condense the graft on the blunted frame by introducing a recirculation zone ahead of the blunt body. To create a recirculation zone we need to incorporate an Aero-spike on the nose of the blunt body.

Different kinds of Aero-spikes are used to reduce the drag. In that we concentrate on some Aero-spike shapes which are efficiently reduces the drag, also we study different length to diameter ratio of Aero-spike, in that we observe which L/D Ratio reduces the drag efficiently, also we study different combinations of Aerospike and compare the results and find which is suitable for real time applications.

IOP Conf. Series: Materials Science and Engineering **623** (2019) 012022 doi:10.1088/1757-899X/623/1/012022

[1] Says that Untried and computational lessons have been ended consuming spines of changed characters riding on a curved blunted bulk at supersonic hurry of Mach numerical of [2]. A series of wind tunnel tests have been performed on an aero-spike-protected missile dome at a Mach number of [5] High-speed tide over a blunt body spawns a bow shock wave which bases high outward gravity and as a result the change of high clean drags.
[6]spike with frontward facing devoted to a hemispherical body pointedly alter its flow field aerodynamic drag and heat flux in wall of high speed flow.
[7]A advancing incrustation aero-spike involved to a unsharpened frame of a bullet and re-entry typical fluctuations its f-drag reduction.
[8]High surface pressure origin by Bow shock wave influence high aerodynamic drags desired governor outsides at hypersonic hustles. In the modern year's interstellar truck like speeds, re-entry means of transportation irrespective their exclusive designs desired governor outsides at hypersonic hustles.

2. Methodology

Model is designed using CATIA V5 Designed model meshed using ANSYS ICEM CFD 15 Meshed model analysed using ANSYS FLUENT 15 Results were taken using FLUENT 15 and CFD Post processing Finally results were compared and graph plotted

2.1 Model geometry

Diameter of base model-300 mm.
Length of base model- 750 mm.
Diameter of spike-30 mm.
Diameter of disk- 60 mm .
Length of spike- 300 mm (L/D_1) 600mm(L/D_2).

2.2 Mesh details:

Number of Elements- 45216.
Number of Nodes- 46819.
Relevance centre- Fine Smoothing- Medium ,Min size- 0.3 mm Max size- 10 mm ,Growth rate - 1.20.

2.3 Design and mesh:

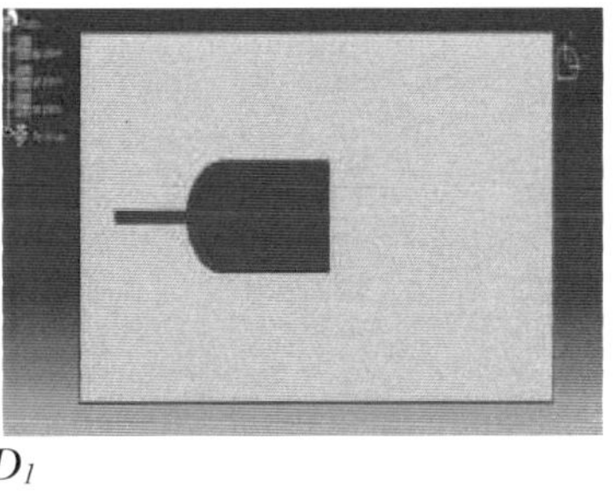

Simple spike L/D_1

Extended Hemisphere L/D of 1

IOP Conf. Series: Materials Science and Engineering **623** (2019) 012022 doi:10.1088/1757-899X/623/1/012022

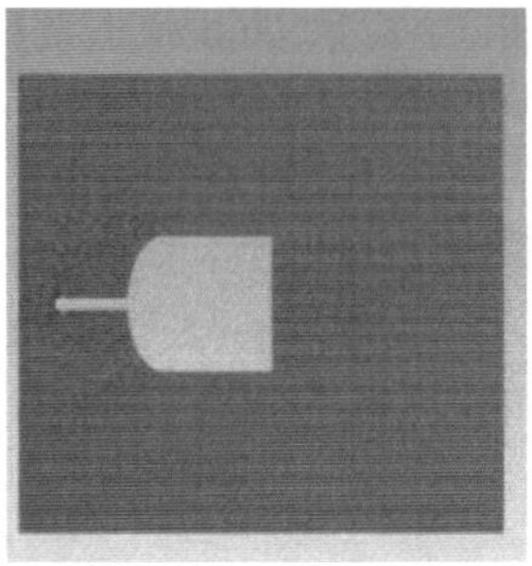

Truncated hemisphere L/D of 1

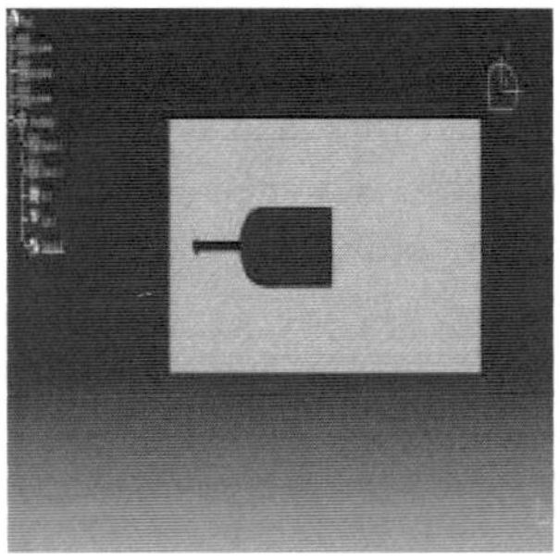
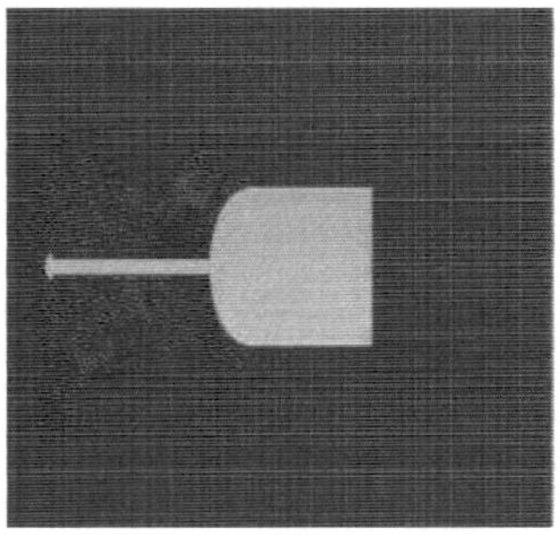

Flat faced cone L/D of 1

2.4 Viscous model:

k-ω sst, Energy- on,Material: air
Density- ideal gas kg/m3,
Specific heat Cp- 1006.43 j/kg-k,
Thermal conductivity- 0.0242 w/m-k,
Viscosity: southerland kg/m-s (three coefficient method),
Reference viscosity- 1.716e-05 kg/m-s,
Reference temperature- 273.11 k,
Effective temperature- 110.56 k

2.5 boundary condition

All the four sides of domain- pressure far field
Mach no- 2.
Pressure- 101325 pa
Blunt body: wall (standard wall condition)

3. Results

Blunt Body with Simple Spike L/D Of 1/ 0 Deg AOA

IOP Conf. Series: Materials Science and Engineering **623** (2019) 012022 doi:10.1088/1757-899X/623/1/012022

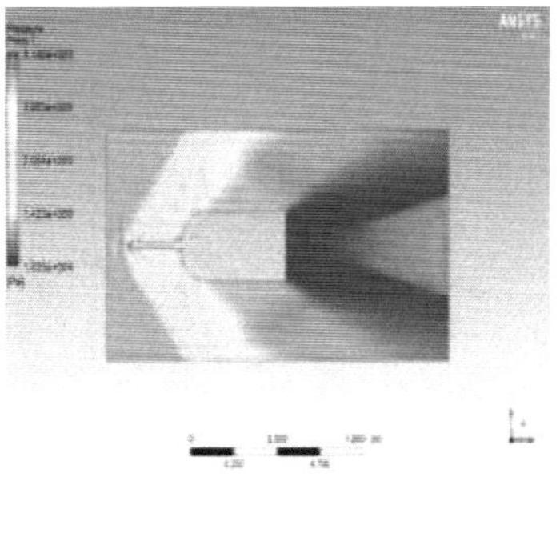

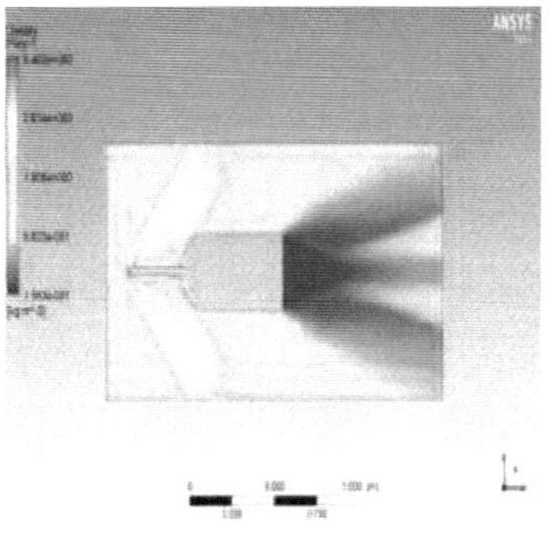

Pressure contour Density contours

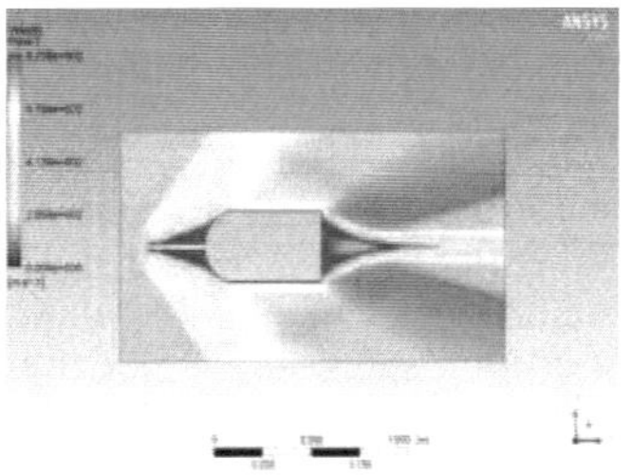

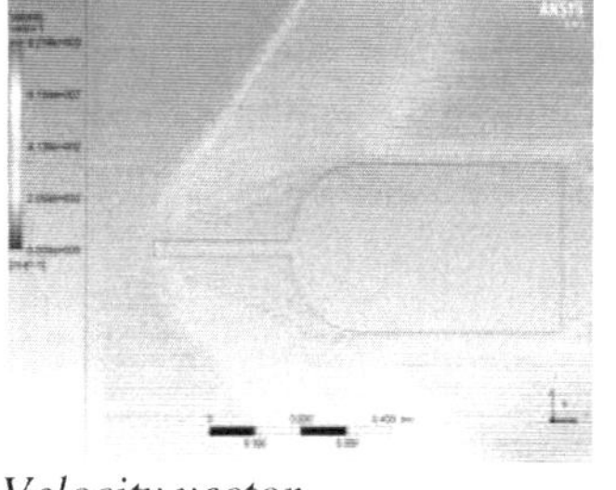

Velocity contour *Velocity vector*

3.1 Graphs

Cd Value of various shapes of aero spikes for four different angles

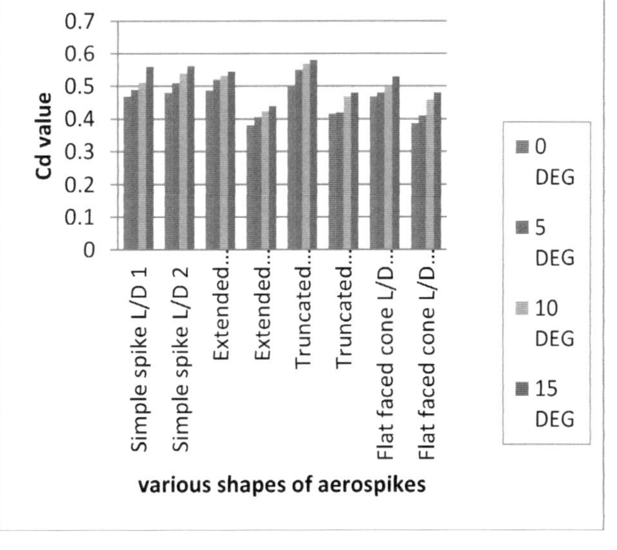

3.2 Discussions

Aero spikes with three disk shapes namely truncated hemisphere, extended hemisphere, flat faced cone and one simple aero-spike without any disk have been analysed at Mach no 2 and at different angles of attack. We noticed increment of spike length increased the recirculation zone; recirculation zone dramas a main person in the drop of coefficient of drag. Larger recirculation zones acts as a more stream lined body and reduce the drag. The position of outbreak increase the grind similarly increased. Spike with L/D share of 2 reduces the grind ably.

4. Conclusion

From all the aero-spike reproductions which were investigated it can be publicised that the competent drag saving given by spike with L/D fraction of 2 and 0 degree position of spasm for all the prototypes principally Flat faced cone and stretched hemisphere shrinks the quantity of drag well 0.39 and 0.38 respectively.

5. References

[1]. B. Kaleeswaran, S.Ranjith Kumar, Jeniwer Bimro.N, An Aerodynamic Optimization of Supersonic Flow over the Nose Section of Missiles. International Journal of Engineering Research & Technology (IJERT)Vol. 2 Issue 4, April – 2013 ISSN: 2278-0181.

[2.] Hemanth Raghav.P, Dinesh Kumar.G, Flow Analysis on Various Bluff Bodies, IJISET - International Journal of Innovative Science, Engineering & Technology, Vol. 2 Issue 4, April 2015. ISSN 2348 – 7968.

[3]. M. Venkatesan, N.Kannamanimuthu, S. Das, P.Kumar and J. K. Prasad, Flow field investigation over a hemispherical blunt body with different spikes at supersonic speed, Proceedings of the 38th National Conference on Fluid Mechanics and Fluid Power December 15-17-2011, MANIT,Bhopal Paper Code: AM-02.

[4]. Md Akhtar Khan, Karrothu Vigneshwara, Suresh Kukutla, Avinash Gupta Effect of spikes integrated to airfoil at supersonic speed, International Journal of Research in Engineering and Technology eISSN: 2319-1163 pISSN: 2321-7308.

[5]. N.Kannamanimuthu, S. Das, P.Kumar and J. K. Prasad, Flow field around a blunt body with spike at Supersonic speed, Symposium on Applied Aerodynamics and Design of Aerospace Vehicle (SAROD 2011) November 16-18, 2011, Bangalore, India. 61

[6]. N. Tulasi Radha, K. Dorathi, Modelling and CFD analysis over a missile blunt body with spherical aero-spike, International journal of advance research in science and engineering ISSN 2349-8354, Vol.no.5, issue no.08, august 2016.

[7]. Rajesh Yadav, Ugur Guven, V K Patidar and Ramesh Repaka, Sudhir joshi, Effect of double disk aero-spikes on aerothermodynamics of blunt body at mach 6.2, International Journal of Applied Engineering Research ISSN 0973-4562 Volume 11, Number 1 (2016) pp 366-376.

Institute of Physics
Dirac House, Temple Back
Bristol BS1 6BE UK

ISSN: 1757-8981
ISBN 978-1-5108-9827-1

9 781510 898271